普通高等教育机电类系列教材
"十三五"江苏省高等学校重点教材（2019-2-260）

Electromechanical Control Engineering
机电控制工程
（双语教材）

英汉对照

Editors in Chief: Dai Min, Zeng Li
Auditors in Chief: Miguel A. Salido Adriana Giret

主编 戴 敏 曾 励
主审 Miguel A. Salido Adriana Giret

机械工业出版社

本书主要介绍经典控制理论的基本概念、原理和分析方法。全书共七章，主要包括机电控制工程概论、控制系统的数学模型、控制系统的时域分析、控制系统的频率特性分析、控制系统的性能分析、控制系统的综合校正、基于 MATLAB 的控制系统分析与校正。

本书以双语形式出版，以便于读者的阅读和学习。本书既可作为机械专业本科生的教材，也可作为工程技术人员的参考书。对于专科生和少学时专业，可适当调整学时数。

This book mainly introduces the basic concepts, principles and analysis methods of classical control theory, consisting of seven chapters, including introduction to electromechanical control engineering, mathematical models of control systems, time-domain analysis of control systems, frequency characteristic analysis of control systems, performance analysis of control systems, comprehensive compensation of control systems, and analysis and compensation of control systems based on MATLAB.

This book is published in a bilingual form for the convenience of readers. It could be used as a textbook for mechanical engineering undergraduate students, as well as a reference book for the engineers and technicians. For junior college students and majors with less class hours, the class hours can be adjusted appropriately.

图书在版编目（CIP）数据

机电控制工程：双语教材：英汉对照/戴敏，曾励主编. —北京：机械工业出版社，2021.11
普通高等教育机电类系列教材
ISBN 978-7-111-69335-2

Ⅰ. ①机… Ⅱ. ①戴… ②曾… Ⅲ. ①机电一体化-控制系统-双语教学-高等学校-教材-英、汉 Ⅳ. ①TH-39

中国版本图书馆 CIP 数据核字（2021）第 204144 号

机械工业出版社（北京市百万庄大街 22 号　邮政编码 100037）
策划编辑：徐鲁融　　责任编辑：徐鲁融
责任校对：王勇哲　　封面设计：陈　沛
责任印制：李　昂
北京捷迅佳彩印刷有限公司印刷
2022 年 1 月第 1 版第 1 次印刷
184mm×260mm · 23.5 印张 · 573 千字
标准书号：ISBN 978-7-111-69335-2
定价：69.00 元

电话服务　　　　　　　　　　网络服务
客服电话：010-88361066　　　机　工　官　网：www.cmpbook.com
　　　　　010-88379833　　　机　工　官　博：weibo.com/cmp1952
　　　　　010-68326294　　　金　书　网：www.golden-book.com
封底无防伪标均为盗版　　　　机工教育服务网：www.cmpedu.com

Preface

At present, domestic education has entered a new stage of comprehensively deepening reform. The trend of current engineering education reform is to improve the international competitiveness of higher education and promote the construciton of world-class universities. Building a bilingual model course using international advanced engineering teaching concepts and outstanding domestic teaching models, ideas, and systems is reasonable and feasible. Bilingual teaching is one of the important measures to improve the quality of higher education in China.

"Fundamentals of control engineering" is an important professional basic course for mechanical engineering students, which is taught in Chinese in most of higher education institutions. With the further development of "undergraduate teaching reform and teaching quality project" initiated by the Ministry of Education, the construction of bilingual model courses has been gradually piloted in some colleges and universities, and good results have been obtained. Because of this, based on summarizing the teaching practice and teaching reform experience in recent years and referring to similar textbooks and works at home and abroad, the compliers accomplish this book. This book mainly introduces the basic concepts, principles and analysis methods of classical control theory, consisting of seven chapters, including introduction to electromechanical control engineering, mathematical models of control systems, time-domain analysis of control systems, frequency characteristic analysis of control systems, performance analysis of control systems, comprehensive compensation of control systems, and analysis and compensation of control systems based on MATLAB. This book tries to simplify the content and make it easy to understand. It not only considers the systematization and integrity of the classical control theory, but also strives to give prominence to key points and contacts theory with practice, which gradually cultivates students' ability to solve complex engineering problems.

This book could be used as a textbook for mechanical engineering undergraduate students, as well as a reference book for engineers and technicians. For junior college students and majors with less class hours, the class hours can be adjusted appropriately. Through the study of this course, readers can not only master the basic analysis and comprehensive methods of classical control theory, but also basically analyze and design engineering control systems by using MATLAB software.

Dai Min and Zeng Li from the School of Mechanical Engineering, Yangzhou University are the editors in Chief of this book. Participating editors are Zhang Fan, Zhu Zhida, and Wang Lixing. At the same time, the book is reviewed by Professors Miguel A. Salido and Adriana Giret from Universidad Politécnica de Valencia, Spain. We would like to thank two foreign experts for their strong support and valuable suggestions.

In the process of compiling this book, we have been supported by the Publication Fund of Yangzhou University, and many similar textbooks and works have been referred to. We would like to express our deep gratitude to the personnel concerned. Due to the limited knowledge of the compilers, inaccuracies and omissions are unavoidable, and any criticism and correction would be much appreciated.

Compliers
Yangzhou University

前 言

目前,我国教育已进入全面深化改革的新阶段,提高高等教育的国际竞争力,推动世界一流大学建设,是当前工程教育改革的趋势所在。借鉴国外先进的工科教学理念,兼顾国内优秀的教学模式、思想和体系,开展双语教学示范课程建设是提升我国高等教育质量的重要举措之一。

控制工程基础是机械工程类学生一门重要的专业基础课,开设该课程的高等院校基本采用中文授课的教学形式。随着教育部"高等学校本科教学改革与教学质量工程"的深入开展,双语教学示范课程建设工程逐步在一些高校进行了试点,并取得了良好的效果。鉴于此,编者在总结了近年来的教学实践和教学改革经验,参考了国内、外同类教材和相关著作的基础上,以学生能力提高为导向,以加强学习者的控制工程基础为目的,编写本书。本书主要介绍经典控制理论中系统分析与综合的基本方法。本书共七章,包括机电控制工程概论、控制系统的数学模型、控制系统的时域分析、控制系统的频率特性分析、控制系统的性能分析、控制系统的综合校正、基于 MATLAB 的控制系统分析与校正。本书在阐述上力求内容精简、浅显易懂,既考虑了经典控制理论体系的系统性和完整性,又尽量做到重点突出、理论联系实际,逐步培养学生解决复杂工程问题的能力。

本书主要面向机械类本科生教学,也可供工程技术人员参考。对于专科生和少学时专业,使用本书时可适当调整学时数。读者通过对本书的学习,不仅能掌握经典控制理论的基本分析和综合方法,而且能应用 MATLAB 软件分析和设计工程控制系统。

本书由扬州大学机械工程学院戴敏、曾励任主编,参与编写的还有张帆、竺志大和王礼星。全书由西班牙瓦伦西亚科技大学(Universidad Politécnica de Valencia)米格尔·萨利多(Miguel A. Salido)教授和阿德里安娜·吉雷(Adriana Giret)教授审核。编者对两位教授所给予的大力支持和提出的宝贵建议深表感谢。

本书在编写过程中得到了扬州大学出版基金的资助,并参考了许多同类教材和相关著作,在此对有关人员表示深深的谢意。限于编者的水平,书中错误和疏漏之处在所难免,恳请广大读者批评指正。

编 者
于扬州大学

CONTENTS
目 录

Preface　前言

Chapter1　Introduction to Electromechanical Control Engineering　机电控制工程概论 ⋯ 1
  1.1　Introduction　概述 ⋯ 2/3
  1.2　Structures of Control Systems　控制系统的结构 ⋯ 4/3
    1.2.1　Open-loop control systems　开环控制系统 ⋯ 4/3
    1.2.2　Closed-loop control systems　闭环控制系统 ⋯ 4/5
    1.2.3　The composition of feedback control systems　反馈控制系统的组成 ⋯ 8/9
  1.3　The Classification of Control Systems　控制系统的分类 ⋯ 10/11
    1.3.1　The classification in terms of the characteristics of input signals　按输入信号的特征分类 ⋯ 10/11
    1.3.2　The classification in terms of realization ways of controllers　按控制器的实现方式分类 ⋯ 12/13
    1.3.3　The classification in terms of error　按误差分类 ⋯ 12/13
  1.4　Basic Requirements of Control Systems　控制系统的基本要求 ⋯ 14/13
    1.4.1　Stability　稳定性 ⋯ 14/13
    1.4.2　Rapidity　快速性 ⋯ 14/15
    1.4.3　Accuracy　准确性 ⋯ 16/15

Chapter2　Mathematical Models of Control Systems　控制系统的数学模型 ⋯ 18
  2.1　Differential Equations of Control Systems　控制系统的微分方程 ⋯ 20/21
    2.1.1　Linear systems and nonlinear systems　线性系统与非线性系统 ⋯ 20/21
    2.1.2　The establishment of differential equations　微分方程的建立 ⋯ 24/25
    2.1.3　Solution of differential equations　微分方程的求解 ⋯ 28/29
  2.2　The Laplace Transform　拉普拉斯变换 ⋯ 30/31
    2.2.1　Definition of the Laplace transform　拉普拉斯变换的定义 ⋯ 30/31
    2.2.2　Laplace transforms of typical functions　典型函数的拉普拉斯变换 ⋯ 30/31
    2.2.3　Fundamental theorems of the Laplace transform　拉普拉斯变换的基本定理 ⋯ 32/33
    2.2.4　The inverse Laplace transform　拉普拉斯反变换 ⋯ 38/39
  2.3　The Transfer Function of Control Systems　控制系统的传递函数 ⋯ 42/45
    2.3.1　Definition of the transfer function　传递函数的定义 ⋯ 44/45
    2.3.2　Characteristics of the transfer function　传递函数的特点 ⋯ 44/45
    2.3.3　Forms of the transfer function　传递函数的形式 ⋯ 46/47
  2.4　Transfer Functions of Typical Elements　典型环节的传递函数 ⋯ 48/47
    2.4.1　Proportion element　比例环节 ⋯ 48/49
    2.4.2　Differential element　微分环节 ⋯ 48/49
    2.4.3　Integral element　积分环节 ⋯ 52/55
    2.4.4　Inertial element　惯性环节 ⋯ 54/55

  2.4.5 First-order differential element 一阶微分环节 ·················· 56/57
  2.4.6 Second-order oscillation element 二阶振荡环节 ·················· 56/57
  2.4.7 Second-order differential element 二阶微分环节 ·················· 58/61
  2.4.8 Delay element 延时环节 ·· 60/61
 2.5 Function Block Diagrams of Control Systems 控制系统的框图模型 ······ 60/63
  2.5.1 Basic connection modes of control systems 控制系统的基本连接方式 ······ 62/63
  2.5.2 The closed-loop control system under disturbances 扰动作用下的闭环控制系统 ······ 66/67
  2.5.3 The drawing of block diagrams 框图的绘制 ···················· 68/69
  2.5.4 The simplification of block diagrams 框图的简化 ················ 70/69
 2.6 Mathematical Models of Typical Control Systems 典型控制系统的数学模型 ······ 74/75
  2.6.1 Mechanical systems 机械系统 ···································· 74/77
  2.6.2 Electrical systems 电气系统 ····································· 80/81

Chapter3 Time-Domain Analysis of Control Systems 控制系统的时域分析 ······ 85

 3.1 Transient Response of Control Systems 控制系统的瞬态响应 ··········· 86/87
  3.1.1 Typical input signals 典型输入信号 ······························· 86/87
  3.1.2 Time response 时间响应 ··· 92/93
  3.1.3 Time-domain performance indicators of control systems 控制系统的时域性能指标 ······ 94/95
 3.2 Mathematical Models and Time Response of First-Order Systems 一阶系统的数学模型和时间
   响应 ··· 98/97
  3.2.1 Mathematical models of first-order systems 一阶系统的数学模型 ······ 98/97
  3.2.2 Time response of first-order systems 一阶系统的时间响应 ········· 98/99
 3.3 Mathematical Models and Time Response of Second-Order Systems 二阶系统的数学模型和时间
   响应 ·· 106/107
  3.3.1 Mathematical models of second-order systems 二阶系统的数学模型 ······ 108/107
  3.3.2 Time response of second-order systems 二阶系统的时间响应 ······· 110/111
  3.3.3 Time-domain performance indicators of second-order systems 二阶系统的时域性能
    指标 ·· 116/119
 3.4 Time Response and Performance Analysis of Higher-Order Systems 高阶系统的时间响应与性能
   分析 ·· 126/127
  3.4.1 Time response of higher-order systems 高阶系统的时间响应 ········ 126/127
  3.4.2 Performance analysis of higher-order systems 高阶系统的性能分析 ···· 130/131

Chapter4 Frequency Characteristic Analysis of Control Systems 控制系统的频率
   特性分析 ··· 138

 4.1 Basic Concepts of Frequency Characteristics 频率特性的基本概念 ········ 140/141
  4.1.1 Frequency response 频率响应 ··································· 140/141
  4.1.2 Frequency characteristics and solving methods 频率特性及其求取方法 ··· 142/143
  4.1.3 Graphical representation of frequency characteristics 频率特性的图形表示法 ··· 146/145
 4.2 Frequency Characteristics of Typical Elements 典型环节的频率特性 ······ 148/147
  4.2.1 Proportion element 比例环节 ···································· 148/147
  4.2.2 Integral element and differential element 积分环节和微分环节 ········ 148/149

  4.2.3 Inertial element and first-order differential element 惯性环节和一阶微分环节 ·············· 150/151

  4.2.4 Oscillation element and second-order differential element 振荡环节和二阶微分环节 ·········· 154/157

  4.2.5 Time delay element 延时环节 ·· 160/161

 4.3 Open-loop Frequency Characteristics of Control Systems 控制系统的开环频率特性 ················ 160/161

  4.3.1 The open-loop polar plot of control systems 控制系统的开环极坐标图 ···························· 160/161

  4.3.2 The open-loop Bode plot of control systems 控制系统的开环伯德图 ································ 174/175

 4.4 Closed-loop Frequency Characteristics of Control Systems 控制系统的闭环频率特性 ············· 182/185

  4.4.1 Closed-loop frequency characteristics of control systems 控制系统的闭环频率特性 ·········· 182/185

  4.4.2 Estimation of closed-loop frequency characteristics using open-loop frequency characteristics

    利用开环频率特性估计闭环频率特性 ··· 182/185

  4.4.3 Closed-loop frequency domain performance indicators of control systems 控制系统的闭环

    频域性能指标 ·· 186/189

Chapter5 Performance Analysis of Control Systems 控制系统的性能分析 ·················· 193

 5.1 Stability Analysis of Control Systems 控制系统的稳定性分析 ·· 194/195

  5.1.1 Introduction 概述 ··· 194/195

  5.1.2 The algebraic stability criterion 代数稳定性判据 ··· 198/199

  5.1.3 The stability criterion of control systems in the frequency domain 控制系统的频域稳定性

    判据 ··· 210/211

  5.1.4 Relative stability analysis of control systems 控制系统的相对稳定性分析 ····················· 222/223

 5.2 Error Analysis of Control Systems 控制系统的误差分析 ··· 232/235

  5.2.1 System error and error transfer function 系统误差和误差传递函数 ························· 234/235

  5.2.2 Analysis and calculation of system static errors 系统静态误差分析与计算 ···················· 240/239

  5.2.3 Analysis and calculation of system dynamic errors 系统动态误差分析与计算 ·············· 248/251

  5.2.4 The relationship between open-loop characteristics and steady-state error 开环特性与稳态

    误差的关系 ··· 254/255

 5.3 Dynamic Performance Analysis of Control Systems 控制系统的动态性能分析 ······················ 258/259

  5.3.1 The relationship between dynamic performance and open-loop frequency characteristics

    动态性能与开环频率特性的关系 ·· 258/259

  5.3.2 The relationship between dynamic performance and closed-loop frequency characteristics

    动态性能与闭环频率特性的关系 ·· 262/262

Chapter6 Comprehensive Compensation of Control Systems 控制系统的综合校正 ······· 269

 6.1 Introduction 概述 ··· 270/271

 6.2 Cascade Compensation of Control Systems 控制系统的串联校正 ······································· 272/271

  6.2.1 Phase-lead compensation 相位超前校正 ·· 272/273

  6.2.2 Phase-lag compensation 相位滞后校正 ··· 280/281

  6.2.3 Phase lag-lead compensation 相位滞后-超前校正 ·· 288/287

  6.2.4 PID compensation PID 校正 ··· 294/293

 6.3 Parallel Compensation of Control Systems 控制系统的并联校正 ·· 300/299

  6.3.1 Feedback compensation 反馈校正 ·· 300/299

  6.3.2 Compound compensation 复合校正 ··· 302/301

Chapter 7 Analysis and Compensation of Control Systems based on MATLAB 基于 MATLAB 的控制系统分析与校正 ………………………………… /305

7.1 Introduction 概述 …………………………………………………………… 306/307
 7.1.1 Basic operations and commands 基本操作及命令 ………………………… 306/307
 7.1.2 MATLAB functions MATLAB 函数 ……………………………………… 310/311
 7.1.3 Plotting response curves 绘制响应曲线 ………………………………… 310/311
7.2 Description of Mathematical Models based on MATLAB 基于 MATLAB 的数学模型描述 …… 316/317
 7.2.1 Description of mathematical models of continuous systems 连续系统数学模型的描述 …… 316/317
 7.2.2 Transformations between various mathematical models 各种数学模型之间的转换 ……… 324/325
 7.2.3 Control system modeling based on MATLAB 基于 MATLAB 的控制系统建模 ………… 326/325
 7.2.4 Modeling method based on Simulink 基于 Simulink 的建模方法 ……………… 328/327
7.3 Performance Analysis of Control Systems based on MATLAB 基于 MATLAB 的控制系统性能分析 …………………………………………………………………… 330/329
 7.3.1 Time-domain analysis of control systems 控制系统的时域分析 ……………… 330/329
 7.3.2 Frequency-domain analysis of control systems 控制系统的频域分析 …………… 336/337
 7.3.3 Stability analysis of control systems 控制系统的稳定性分析 ………………… 340/341
7.4 Compensation of Control Systems based on MATLAB 基于 MATLAB 的控制系统校正 …… 346/345

Appendix A Exercises 课后练习 ……………………………………………………… 351

References 参考文献 ……………………………………………………………… 368

Chapter 1   Introduction to Electromechanical Control Engineering

第1章　机电控制工程概论

1.1 Introduction

Electromechanical control engineering is an emerging technical discipline, which is about the basis of control theory and mainly focuses on the theory of automatic control with application to engineering. The basic knowledge of control engineering is necessary for lots of fields for example, CNC machine tools, industrial robots, etc. The so-called automatic control means that certain physical variables of the production process or the controlled object are accurately changed based on the expected rule without anyone being directly involved. For example, in order to make the generator supply power run smoothly in the production process of a power plant, without being affected by load changes and fluctuations of the prime mover's speed, it is necessary to keep its output voltage constant; when the food factory produces cooked food, it must strictly control the furnace temperature of the oven according to the processing requirements; in the machining process, the parts can be produced with high-precision only if the position of the machine table and the tool holder accurately follow the feeding command. All of these systems have one thing in common that they are controlled by physical variables that vary according to the changes of given variables, which could be specific physical variables (like voltage, displacement, angle, etc) or digital variables. In general, the basic task of control systems is to make the controlled variables change along with the given variables.

Based on the content of control theory and the different development stages, there are two main types of control theory: one is classical control theory, and the other is modern control theory.

Classical control theory is a type of single-loop linear control theory, which is only suitable for single-input and single-output control systems. Its main research object is single variable linear systems with constant coefficients. The mathematical model of the system is simple, and the basic analysis method is based on the frequency method and the graphic method. The research objects, mathematical methods and calculation methods of the classical control theory are closely related to the contemporary needs of the society and level of technological development. Nevertheless, classical control theory promoted the development and popularization of automation technology, and it is applied in many engineering and technical fields nowadays as well.

Modern control theory focuses on the analysis and design of control systems regarding to multiple inputs, multiple-outputs, variable parameters, nonlinearity, high-precision, high-efficiency, and so on. Its main content is on the basis of the state-space approach, and theories of optimal control, optimal filtering, system identification, and adaptive control are its main research topics in the field. In addition, due to the fast development of computer technology and modern applied mathematics in recent years, modern control theory has made great progress in large system theory and artificial intelligence control.

Throughout the development history of electromechanical control engineering, it is closely related to the development of control theory, computer technology, modern applied mathematics and so on. At present, control theory has been constantly developing along with fuzzy mathematics, genetic

1.1 概述

机电控制工程是一门新兴学科，它以控制理论为基础，聚焦有关自动控制的理论及其在工程中的应用。例如，数控机床、工业机器人等都需要用到控制工程的基础知识。所谓自动控制，是在没有人直接参与的情况下，使生产过程或被控对象的某些物理量准确地按照预期的规律变化。例如，在发电厂的生产过程中，要想使发电机不受负载变化和原动机转速波动的影响而正常供电，就必须保持其输出电压恒定；食品厂在生产熟食时，必须按照加工要求严格控制烘炉的温度；在机械加工的过程中，只有机床工作台和刀架的位置准确地跟随进给指令，才能加工出高精度的零件。所有这些系统都有一个共同点，即其中都有一些被控制的物理量，它们按照给定量的变化而变化，给定量可以是具体的物理量，如电压、位移、角度等，也可以是数字量。一般来说，如何使被控量按照给定量的变化规律而变化，这就是控制系统所要解决的基本问题。

根据控制理论的内容和发展的不同阶段，控制理论可分为"经典控制理论"和"现代控制理论"两大阶段。

经典控制理论是一种单回路线性控制理论，只适用于单输入单输出控制系统。主要研究对象是单变量常系数线性系统，系统的数学模型简单，分析方法是频率法和图解法。经典控制理论的研究对象、数学方法和计算手段与当时的社会需要和技术水平密切相关。尽管如此，经典控制理论不仅推动了当时自动化技术的发展和普及，还在今天许多工程和技术领域中继续得到应用。

现代控制理论聚焦于多输入、多输出、变参数、非线性、高精度、高效能的控制系统的分析和设计。它的主要内容是以状态空间法为基础，最优控制、最佳滤波、系统辨识、自适应控制等理论都是这一领域的主要研究主题。特别是近年来，计算机技术和现代应用数学迅速发展，使现代控制理论在大系统理论和人工智能控制等方面取得了很大发展。

纵观控制工程的发展历程，它是与控制理论、计算机技术、现代应用数学等的发展息息相关的。目前，控制理论正在与模糊数学、遗传算法、神经网络等学科的交叉渗透中不断向前发展着。

1.2 控制系统的结构

控制系统一般由被控对象和控制器两部分组成。其中，被控对象是指系统中需要加以控制的机器、设备或生产过程；控制器是指能够对被控对象进行控制的设备的集合。控制系统的任务就是使生产过程或生产设备中的某些物理量保持恒定，或者让它们按照一定的规律变化。为完成控制系统的分析和设计，首先必须对控制系统结构有明确的了解。一般，可将控制系统分为两种基本形式：开环控制系统和闭环（反馈）控制系统。

1.2.1 开环控制系统

若控制系统的输出端和输入端之间没有反馈回路，输出量对系统的控制作用没有影响，即在控制器和控制对象间只有正向控制作用的系统称为开环控制系统，如图1-1所示。

algorithms, neural networks and other disciplines.

1.2 Structures of Control Systems

A control system generally has two parts: the controlled object and the controller. The controlled object is the machine, equipment or production process which needs to be controlled in the system; the controller is an organized collection of devices that can control the controlled object. The task of the control system is to keep certain physical variables stable in production process or production equipment constant, or to change them according to the given rules. In order to complete the analysis and design of the control system, we must have a clear understanding of the control system structure firstly. In general, control systems can be divided into two basic forms: open-loop control systems and closed-loop (or feedback) control systems.

1.2.1 Open-loop control systems

When there is no feedback loop between the output and the input of the control system, and the output has no effect on the control of the system, i.e., there is only forward control between the controller and the controlled object, the system is called open-loop control system (see Fig. 1-1).

For example, as is shown in Fig. 1-2, a manually controlled thermostat is a typical open-loop control system. In the system, the thermostat is the controlled object. The thermostat temperature is the controlled variable, and it is expected to be kept within the acceptable deviation. The temperature control process of the thermostat is described as follows: according to the temperature value required in the thermostat (i.e., the given value of the controlled variable), the voltage regulator is adjusted so that the resistive heater is heated to achieve the intended temperature. Unfortunately, it is an inaccurate control system. The temperature in the thermostat will deviate from the original value (i.e., the given value) with interference. Moreover, the deviation will not be corrected by the system once the allowable deviation is exceeded.

In summary, open-loop control systems get output values related to input values. If the output value is deviated from the original one due to some interference, the system does not have the ability to automatically correct the deviation. If the deviation is to be compensated, the input value has to be changed manually. Obviously, the accuracy of the open-loop control system is not high. However, if the component characteristics and parameter values of the system are relatively stable and the external interference is relatively small, the control system accuracy could be ensured. Therefore, the greatest advantage of the open-loop control system is the simplicity of its structure. In general, it works stably and reliably, and can be applied to systems with low accuracy requirements.

1.2.2 Closed-loop control systems

When there is a feedback loop between the output and the input of the control system, and the output has a direct impact on the control of the system, the system is called closed-loop control sys-

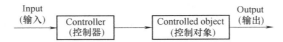

Fig. 1-1　Open-loop controlled system（开环控制系统）

如图 1-2 所示，人工控制的恒温箱是一个典型的开环控制系统。在这个控制系统中，恒温箱是被控对象，恒温箱内的温度是被控量，控制要求是温度保持在允许的偏差范围内。恒温箱温度控制过程如下：根据要求的温度值（被控制量的给定值）调节调压器，使加热电阻丝发热，以达到预定温度。但这是个不精确的控制系统，在干扰作用下，恒温箱内的温度将偏离原定值（给定值），且一旦超过允许的范围，系统将无法纠正偏差。

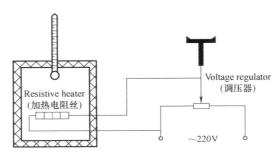

Fig. 1-2　A manually controlled thermostat
（人工控制的恒温箱）

综上，开环控制系统用一定输入值产生一定输出值，若某种干扰作用使输出值偏离原始值，这种系统没有自动纠正偏差的能力。如果要进行补偿，则必须借助人工改变输入值。显然，开环控制系统的精度不高。但是如果组成系统的元件特性和参数值比较稳定，且外界干扰比较小，则这种控制系统也可以保证一定的精度。开环控制系统最大的优点是其结构比较简单，一般都能稳定可靠地工作，适用于对精度要求不高的系统。

1.2.2　闭环控制系统

若控制系统的输出端和输入端之间有反馈回路，输出量对系统的控制作用有直接影响，这种系统称为闭环控制系统。这里，闭环的作用就是利用反馈来减少偏差，因此，反馈控制系统必定是闭环系统。闭环控制系统在控制器和被控对象之间不仅存在正向作用，还存在反向作用。将检测出来的输出量送回到系统的输入端，并与输入信号比较，称为反馈。因此，闭环控制又称为反馈控制，其结构如图 1-3 所示。其中，⊗表示比较器，→表示信号流的方向。在该结构下，控制器和控制对象共同构成了系统的前向通道，而反馈装置构成了系统的反馈通道。

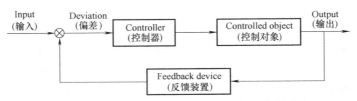

Fig. 1-3　Closed-loop controlled system（闭环控制系统）

tem. The role of closed loop is to reduce the deviation by using feedback. Thus, a feedback control system must be a closed-loop system. In closed-loop control system, there is not only forward control between the controller and the controlled object, but also backward control effect between them. The detected output signal is sent back to the input of the system and compared with the input signal, which is called feedback. Therefore, the closed-loop control is also called the feedback control, and its structure is shown in Fig. 1-3. In the figure, ⊗ represents a comparator, and→indicates the direction of signal flow. In such a framework, the controller and controlled object constitute the forward channel of the system, and the feedback device constitutes the feedback channel of the system.

In terms of control system, the concept of feedback is very important. In the system as is shown in Fig. 1-3, if the actual output signal obtained by the feedback device is subtracted from the input signal, and the resulting value is fed into the controller to control the controlled object, the process is called negative feedback. Conversely, if the input signal and the feedback signal are added as the input of the controller, it is called positive feedback.

In fact, a system with positive feedback generally cannot improve the system performance, but tends to deteriorate the performance of the system. So it is rarely used. A system with negative feedback automatically corrects the deviation so that the output of the system reaches a desired value. Moreover, it suppresses the effects of internal disturbances and external disturbances in the system loop, and finally achieves the purpose of automatic control. Feedback control usually refers to negative feedback control.

Compared to the open-loop control system, the most important characteristic of the closed-loop control system is that it can detect and correct deviations. Firstly, from the perspective of system structure, the closed-loop system has a backward channel such as feedback. Secondly, from the perspective of system function, the control precision of the system is higher due to the existence of feedback element. At the same time, it can better improve the dynamic performance of the system.

Of course, if an inappropriate feedback is introduced (such as positive feedback, or unreasonable parameter selection), it could make a stable system become unstable, instead of improving the system performance.

Although there are different forms of feedback control in actual systems, they generally can be depicted as the form shown in Fig. 1-3. To illustrate the working principle of a closed-loop control system, an example about an automatic control system of a thermostat is given as follows.

The automatic control system in the thermostat is shown in Fig. 1-4. The task of the system is to overcome external disturbances (like voltage fluctuations, external environmental changes, etc.) and keep the temperature of the thermostat constant. In the automatic control system, the temperature of the thermostat is given by the voltage signal u_1. When the temperature changes due to external conditions, the thermocouple, as a measuring element, converts the temperature into the corresponding voltage signal u_2. The voltage signal u_2 is fed back and compared with u_1, and the difference value is the temperature deviation signal $\Delta u = u_1 - u_2$. After the deviation signal is amplified by voltage amplifier and power amplifier, it is used to change the rotation speed and rotation direction

对控制系统而言，反馈的概念非常重要。在如图 1-3 所示系统中，如果将反馈装置获得的实际输出信号从输入信号中减去，并将其差值输入到控制器中去控制被控对象，则称这样的反馈为负反馈；反之，若由输入量和反馈量相加作为控制器的输入，则称为正反馈。

实际上，具有正反馈形式的系统一般是不能改进系统性能的，而且容易使系统的性能变差，因此很少采用。而具有负反馈形式的系统，它通过自动修正偏差值，使系统输出趋向于给定值，并抑制系统回路中存在的内扰和外扰的影响，最终达到自动控制的目的。通常反馈控制就是指负反馈控制。

与开环控制系统相比较，闭环控制系统最重要的特征是具有检测偏差、纠正偏差的能力。首先，从系统结构上看，闭环系统具有反向通道，即反馈；其次，从系统功能上看，由于反馈环节的存在，系统的控制精度更高；同时，可以较好地改善系统的动态性能。

当然，如果引入不适当的反馈（如正反馈，或者参数选择不合理），则达不到改善系统性能的目的，甚至会导致一个稳定系统变为不稳定系统。

实际系统虽然有不同形式的反馈控制，但一般均可简化为如图 1-3 所示的形式。下面以恒温箱自动控制系统为例说明闭环控制系统的工作原理。

恒温箱自动控制系统如图 1-4 所示，系统的任务是克服外部干扰（电源电压波动、外部环境变化等），使恒温箱内的温度保持恒定。在这一自动控制系统中，恒温箱的温度是由电压信号 u_1 给定，当外界条件引起箱内温度变化时，作为测量元件的热电偶把温度转换成对应的电压信号 u_2，电压信号 u_2 作为反馈信号而与 u_1 相比较，该差值即为温度偏差信号 $\Delta u = u_1 - u_2$。偏差信号经过电压、功率放大后，用以改变电动机的转速和转动方向，并通过传动装置拖动调压器的调节触头。当温度偏高时，动触头向着减小电流的方向运动；反之，加大电流，直到温度达到给定值为止。偏差信号 $\Delta u = 0$ 时，电动机停转，此时控制调节过程结束，系统达到新的稳定状态。由于干扰因素是经常出现的，因而调控过程是不断进行的。

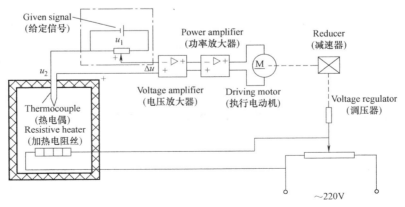

Fig. 1-4　The automatic control system of a thermostat
（恒温箱自动控制系统）

从上述恒温箱工作过程的分析中可以看出，自动控制系统就是要先检测偏差，再用检测

of the motor, and trigger the moving contact of the voltage regulator assisted by the transmission device. When the temperature is high, the moving contact moves towards the direction of decreasing the current; otherwise, the current is increased until the temperature reaches the given value. When the deviation signal $\Delta u = 0$, the motor stops. At this time, the control adjustment process ends, and the system reaches a new stable state. Since interference factors often occur, the control process is carried out continuously.

According to above analysis of the working process of the thermostat, it can be seen that the automatic control system is to detect the deviation firstly, and then use the detected deviation to correct the deviation. Therefore, if there is no deviation, there is no control process.

In a control system, the given value is the input of the system, and the controlled variable is the output of the system. The backward process of the output is feedback process, which means that part or all of the signals from the output are returned to the input assisted by a measuring device and compared with the input. The result of the comparison is called deviation. In an automatic control system, the deviation is generated by the controller through comparison and calculation. Therefore, the principle of "detecting deviation to correct deviation" based on feedback is also called the feedback control principle; a system based on feedback control principle is called feedback control system.

For example, the automatic control system of the temperature in the thermostat can be represented by the block diagram of the control system shown in Fig. 1-5. The basic principle of feedback control system can be observed from the figure, the output of each element is controlled by the input. It is worth noting that while there may be a variety of devices for achieving automatic control, the principle of feedback control can be the same. Feedback control is the most basic method to achieve the goal of automatic control.

Besides the open-loop control system and the closed-loop control system described above, there is also a type of semi-closed loop control system. If the feedback signal of the control system is not taken directly from the output of the system, but indirectly from an intermediate measuring element, the system is called a semi-closed loop control system. For example, in the feed servo system of CNC machine, if the position detection device is mounted at the end of the drive screw, the actual displacement of the table is obtained indirectly. It is a semi-closed loop control system. The semi-closed loop control system can achieve higher control accuracy than the open-loop control system, but lower control accuracy than the closed-loop control system.

1.2.3 The composition of feedback control systems

The above closed-loop system is only the basic form of the closed-loop control system. In order to obtain the desired control effect, other related elements should be added. A typical feedback control system is shown in Fig. 1-6. It should include a given element, a feedback element, comparators, amplifying elements, an actuating element, a controlled object, compensating elements, etc.

1. Given element

A given element is used to generate a given signal or input signal, such as a given potentiome-

到的偏差纠正偏差。因此,可以说,若没有偏差的存在,就没有控制过程。

在控制系统中,给定值即为系统的输入量,被控量为系统的输出量。输出的返回过程即为反馈过程,它表示输出量通过测量装置将信号的部分或全部返回输入端,使之与输入量进行比较,比较的结果称为偏差。在自动控制系统中,偏差是由控制器进行比较、计算产生的。因此,基于反馈的"检测偏差用以纠正偏差"的原理又称为反馈控制原理;利用反馈控制原理组成的系统称为反馈控制系统。

例如,恒温箱温度的自动控制系统可由如图1-5所示的框图表示。从图1-5中可以看到反馈控制系统的基本原理,每个环节的输出都受到输入的控制。值得注意的是,虽然实现自动控制的装置可能多种多样,但反馈控制的原理却可以是相同的,反馈控制是实现自动控制最基本的方法。

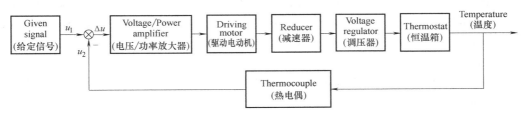

Fig. 1-5 The block diagram of automatic control system of a thermostat
(恒温箱温度自动控制系统框图)

除上述的开环控制系统和闭环控制系统外,还有半闭环控制系统。如果控制系统的反馈信号不是直接从系统的输出端引出,而是间接地从中间的测量元件获得,则这种系统称为半闭环控制系统。例如,在数控机床的进给伺服系统中,若将位置检测装置安装在传动丝杠的端部,工作台的实际位移为间接测量获得的,其为半闭环控制系统。半闭环控制系统可以获得比开环控制系统更高的控制精度,但比闭环控制系统的控制精度低。

1.2.3 反馈控制系统的组成

上述的闭环系统,只是闭环控制系统的基本组成形式,要想获得理想的控制效果,还应增加其他有关元件。一个典型的反馈控制系统如图1-6所示,它应该包括给定元件、反馈元件、比较元件、放大元件、执行元件、控制对象及校正元件等。

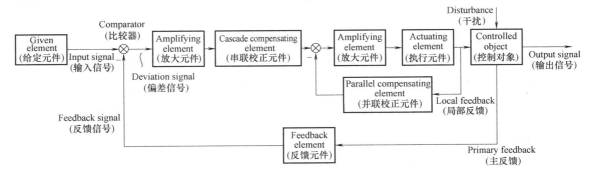

Fig. 1-6 Block diagram of a typical closed-loop control system
(典型闭环控制系统框图)

ter in a speed control system.

2. Feedback element

A feedback element is used to measure the controlled variable or output to generate a feedback signal, and there is a defined functional relationship between the signal and the output. Generally, for the convenience of transmission, the feedback signal is usually an electrical signal.

3. Comparing element (or comparator)

A comparing element (or a comparator) is used to compare the deviation of the feedback signal from the input signal. It can be a physical comparator (like a rotating transformer) or a differential circuit.

4. Amplifying element

An amplifying element is used to amplify a deviation signal, such as a servo power amplifier, a electro-hydraulic servo valve, etc. The output of an amplifying element must be energetic enough, so that it can drive the actuators and achieve the control function.

5. Actuating element

An actuating element is an element that directly operates on the controlled object, such as an electric motor, a hydraulic motor, etc.

6. Controlled object

A controlled object is the object that the control system manipulates, of which the output is the controlled variable of the system, such as a machine table.

7. Compensating element

A compensating element is an element that makes the system keep good static and dynamic performance to ensure the control quality. The compensating element has two forms: cascade compensating element and parallel compensating element.

Except for the controlled object, the above-mentioned given element, the feedback element, the comparing element, the amplifying element, the actuating element and the compensating element constitute the control part of the control system jointly. So, the control system is composed of two parts: the control part and the controlled object.

1.3 The Classification of Control Systems

There are many types of control systems. In order to study the problem conveniently, they are often classified according to their structural characteristics, input signal characteristics, realization ways, etc. The common classifications are introduced as follows.

1.3.1 The classification in terms of the characteristics of input signals

1. Constant value control system

The input of such a system is a constant value. The basic task of the system is to eliminate the influence of various disturbance factors, so that the controlled variable maintains the desired value with a certain precision. For example, control systems that require constant values such as speed,

1. 给定元件
给定元件用于产生给定信号或输入信号，如调速系统中的给定电位计。

2. 反馈元件
反馈元件用于测量被控量或输出量来产生反馈信号，该信号与输出量之间存在确定的函数关系。一般来说，为便于传输，反馈信号通常为电信号。

3. 比较元件（比较器）
比较元件（比较器）用来比较输入信号与反馈信号之间的偏差，它可以是物理比较元件（如旋转变压器），也可以是差接电路。

4. 放大元件
放大元件用来对偏差信号进行放大，如伺服功率放大器、电液伺服阀等。放大元件的输出一定要有足够的能量，这样才能驱动执行元件、实现控制功能。

5. 执行元件
执行元件是直接对控制对象进行操作的元件，如执行电动机、液压马达等。

6. 控制对象
控制对象是控制系统操纵的对象，它的输出量是系统的被控量，如机床工作台。

7. 校正元件
校正元件是为保证控制质量，使系统具有良好的静、动态性能的元件。校正元件有串联校正和并联校正两种形式。

除被控对象外，上述的给定元件、反馈元件、比较元件、放大元件、执行元件、校正元件一起组成了控制系统的控制部分。因此，控制系统是由控制部分和被控对象两大部分组成的。

1.3 控制系统的分类

控制系统种类繁多，为便于研究问题，常按照它们的结构特征、输入信号特征、实现方式等内容进行分类。下面列举常见的分类方式。

1.3.1 按输入信号的特征分类

1. 恒值控制系统
这类系统的输入量是一个恒值，系统的基本任务是排除各种干扰因素的影响，使被控制量以一定精度保持期望值。例如，工业生产中要求速度、压力、流量等数值恒定的控制系统都属恒值控制系统。

2. 伺服跟踪系统
这类系统的输入是变化的，且随时间的变化规律不能预先确定。当输入量发生变化时，要求输出量排除各种干扰因素的影响，快速、平稳地响应变化，准确地重现输入信号的变化规律。伺服跟踪系统的一个例子是武器装备中的火炮自动跟踪系统。

3. 程序控制系统
这类系统的输入量是变化的，且随时间的变化规律是能预先确定的。系统预先将输入的变化规律编写为程序，由该程序发出控制指令，然后，控制指令将在输入装置中被转换成控

pressure, and flow rate in industrial production are constant value control systems.

2. Servo tracking system

The input of such a system is variable, and its changing law with time cannot be predetermined. When the input value changes, the output value is required to eliminate the influence of various disturbance factors, respond to the change rapidly and smoothly, and reproduce the changing law of the input signal accurately. An example of servo tracking system is the artillery automatic tracking system in weaponry.

3. Program control system

The input of such a system is variable, and its changing law with time can be predetermined. The system pre-programs the changing law of the input into a program, and the control command is issued by the program. Then, the control command is converted into a control signal in the input device, and the signal is fed into the controller to let the controlled object work in accordance with the requirements of the command.

1.3.2 The classification in terms of realization ways of controllers

1. Continuous analog control system

The signals of continuous analog control systems continuously change. Most closed-loop control systems, such as centrifugal governors and hydraulic servo systems, fall into this category. Continuous control systems can be further divided into two categories: linear systems and nonlinear systems. A system that can be described by linear differential equations is called linear system; a system that cannot be described by linear differential equations, or has nonlinear components, is called nonlinear system.

2. Discrete digital control system

The signal composition in discrete digital control system is generally complex, and it may include continuous analog signals, discrete signals, digital signals, etc. Among these signals, the digital signal plays a direct role in the control process.

1.3.3 The classification in terms of error

A closed-loop control system does control according to the deviation. If the controlled variable of a system deviates from the steady-state value and generates an error due to disturbance factors, the system will perform a dynamic adjustment process in which the deviation is detected and eliminated to reach a steady state. Therefore, according to the difference between the controlled variable and the expected value, the closed-loop control system can be divided into the following two forms.

1. Non-error system

If a control system is stabilized by dynamic adjusting, and the controlled variable can be restored to the original value, which means the controlled variable is equal to the expected value and the error is zero, the system is called a non-error system.

2. Error system

If a control system is adjusted by dynamic adjusting, and the controlled variable can be close to

制信号,并被输入到控制器中,使被控对象按照指令的要求进行工作。

1.3.2 按控制器的实现方式分类

1. 连续模拟式控制系统

连续模拟式控制系统中的信号为连续变化的模拟信号。离心调速器、液压伺服系统等大多数的闭环控制系统都属于此类。连续控制系统可以进一步分为线性系统和非线性系统两大类,其中,能用线性微分方程描述的系统称为线性系统;不能用线性微分方程描述、存在非线性部件的系统称为非线性系统。

2. 离散数字式控制系统

离散数字式控制系统中的信号成分一般比较复杂,可能包含连续模拟信号、离散信号、数字信号等。这些信号中,数字信号在控制过程中起直接作用。

1.3.3 按误差分类

闭环控制系统是按偏差进行调节的。若系统的被控量受到干扰因素的影响,偏离稳态值而产生误差,那么系统就会通过检测偏差、纠正偏差的动态调节过程再次达到稳定状态。因此,闭环控制系统按照被控量与期望值相比的差值,可分为以下两种形式。

1. 无差系统

若控制系统通过动态调节保持稳定,且被控量能恢复原值,即被控量与期望值一样,误差为零,则称这种系统为无差系统。

2. 有差系统

若控制系统能够进行动态调节,但被控量可以接近却不能恢复原值,即被控量与期望值之间存在误差,则称这种系统为有差系统。能否消除偏差,取决于闭环控制系统的结构和参数。

此外,控制系统还有多种分类方式。例如,按照控制方式,可分为开环系统、闭环系统和半闭环系统;按照稳定性,可分为稳定系统和不稳定系统;按照系统的数学描述,可分为线性系统和非线性系统;按照系统部件的物理属性,可分为机械控制系统、电气控制系统、机电控制系统、液压控制系统、气动控制系统等。

1.4 控制系统的基本要求

不同的控制系统,由于工作场合及要完成的任务的差异,其性能指标也各不相同。但对所有的控制系统来说,控制目标是一致的。简言之,就是系统的被控量应能迅速、准确地跟踪给定量(或希望值)的变化。被控量与给定值之间应具有确定的函数关系,并且这种关系应最大程度地不受各种干扰因素的影响。具体来说,控制系统应满足稳定性、快速性及准确性三方面的要求。

1.4.1 稳定性

为了确保控制系统能正常工作,稳定性是必须具备的先决条件。一般情况下,系统的输出量在没有外作用时处在某一稳定状态,当系统受到外作用(输入量或扰动量)后,其输

but not reach the original value, which means there is an error between the controlled variable and the expected value, the system is called an error system. Whether the deviation can be eliminated or not depends on the structure and parameters of the closed-loop control system.

Besides, there are many other classification methods for control systems. For example, according to the control mode, they can be classified into open-loop systems, closed-loop systems and semi-closed loop systems; according to the stability, they can be classified into stable systems and unstable systems; according to the mathematical description of the system, they can be classified into linear systems and nonlinear systems; according to the physical properties of system components, they can be classified into mechanical control systems, electrical control systems, electromechanical control systems, hydraulic control systems, pneumatic control systems, etc.

1.4 Basic Requirements of Control Systems

Due to differences in their working situations and tasks to be completed, different control systems have different performance indicators. However, in any control system, the goals to be achieved by the control process should be the same. In short, it is required that the controlled variable of a system should be able to track the change of a given value (or desired value) quickly and accurately. There should be a certain functional relationship between the controlled variable and the given value, and the relationship ought to avoid being affected by various disturbance factors to the greatest extent. Specifically, a control system should meet the requirements of stability, rapidity and accuracy.

1.4.1 Stability

In order to ensure a control system to work well, stability is the first need to be met. In general, the output of the system is in a stable state when there is no external influence. Once the system is affected by external influence, such as input or disturbance, its output will deviate from the original steady state. If the output of the system can converge with time after deviates from the steady state and return to an equilibrium state, the system is a stable system, as is shown in see Fig. 1-7.

Conversely, if the output of the system cannot return to an equilibrium state but is in continuous oscillation or divergence state, the system is unstable, as shown in Fig. 1-8. Since an unstable system can never reach a stable state, instead of working well, it cannot complete its control tasks, and it may even destroy equipment and cause accidents.

1.4.2 Rapidity

For an stable system, rapidity refers to the speed of which the deviation is eliminated when there is a deviation between the output and the given input. It takes a certain amount of time for the system to move from one equilibrium state to another. In other words, there exists a transition process. Therefore, rapidity has two meanings: one is the rapidity of the transition process, which is expressed as the speed of which the output tracks the input after the input is applied; the second

出量将偏离原来的稳定状态。若系统输出量在偏离稳定状态后，能随着时间收敛，并重新回到平衡状态，那么该系统是稳定的，如图 1-7 所示。

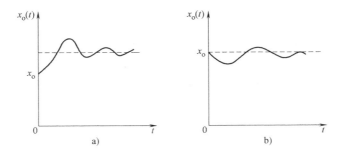

Fig. 1-7　The time response curve of a stable system（稳定系统的时间响应曲线）
　　a）The response curve under the input variable（输入量作用下的响应曲线）
　　b）The response curve under the disturbance（扰动作用下的响应曲线）

反之，若系统的输出量不能重新回到平衡状态，而呈持续振荡或发散振荡状态，则系统是不稳定的，如图 1-8 所示。由于不稳定系统无法进入稳定状态、不能正常工作，因此也不能完成控制任务，甚至会毁坏设备、造成事故。

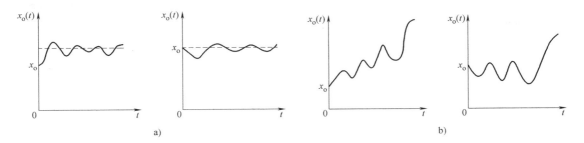

Fig. 1-8　The time response curve of an unstable system（不稳定系统的时间响应曲线）
　　a）Persistent oscillation（等幅振荡）　b）Divergent oscillation（发散振荡）

1.4.2　快速性

对于稳定系统快速性是指当系统的输出量与给定的输入量之间存在偏差时，消除这种偏差的快慢程度。系统从一个平衡状态过渡到另一个平衡状态都需要一定的时间，换言之，存在一个过渡过程。因此，快速性包含两方面的含义：一是过渡过程的快速性，它表现为施加输入量后，输出量跟随输入量变化的快慢程度；二是过渡过程的平稳性，它表现为瞬态响应过程结束后，输出量跟随输入量变化的快慢程度。

1.4.3　准确性

准确性是指系统响应的动态过程结束后，其被控量与期望值之间的差值，即稳态误差的大小，它反映系统稳态精度的高低。稳态精度也是衡量系统性能的一个重要指标。对控制系统来说，稳态精度越高越好，即系统的稳态输出越接近期望值越好，稳态误差越小越好。

is the smoothness of the transition process, which is expressed as the speed of which the output follows the input after the end of the transient response process.

1.4.3 Accuracy

Accuracy refers to the difference between the controlled variable and the desired value after the dynamic process of the system response As known as the magnitude of the steady-state error, it reflects the steady-state accuracy of the system. The steady-state accuracy is also an important indicator of system performance. For a control system, the higher the steady-state accuracy is, the better it is. In other words, the closer the steady-state output of the system is to the desired value, and the smaller the steady-state error is, the better.

Due to the differences among the controlled objects, different control systems have different requirements for the three aspects mentioned above. For example, program control systems have high requirements for the accuracy of the response; servo systems have high requirements for rapidity; constant temperature systems have strict requirements for stability. Furthermore, the requirements of these three aspects are often contradictory in the same control system. Increasing the rapidity may cause the system to oscillate strongly and make the stability lower; on the contrary, if the stability is highlighted excessively, it may affect the rapidity and accuracy of the system. Therefore, when we design a control system, it is necessary to make overall considerations in terms of the requirements of the specific controlled object.

由于控制对象间存在差异，不同的控制系统对"稳、快、准"三个方面的要求也各不相同。例如，程序控制系统对响应的准确性要求较高，伺服系统对快速性要求较高，而恒温系统对稳定性的要求又更为严格。另外，这三个方面的要求体现在同一个控制系统中时，常常是相互矛盾的。提高快速性，可能引起系统强烈振荡，而降低稳定性；相反，若过分讲究稳定性，又有可能影响系统的快速性和准确性。因此，在设计控制系统时，需要根据具体被控对象提出的要求，统筹兼顾，有所侧重。

Chapter 2　Mathematical Models of Control Systems

第2章　控制系统的数学模型

In order to theoretically analyze the performance of control systems, mathematical models of the systems should be established. The so-called mathematical model refers to the mathematical expression that describes the relations among the variables of a system. It reveals the inherent relationships among structure, parameters and dynamic performance of the system. The mathematical models of the system can take many forms, depending on the choice of variables and coordinates. In the time domain, the form of differential equations is usually used; in the complex number field, the form of the transfer function is used; in the frequency domain, the form of the frequency characteristic is used.

It should be pointed out that the establishment of a reasonable mathematical model is extremely important for the analysis of the system. Whether it is a mechanical, electrical system, or other systems, mathematical models can be used to describe its motion characteristics as long as it is a deterministic system. However, it is not an easy task to establish a reasonable mathematical model of the system, which requires a sufficient understanding of the construction principles and working conditions of the system. In engineering, to establish a relatively reasonable and accurate mathematical model, it is often necessary to make some assumptions and simplifications, ignoring the factors that have little influence on system characteristics and linearizing some nonlinear relations.

2.1 Differential Equations of Control Systems

A differential equation is a mathematical model that describes the dynamics of a system in the time domain. Meanwhile, the other forms of mathematical models that describe the dynamics of the system can be obtained by using the differential equation. If the differential equation of the system can be solved, the dynamic process of the system's output change with time can be obtained.

2.1.1 Linear systems and nonlinear systems

1. Linear systems

When the mathematical model of a system can be described by a linear differential equation, the system is called a linear system. If the coefficients of the differential equation are constant, the system is called a linear time-invariant system; if the coefficients of the differential equation are functions of time, the system is called a linear time-variant system.

An important characteristic of a linear system is that the superposition principle can be used. The superposition principle includes superposition and homogeneity. The so-called superposition means that the total output response $c(t)$ of multiple input signals applied on a linear system is equal to the algebraic sum of the output response ($c_1(t)$, $c_2(t)$, $\cdots$, $c_n(t)$) of each input signal applied on the linear system separately, i.e., $c(t) = c_1(t) + c_2(t) + \cdots + c_n(t)$. The so-called homogeneity means that if the output response of the input signal $r(t)$ applied on a linear system is $c(t)$, the output response of the linear system becomes $k \cdot c(t)$ when a signal $k \cdot r(t)$ is input. (k is a constant). In short, when the system has multiple inputs simultaneously, each input can be considered separately to get the corresponding output response, and then these responses are su-

为了从理论上对控制系统进行性能分析,应建立系统的数学模型。所谓数学模型,是指描述系统内各变量之间关系的数学表达式,它揭示了系统结构、参数与其动态性能之间的内在关系。系统的数学模型有多种形式,这取决于变量与坐标的选择。在时间域,通常采用微分方程或微分方程组的形式;在复数域,则采用传递函数形式;在频率域,则采用频率特性形式。

应当指出,建立合理的数学模型,对于系统的分析极为重要。无论是机械、电气系统,还是其他系统,只要是确定的系统,都可以用数学模型描述其运动特性。但是,要建立一个系统的合理的数学模型并非易事,这需要对系统的构造原理、工作情况有足够的了解。在工程上,为建立一个比较合理而准确的数学模型,常常做一些必要的假设和简化,忽略对系统特性影响小的因素,并对一些非线性关系进行线性化。

2.1 控制系统的微分方程

微分方程是在时域中描述系统动态特性的数学模型。同时,可以利用微分方程得到描述系统动态特性的其他形式的数学模型。如果能对系统的微分方程加以求解,则可以得到系统的输出随时间变化的动态过程。

2.1.1 线性系统与非线性系统

1. 线性系统

当系统的数学模型能用线性微分方程描述时,该系统称为线性系统。如果微分方程的系数为常数,则该系统称为线性定常系统;如果微分方程的系数为时间的函数,则称为线性时变系统。

线性系统的重要特性之一是可以运用叠加原理。叠加原理包括叠加性和齐次性。所谓叠加性是指作用于线性系统的多个输入信号的总输出响应 $c(t)$ 等于各个输入信号单独作用时产生的输出响应 $c_1(t)$,$c_2(t)$,…,$c_n(t)$ 的代数和,即 $c(t)=c_1(t)+c_2(t)+\cdots+c_n(t)$。所谓齐次性是指若输入信号 $r(t)$ 作用于线性系统引起的输出响应为 $c(t)$,则在 $kr(t)$ 作用下,该线性系统的输出响应变为 $kc(t)$(k 为常数)。简言之,当系统同时有多个输入时,可以对每个输入分别进行考虑以得到相应的输出响应,然后将这些响应叠加起来,就得到系统的最终输出响应。

线性定常系统除满足叠加原理外,还有一个重要的特性,就是系统对某输入信号的导数的时间响应等于该输入信号时间响应的导数。根据这一特点,在测试系统时,可以用一种输入信号推断出相应信号的响应结果,而线性时变系统和非线性系统都不具备这种特性。

2. 非线性系统

当系统的数学模型能用非线性微分方程描述时,该系统就称为非线性系统。虽然许多物理系统常以线性方程来表示,但是在大多数情况下,实际的函数关系并不是线性的。事实上,对物理系统仔细研究后可以发现系统或元件都有不同程度的非线性,即输入与输出之间的关系不是线性的。下面是一些在机电系统中常见的非线性特性的例子。

(1) 传动间隙非线性 在机床进给传动系统(主要由齿轮及丝杠副组成)中,经常存在传动间隙 Δ(如图 2-1 所示),它会使输入转角 x_i 和输出位移 x_o 间有滞环关系。若把传动

perimposed to obtain the final output response of the system.

In addition to satisfying the superposition principle, a linear constant system has another important characteristic, which is the system's time domain response to the derivative of an input signal is equal to the derivative of the time domain response of the input signal. According to this characteristic, the response results of several corresponding signals can be inferred from one signal input when testing a system. However, neither a linear time-variant system nor nonlinear system has such characteristic.

2. Nonlinear system

When the mathematical model of a system can be described by nonlinear differential equations, the system is called a nonlinear system. Although many physical systems are often represented by linear equations, in most cases the actual functional relationship is not linear. In fact, after studying the physical system carefully, it can be seen that any system or component has different degrees of nonlinearity, that is, the relation between input and output is not linear. Some examples of the nonlinear characteristics commonly found in electromechanical systems are as follows.

(1) Transmission clearance nonlinearity In feed driving systems (mainly consisting of gears and lead screw pairs) of a machine tool, there often is a transmission clearance Δ (as is shown in Fig. 2-1), which results in a hysteresis relation between the input angle x_i and output displacement x_o. If the transmission clearance is eliminated, there is a linear relation between x_i and x_o.

(2) Dead zone nonlinearity In the dead zone, there is an input but no output of the system, as shown in Fig. 2-2. An Example of the dead zone is a hydraulic servo valve.

(3) Friction nonlinearity Sliding pairs, such as the kinematic pair in a machine tool sliding guide, the kinematic pair in a spindle quill, the kinematic pair in a piston hydraulic cylinder and etc., have friction in motion. Assuming that the friction is dry friction, the friction nonlinearity is depicted as shown in Fig. 2-3a. The magnitude of the friction is f, and its direction is always opposite to the direction of speed. Assuming that the friction is viscous friction, the friction nonlinearity is depicted as shown in Fig. 2-3b.

(4) Saturation nonlinearity Under the input of a large signal, the output of a component reaches saturation, as shown in Fig. 2-4.

(5) Square-law nonlinearity There is a square-law relation between the input and output of the component, as in shown in Fig. 2-5. For example, the flow and the pressure of a servo valve in a hydraulic system has a square-law nonlinearity relation in the tiny opening condition.

Since the superposition principle is not applicable to a nonlinear system, the solving process is usually very complicated when the problem involves a nonlinear system. Therefore, it is necessary to deal with the nonlinear system properly to avoid the mathematic difficulty caused by the system.

If a system has severe nonlinear properties near an operating point, such as discontinuity, jump, broken line, non-single-value relation etc., it is called an inherently nonlinear system. In order to obtain a linear equation, these properties have to be neglected when we establish a mathematical model, and an approximate solution can be obtained consequently. If the error caused by the neglect is large, we can only use complex nonlinear processing methods to deal with the system.

间隙消除，则 x_i 和 x_o 间有线性关系。

（2）死区非线性 在死区范围内，系统有输入而无输出，如图 2-2 所示。例如，液压伺服阀就具有死区。

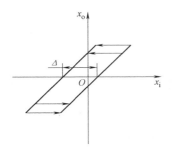
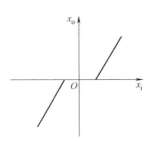

Fig. 2-1　Transmission clearance nonlinearity
（传动间隙非线性）

Fig. 2-2　Dead zone nonlinearity
（死区非线性）

（3）摩擦力非线性 滑动副，如机床滑动导轨中的运动副，主轴套筒中的运动副，活塞液压缸中的运动副等，在运动中都存在摩擦力。若为干摩擦力，则非线性特性如图 2-3a 所示，其大小为 f，方向总是与速度的方向相反；若为黏性摩擦力，则非线性特性如图 2-3b 所示。

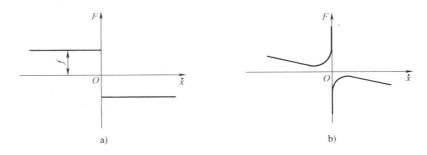

Fig. 2-3　Friction nonlinearity（摩擦力非线性）
a）Dry friction（干摩擦力）　b）Viscous friction（黏性摩擦力）

（4）饱和非线性 在大输入信号作用下，元件的输出量达到饱和，如图 2-4 所示。

（5）平方律非线性 元件的输入与输出之间存在平方律关系，如图 2-5 所示。例如，液压系统中伺服阀在微小开口条件下流量与压力之间就存在平方律非线性关系。

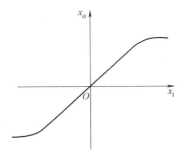
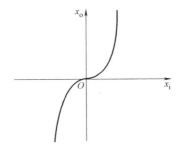

Fig. 2-4　Saturation nonlinearity（饱和非线性）

Fig. 2-5　Square law nonlinearity（平方律非线性）

If a system has others nonlinear properties that are not severe nonlinear properties as described above, it is called a non-inherently nonlinear system. For this type of nonlinear system, it can be linearized by the tangent method near the operating point. This linearization will not result in a significant error when the variable is slightly changed near its operating point. A nonlinear system can be described by a linear differential equation after linearizing. And then general linear methods can be used to analyze and design the system.

2.1.2 The establishment of differential equations

Assuming that the input is $x_i(t)$ and the output $x_o(t)$ is for a linear time-invariant system, a differential equation describing the input-output relation of the system is

$$a_n x_o^{(n)}(t) + a_{n-1} x_o^{(n-1)}(t) + \cdots + a_1 \dot{x}_o(t) + a_0 x_o(t)$$
$$= b_m x_i^{(m)}(t) + b_{m-1} x_i^{(m-1)}(t) + \cdots + b_1 \dot{x}_i(t) + b_0 x_i(t) \tag{2-1}$$

where a_0, a_1, $\cdots$, a_n and b_0, b_1, $\cdots$, b_m are coefficients that are dependent of the system structure and its parameters ($n \geq m$).

When the input signal becomes zero (i.e., zero input state) during the working process, the system is in the state of free motion, and the changing law of its output signal is called the free motion mode of the system. It characterizes the inherent features of the system. Meanwhile, the equation of free motion for the system is a homogeneous differential equation, i.e.

$$a_n x_o^{(n)}(t) + a_{n-1} x_o^{(n-1)}(t) + \cdots + a_1 \dot{x}_o(t) + a_0 x_o(t) = 0 \tag{2-2}$$

The purpose of establishing the differential equation of the system is to determine the functional relationship between the output of the system and the given input or disturbance input of the system. As the system is composed of various components, the general steps for listing equations are as follows.

1) Analyze the working principle of the system and the process of signal transmission to determine the input and output of the system.

2) List the dynamic differential equations of each component in terms of the physical laws followed by each variable by ignoring some secondary factors and linearizing the nonlinear parts.

3) Eliminate the intermediate variables of the differential equations listed, and obtain the differential equations describing the relation between the input and output of the system.

4) Arrange the resulting differential equations. Generally, the items related to the output are placed on the left side of the equation, and the items related to the input are placed on the right side of the equation. At the same time, all-order derivatives are arranged in descending order.

1. Establishment of differential equations for linear systems

Example 2-1 As is shown in Fig. 2-6a, a power slide system regarding the mass, viscous damping and stiffness can be simplified as a mass-damping-spring system shown in Fig. 2-6b. Try to write down the differential equation of motion between the external force $f(t)$ and displacement $y(t)$.

Solution: The input of the system is force $f(t)$, and the output of the system is displacement $y(t)$. If the zero point of $y(t)$ is taken as the natural equilibrium position of the equivalent mass

由于叠加原理不适用于非线性系统,当问题涉及非线性系统时,求解过程通常非常复杂,因此为了避免由非线性系统造成的数学难关,需要对非线性系统进行适当处理。

如果一个系统在某个工作点附近具有严重的非线性特性,如不连续、跳跃、折线和非单值关系等,则该系统称为本质非线性系统。为得到线性方程,在建立数学模型时只能忽略这些性质,由此获得近似的解。若这种忽略带来的误差较大,就只能用复杂的非线性处理方法来求解。

如果一个系统具有其他非线性特性,而非上述严重非线性特性的,则该系统称为非本质非线性系统。对于这种非线性系统,可以在工作点附近用切线法进行线性化。当变量在其工作点附近发生轻微变化时,这种线性化不会产生显著误差。非线性系统经线性化处理后,就可以用线性微分方程描述,进而采用普通的线性方法来分析和设计系统。

2.1.2 微分方程的建立

设线性定常系统的输入为 $x_i(t)$,输出为 $x_o(t)$,则描述系统输入输出关系的微分方程为

$$a_n x_o^{(n)}(t) + a_{n-1} x_o^{(n-1)}(t) + \cdots + a_1 \dot{x}_o(t) + a_0 x_o(t) \\ = b_m x_i^{(m)}(t) + b_{m-1} x_i^{(m-1)}(t) + \cdots + b_1 \dot{x}_i(t) + b_0 x_i(t) \tag{2-1}$$

式中,a_0,a_1,$\cdots$,a_n 和 b_0,b_1,$\cdots$,b_m 为取决于系统结构及其参数的系数($n \geq m$)。

当系统在工作过程中输入信号变为零(即零输入状态)时,系统处于自由运动状态,其输出信号的变化规律称为系统的自由运动模态,它表征系统的固有特性。此时系统的自由运动方程为齐次微分方程,即

$$a_n x_o^{(n)}(t) + a_{n-1} x_o^{(n-1)}(t) + \cdots + a_1 \dot{x}_o(t) + a_0 x_o(t) = 0 \tag{2-2}$$

建立系统的微分方程,是为了确定系统的输出量与给定输入量或扰动输入量之间的函数关系。而系统是由各种元件组成的,因此建立方程的一般步骤如下。

1)分析系统的工作原理和信号传递的过程,确定系统的输入量和输出量。

2)通过忽略一些次要因素和线性化非线性项,再依据各变量所遵循的物理学定律,列出各元件的动态微分方程。

3)消除所列各微分方程的中间变量,得到描述系统的输入量与输出量之间关系的微分方程。

4)整理所得微分方程,一般将与输出量有关的各项放在方程的左侧,与输入量有关的各项放在方程的右侧,同时,将各阶导数项按降幂排列。

1. 线性系统微分方程的建立

例 2-1　图 2-6a 所示为考虑质量、黏性阻尼及刚度的动力滑台系统,其可简化成图 2-6b 所示的质量-阻尼-弹簧系统。试求外力 $f(t)$ 与位移 $y(t)$ 之间的运动微分方程。

解: 该系统的输入量为外力 $f(t)$,输出量为位移 $y(t)$,若取 $y(t)$ 的零点为等效质量 m 的自然平衡位置,应用牛顿第二定律,可列出系统原始运动方程为

$$m \frac{d^2 y(t)}{dt^2} = f(t) - c \frac{dy(t)}{dt} - k y(t) \tag{2-3}$$

式中,c 为等效阻尼系数;k 为等效弹簧刚度。式(2-3)可重写为

$$m \frac{d^2 y(t)}{dt^2} + c \frac{dy(t)}{dt} + k y(t) = f(t) \tag{2-4}$$

m, an original motion equation of the system according to Newton's second law can be defined as

$$m\frac{d^2y(t)}{dt^2}=f(t)-c\frac{dy(t)}{dt}-ky(t) \tag{2-3}$$

where c is an equivalent damping coefficient; k is an equivalent spring stiffness. Equation (2-3) can be rewritten as

$$m\frac{d^2y(t)}{dt^2}+c\frac{dy(t)}{dt}+ky(t)=f(t) \tag{2-4}$$

Therefore, the equation (2-4) is the motion differential equation of the system under the influence of external force $f(t)$.

Example 2-2 As is shown in Fig. 2-7, the vibration system consists of mass blocks and springs in series. The input is the external force $f(t)$, and the output is the displacement $y_1(t)$. Try to write down the dynamic equation of the system.

Solution: When m_2 and k_2 don't exist, the system shown in Fig. 2-7 is a single-degree-of-freedom system. Then, the dynamic equation between its input and output is:

$$m\ddot{y}_1(t)+k_1y_1(t)=f(t) \tag{2-5}$$

When m_2 and k_2 are connected to m_1 and k_1, a load effect is generated on m_1 and k_1. Then, the system becomes a two-degree-of-freedom system, and its dynamic equation is

$$\begin{cases} m_1\ddot{y}_1(t)+k_1y_1(t)+k_2[y_1(t)-y_2(t)]=f(t) \\ m_2\ddot{y}_2(t)+k_2y_2(t)=k_2y_1(t) \end{cases} \tag{2-6}$$

By eliminating $y_2(t)$ from the equation (2-5) and equation (2-6), the system dynamic equation with $f(t)$ as the input and $y_1(t)$ as the output is obtained as

$$m_1m_2y_1^{(4)}(t)+(m_1k_2+m_2k_1+m_2k_2)\ddot{y}_1(t)+k_1k_2y_1(t)=m_2\ddot{f}(t)+k_2f(t) \tag{2-7}$$

2. Linear approximations of nonlinear systems

For a non-inherently nonlinear system, a tangent method (or called a small deviation method) can be used to linearize it near the operating point. A system usually has a predetermined operating point when it works normally, where the system is in an equilibrium state. For an automatic regulating system or servo system, as long as the working state of the system deviates from the equilibrium position, the system will immediately respond to it and try to return to the original equilibrium position. Hence, the deviation of the system variables from the predetermined operating point is generally small. As long as each variable in the nonlinear function has a derivative or a partial derivative at the predetermined operating point, the nonlinear function can be expanded into Taylor series in the form of the deviation of its independent variables near the predetermined operating point. When the deviation is small, the higher-order terms of the deviation in the expansion can be ignored, and only one item is remained to achieve the linearization.

Assuming that the nonlinear characteristics of a system is as shown in Fig. 2-8, and its equation of motion is $y=f(x)$. In the system, $y(t)$ is the output and $x(t)$ is the input. If the predetermined operating point of the system is (x_0, y_0) and the system is continuously differentiable at the point, the nonlinear function $y=f(x)$ can be expanded into a Taylor series near the point, i. e.

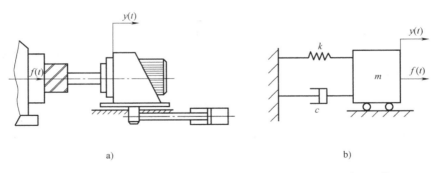

Fig. 2-6　The power slide and its mechanical model（动力滑台及其力学模型）

因此，式（2-4）即为该系统在外力 $f(t)$ 作用下的运动微分方程。

例 2-2　图 2-7 所示为质量块、弹簧串联而成的振动系统。输入为外力 $f(t)$，输出为位移 $y_1(t)$。试求该系统的动力学方程。

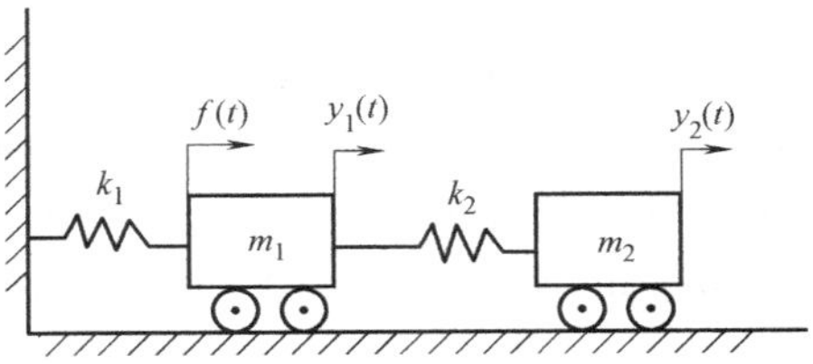

解： 当 m_2 与 k_2 不存在时，图 2-7 所示系统为单自由度系统，其输入与输出之间的动力学方程为

$$m\ddot{y}_1(t)+k_1 y_1(t)=f(t) \tag{2-5}$$

当 m_2 与 k_2 连接到 m_1 与 k_1 上时，便对 m_1 和 k_1 产生了负载效应，此时，系统变成两自由度系统，其动力学方程为

Fig. 2-7　The vibration system（振动系统）

$$\begin{cases} m_1\ddot{y}_1(t)+k_1 y_1(t)+k_2[y_1(t)-y_2(t)]=f(t) \\ m_2\ddot{y}_2(t)+k_2 y_2(t)=k_2 y_1(t) \end{cases} \tag{2-6}$$

从式（2-5）和式（2-6）中消去 $y_2(t)$，则得到的以 $f(t)$ 为输入，$y_1(t)$ 为输出的系统动力学方程为

$$m_1 m_2 y_1^{(4)}(t)+(m_1 k_2+m_2 k_1+m_2 k_2)\ddot{y}_1(t)+k_1 k_2 y_1(t)=m_2\ddot{f}(t)+k_2 f(t) \tag{2-7}$$

2. 非线性系统的线性近似

对于非本质非线性系统，可以在工作点附近用切线法（或称微小偏差法）进行线性化。系统正常工作时通常有一个预定工作点，即系统处于平衡状态的位置。对于自动调节系统或伺服系统，工作状态一旦偏离此平衡位置，系统就会立即做出响应，并试图恢复到原来的平衡位置，因此系统各变量偏离预定工作点的偏差一般很小。只要非线性函数的各变量在预定工作点处存在导数或偏导数，就可在预定工作点附近将此非线性函数以其自变量偏差的形式展成泰勒级数。在偏差很小时，展开式中偏差的高次项可以忽略，只剩下一次项，从而实现线性化。

设某系统的非线性特性如图 2-8 所示，其运动方程为 $y=f(x)$。在该系统中，$y(t)$ 为输出量，$x(t)$ 为输入量。如果系统预定工作点为 (x_0,y_0) 点，且在该点处连续可微，则在这点附近可把非线性函数 $y=f(x)$ 展开成泰勒级数，即

$$y = y_0 + \frac{dy}{dx}\bigg|_{x_0}(x-x_0) + \frac{1}{2!}\frac{d^2y}{dx^2}\bigg|_{x_0}(x-x_0)^2 + \cdots \qquad (2\text{-}8)$$

Since $x-x_0$ is small, the higher-order terms above the second-order in the equation (2-8) could be neglected and we have

$$y - y_0 = \frac{dy}{dx}\bigg|_{x_0}(x-x_0) \quad \text{or} \quad \Delta y = \frac{dy}{dx}\bigg|_{x_0} \cdot \Delta x \qquad (2\text{-}9)$$

Therefore, a linearized equation with increments as variables is obtained, and it is called an increment equation.

$\dfrac{dy}{dx}\bigg|_{x_0}$ is the derivative of the function $y=f(x)$ at the point (x_0, y_0), which is geometrically the tangent slope of the point. In fact, the linearization method is to take a tangent line instead of the curve line in a small range near the operating point, so it is called a tangent method. If the coordinate origin is taken at the equilibrium point, i.e. only to study the changes of the input and output relative to the equilibrium point, the initial condition of the system is equal to zero. It is convenient not only for solving the equation, but also for analyzing the control system.

Using the same method for multivariable nonlinear functions, i.e.

$$y = f(x_1, x_2, \cdots, x_n)$$

a linearized equation can be obtained near the operating point $(x_{10}, x_{20}, \cdots, x_{n0})$, i.e.

$$\Delta y = \frac{\partial y}{\partial x_1}\bigg|_{x_{10},x_{20},\cdots,x_{n0}} \cdot \Delta x_1 + \frac{\partial y}{\partial x_2}\bigg|_{x_{10},x_{20},\cdots,x_{n0}} \cdot \Delta x_2 + \cdots + \frac{\partial y}{\partial x_n}\bigg|_{x_{10},x_{20},\cdots,x_{n0}} \cdot \Delta x_n \qquad (2\text{-}10)$$

Finally, it must be pointed out that the following aspects should be noted in the linearization process.

1) The predetermined operating point of the system must be clarified, because the coefficients of the linearized equations obtained at different operating points are different.

2) The linearization means that a curve line is replaced with a straight line, neglecting the higher-order terms above the second-order in the Taylor series expansion, which is an approximation. If the systems input works within a large range, the established linearized mathematical model will generate a large error. Therefore, the linearization of a nonlinear model is conditional.

3) If the nonlinear function is discontinuous, it cannot be expanded into Taylor series near the discontinuous point, and then it cannot be linearized.

4) The linearized differential equation is an increment equation based on increments.

2.1.3 Solution of differential equations

After establishing the differential equation of a system, the implicit functional relationship between the output and the given input (or disturbance input) of the system can be determined. Solving the differential equation of the system directly to obtain the explicit function relationship of the output is the basic method of analyzing the system. The solution of the equation is the output response of the system. According to the expression of the equation solution, the dynamic characteristics of the system can be analyzed, and the response curve of the output that visually reflects the dynamic

$$y = y_0 + \frac{dy}{dx}\bigg|_{x_0}(x-x_0) + \frac{1}{2!}\frac{d^2y}{dx^2}\bigg|_{x_0}(x-x_0)^2 + \cdots \quad (2\text{-}8)$$

由于 $x-x_0$ 很小，略去式（2-8）中二阶以上的高阶项，得

$$y - y_0 = \frac{dy}{dx}\bigg|_{x_0}(x-x_0) \text{ 或 } \Delta y = \frac{dy}{dx}\bigg|_{x_0} \cdot \Delta x \quad (2\text{-}9)$$

这样就得到了一个以增量为变量的线性化方程，称为增量方程式。

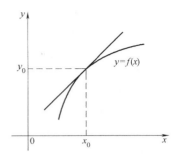

Fig. 2-8　The linearization diagram of tangent method
（切线法线性化示意图）

$\dfrac{dy}{dx}\bigg|_{x_0}$ 是函数 $y=f(x)$ 在点 (x_0,y_0) 处的导数，从几何意义上讲，就是该点的切线斜率。实际上，这种线性化方法就是在工作点附近的小范围内，用切线来代替曲线，故称为切线法。如果把坐标原点取在平衡点处，即只研究输入、输出相对于平衡点的变化，则系统的初始条件为零。这样不但便于求解方程式，而且便于分析控制系统。

采用同样的方法，对于多变量非线性函数

$$y = f(x_1, x_2, \cdots, x_n)$$

在工作点 $(x_{10}, x_{20}, \cdots, x_{n0})$ 附近，可以得到线性化方程为

$$\Delta y = \frac{\partial y}{\partial x_1}\bigg|_{x_{10},x_{20},\cdots,x_{n0}} \cdot \Delta x_1 + \frac{\partial y}{\partial x_2}\bigg|_{x_{10},x_{20},\cdots,x_{n0}} \cdot \Delta x_2 + \cdots + \frac{\partial y}{\partial x_n}\bigg|_{x_{10},x_{20},\cdots,x_{n0}} \cdot \Delta x_n \quad (2\text{-}10)$$

最后，必须指出的是，在线性化处理过程中应注意以下几个方面。

1）必须明确系统的预定工作点，因为在不同工作点处所得线性化方程的系数不同。

2）线性化意味着是以直线代替曲线，省略泰勒级数展开式中二阶以上的高阶项，是一种近似处理。如果系统输入量工作在较大范围内，那么所建立的线性化数学模型势必产生较大的误差。所以，非线性模型线性化是有条件的。

3）非线性函数若为不连续的，则在不连续点附近不能展开为泰勒级数，因而不能线性化。

4）线性化后的微分方程是一个以增量为基础的增量方程。

2.1.3　微分方程的求解

建立了系统的微分方程后，就可以确定系统的输出量与给定输入量（或扰动输入量）之间的隐函数关系。直接求解系统微分方程获得输出量的显函数关系是分析系统的基本方法。方程的解就是系统的输出响应。根据方程解的表达式，可以分析系统的动态特性，也可以绘出能直观地反映系统的动态过程的输出响应曲线。

但微分方程的一般求解过程较为繁琐。微分方程的解包含通解和特解两个部分。通解可由式（2-2）求得，它完全由初始条件生成，反映瞬态过程，工程上称为自然响应。特解只由输入决定，反映系统由输入引起的输出，工程上称为强迫响应。显然，一般求解方法复杂费时，而且难以直接利用微分方程来研究和判断系统的动态性能。由此可见，这种方法有很大的局限性。

此外，可以用拉普拉斯变换来求解线性微分方程。它可将经典数学中的微积分运算转化

process of the system can be drawn.

However, the general solving process of a differential equation is rather complexed. The solution of the differential equation has two parts: general solution and special solution. A general solution can be obtained by equation (2-2). It is completely generated by the initial condition, and reflects the transient process, which is called a natural response in engineering. A special solution is only determined by the input, and reflects the output caused by the input of the system, which is called a forced response in engineering. Obviously, the general solving method is complex and time-consuming. Moreover, it is difficult to directly study and speculate the dynamic performance of the system by using the differential equation. Therefore, the method has great limitations.

In addition, the Laplace transform can be used to solve linear differential equations. It can transform a calculus operation in classical mathematics into an algebraic operation, and directly introduce the influence of the initial condition so that the solution is the full solution. More importantly, through the Laplace transform, the differential equation of the system can be transformed into the transfer function of the system, and various methods based on the transfer function are developed to analyze and design the system.

2.2 The Laplace Transform

The Laplace transform is a mathematical transform in integral operations, and it is often used to solve differential equations in engineering. The complete theory of the Laplace transform is not described in detail in the following, and only some commonly used conclusions are listed as a tool to solve differential equations and for future reference.

2.2.1 Definition of the Laplace transform

Consider a function $f(t)$ with t as a real variable, and the following conditions are satisfied for the function.

1) When $t<0$, $f(t)=0$.
2) When $t \geqslant 0$, $f(t)$ is piecewise continuous at least.
3) An integral of function $f(t)$ as $\int_0^\infty f(t) e^{-st} dt$, exists and converges, where s is a complex variable, and $s = \sigma + j\omega$.

Then, the Laplace transform of function $f(t)$ is defines as

$$F(s) = L[f(t)] = \int_0^\infty f(t) e^{-st} dt \qquad (2\text{-}11)$$

where $f(t)$ is called a primitive function; $F(s)$ is called an image function; s is the Laplacian; L is the symbol for taking the Laplace transform.

2.2.2 Laplace transforms of typical functions

1. Unit step function

The mathematical expression of the unit step function is

为代数运算，并能够直接引入初始条件的影响，求出的解就是全解。更重要的是，通过拉氏变换，可以把系统的微分方程转化为系统的传递函数，并由此发展出基于传递函数来分析和设计系统的各种方法。

2.2 拉普拉斯变换

拉普拉斯变换（简称拉氏变换）是积分运算中的一种数学变换，工程上常利用它来求解微分方程。这里不对拉氏变换的全部理论进行详细描述，而是仅列举一些常用的结论来作为求解微分方程的工具以备查用。

2.2.1 拉普拉斯变换的定义

考虑函数 $f(t)$，其以 t 为实变量，且满足以下条件。

1) 当 $t<0$ 时，$f(t)=0$。
2) 当 $t \geqslant 0$ 时，$f(t)$ 至少是分段连续的。
3) 函数 $f(t)$ 的积分 $\int_0^\infty f(t)\mathrm{e}^{-st}\mathrm{d}t$ 存在且收敛，$s=\sigma+\mathrm{j}\omega$ 为复变量。

此时，则可定义函数 $f(t)$ 的拉氏变换为

$$F(s) = L[f(t)] = \int_0^\infty f(t)\mathrm{e}^{-st}\mathrm{d}t \tag{2-11}$$

式中，$f(t)$ 为原函数；$F(s)$ 为象函数；s 为拉普拉斯算子；L 表示拉氏变换的符号。

2.2.2 典型函数的拉普拉斯变换

1. 单位阶跃函数

单位阶跃函数的数学表达式为

$$1(t) = \begin{cases} 0 & (t<0) \\ 1 & (t \geqslant 0) \end{cases}$$

根据式（2-11），其拉氏变换为

$$F(s) = L[1(t)] = \int_0^\infty 1 \cdot \mathrm{e}^{-st}\mathrm{d}t = \frac{1}{s} \tag{2-12}$$

2. 单位脉冲函数

单位脉冲函数的表达式为

$$\delta(t) = \begin{cases} 0 & (t<0, t>\varepsilon) \\ \lim_{\varepsilon \to 0} \dfrac{1}{\varepsilon} & (0 \leqslant t \leqslant \varepsilon) \end{cases}$$

根据式（2-11），其拉氏变换为

$$F(s) = L[\delta(t)] = \lim_{\varepsilon \to 0}\int_0^\infty \frac{1}{\varepsilon} \cdot \mathrm{e}^{-st}\mathrm{d}t = 1 \tag{2-13}$$

3. 单位斜坡函数

单位斜坡函数的数学表达式为

$$1(t) = \begin{cases} 0 & (t<0) \\ 1 & (t \geq 0) \end{cases}$$

According to equation (2-11), its Laplace transform is

$$F(s) = L[1(t)] = \int_0^\infty 1 \cdot e^{-st} dt = \frac{1}{s} \qquad (2\text{-}12)$$

2. Unit pulse function

The mathematical expression of the unit pulse function is

$$\delta(t) = \begin{cases} 0 & (t<0, t>\varepsilon) \\ \lim_{\varepsilon \to 0} \dfrac{1}{\varepsilon} & (0 \leq t \leq \varepsilon) \end{cases}$$

According to equation (2-11), its Laplace transform is

$$F(s) = L[\delta(t)] = \lim_{\varepsilon \to 0} \int_0^\infty \frac{1}{\varepsilon} \cdot e^{-st} dt = 1 \qquad (2\text{-}13)$$

3. Unit ramp function

The mathematical expression of the unit ramp function is

$$f(t) = \begin{cases} 0 & (t<0) \\ t & (t \geq 0) \end{cases}$$

According to equation (2-11), its Laplace transform is

$$F(s) = L[f(t)] = \int_0^\infty t \cdot e^{-st} dt = \frac{1}{s^2} \qquad (2\text{-}14)$$

4. Acceleration function

The mathematical expression of the acceleration function is

$$f(t) = \begin{cases} 0 & (t<0) \\ \dfrac{1}{2}t^2 & (t \geq 0) \end{cases}$$

According to equation (2-11), its Laplace transform is

$$F(s) = L[f(t)] = \int_0^\infty \frac{1}{2}t^2 \cdot e^{-st} dt = \frac{1}{s^3} \qquad (2\text{-}15)$$

5. Exponential function

The mathematical expression of the exponential function is

$$f(t) = \begin{cases} 0 & (t<0) \\ e^{-at} & (t \geq 0) \end{cases}$$

According to equation (2-11), its Laplace transform is

$$F(s) = L[f(t)] = \int_0^\infty e^{-at} \cdot e^{-st} dt = \int_0^\infty e^{-(s+a)t} dt = \frac{1}{s+a} \qquad (2\text{-}16)$$

Laplace transforms and inverse Laplace transforms of commonly used functions are listed in Table 2-1.

2.2.3 Fundamental theorems of the Laplace transform

This section describes some of the fundamental theorems commonly used for solving differential

$$f(t) = \begin{cases} 0 & (t<0) \\ t & (t\geq 0) \end{cases}$$

根据式（2-11），其拉氏变换为

$$F(s) = L[f(t)] = \int_0^\infty t \cdot e^{-st} dt = \frac{1}{s^2} \tag{2-14}$$

4. 加速度函数

加速度函数的数学表达式为

$$f(t) = \begin{cases} 0 & (t<0) \\ \frac{1}{2}t^2 & (t\geq 0) \end{cases}$$

根据式（2-11），其拉氏变换为

$$F(s) = L[f(t)] = \int_0^\infty \frac{1}{2}t^2 \cdot e^{-st} dt = \frac{1}{s^3} \tag{2-15}$$

5. 指数函数

指数函数的数学表达式为

$$f(t) = \begin{cases} 0 & (t<0) \\ e^{-at} & (t\geq 0) \end{cases}$$

根据式（2-11），其拉氏变换为

$$F(s) = L[f(t)] = \int_0^\infty e^{-at} \cdot e^{-st} dt = \int_0^\infty e^{-(s+a)t} dt = \frac{1}{s+a} \tag{2-16}$$

常用函数的拉氏变换及拉氏反变换对照见表 2-1。

2.2.3 拉普拉斯变换的基本定理

本小节介绍在控制系统中基于拉氏变换求解微分方程时常用的一些基本定理。此处只简单介绍相关定理，不进行证明，详细的证明过程可查阅有关数学课程教材。

1. 叠加定理

如果 $F_1(s) = L[f_1(t)]$，$F_2(s) = L[f_2(t)]$，且 a、b 均为常数，则有

$$L[af_1(t) \pm bf_2(t)] = aL[f_1(t)] \pm bL[f_2(t)] = aF_1(s) \pm bF_2(s) \tag{2-17}$$

2. 微分定理

如果 $F(s) = L[f(t)]$，则有

$$L\left[\frac{df(t)}{dt}\right] = sF(s) - f(0)$$

$$L\left[\frac{d^2 f(t)}{dt^2}\right] = s^2 F(s) - sf(0) - f'(0)$$

$$\vdots$$

$$L\left[\frac{d^n f(t)}{dt^n}\right] = s^n F(s) - s^{n-1} f(0) - s^{n-2} f'(0) - \cdots - f^{(n-1)}(0)$$

equations in a control system based on the Laplace transform. The relevant theorems are introduced in brief, and no proof is given. The detailed proof process can be found in the relevant mathematical textbooks.

1. Superposition theorem

If $F_1(s) = L[f_1(t)]$, $F_2(s) = L[f_2(t)]$, a and b are constant, then

$$L[af_1(t) \pm bf_2(t)] = aL[f_1(t)] \pm bL[f_2(t)] = aF_1(s) \pm bF_2(s) \tag{2-17}$$

2. Differential theorem

If $F(s) = L[f(t)]$, then

$$L\left[\frac{df(t)}{dt}\right] = sF(s) - f(0)$$

$$L\left[\frac{d^2 f(t)}{dt^2}\right] = s^2 F(s) - sf(0) - f'(0)$$

$$\vdots$$

$$L\left[\frac{d^n f(t)}{dt^n}\right] = s^n F(s) - s^{n-1} f(0) - s^{n-2} f'(0) - \cdots - f^{(n-1)}(0)$$

When the initial conditions are equal to zero, which is all initial values for all-order derivatives of $f(t)$ and itself are zero at $t = 0$, the above expressions can be written as

$$\begin{cases} L\left[\dfrac{df(t)}{dt^n}\right] = sF(s) \\ L\left[\dfrac{d^2 f(t)}{dt^n}\right] = s^2 F(s) \\ \vdots \\ L\left[\dfrac{d^n f(t)}{dt^n}\right] = s^n F(s) \end{cases} \tag{2-18}$$

3. Integral theorem

If $F(s) = L[f(t)]$, then

$$L\left[\int f(t)\,dt\right] = \frac{1}{s} F(s) + \frac{1}{s} f^{(-1)}(0)$$

$$L\left[\iint f(t)\,dt^2\right] = \frac{1}{s^2} F(s) + \frac{1}{s^2} f^{(-1)}(0) + \frac{1}{s} f^{(-2)}(0)$$

$$\vdots$$

$$L\left[\underbrace{\int \cdots \int}_{n} f(t)\,dt^n\right] = \frac{1}{s^n} F(s) + \frac{1}{s^n} f^{(-1)}(0) + \cdots + \frac{1}{s} f^{(-n)}(0)$$

In the same way, when all initial values for all-order integrals of $f(t)$ and itself are zero at $t = 0$, the above expressions can be simplified as

Table 2-1 Table of Laplace transforms of commonly used functions（常用函数拉普拉斯变换表）

Number(编号)	Primitive function $f(t)$（原函数 $f(t)$）	Image function $F(s)$（象函数 $F(s)$）
1	$\delta(t)$	1
2	$1(t)$	$\dfrac{1}{s}$
3	e^{-at}	$\dfrac{1}{s+a}$
4	t^n	$\dfrac{n!}{s^{n+1}}$
5	te^{-at}	$\dfrac{1}{(s+a)^2}$
6	$t^n e^{-at}$	$\dfrac{n!}{(s+a)^{n+1}}$
7	$\sin\omega t$	$\dfrac{\omega}{s^2+\omega^2}$
8	$\cos\omega t$	$\dfrac{s}{s^2+\omega^2}$
9	$\dfrac{1}{b-a}(e^{-at}-e^{-bt})$	$\dfrac{1}{(s+a)(s+b)}$
10	$\dfrac{1}{b-a}(be^{-bt}-ae^{-at})$	$\dfrac{s}{(s+a)(s+b)}$
11	$\dfrac{1}{a}(1-e^{-at})$	$\dfrac{1}{s(s+a)}$
12	$\dfrac{1}{ab}\left[1+\dfrac{1}{a-b}(be^{-at}-ae^{-bt})\right]$	$\dfrac{1}{s(s+a)(s+b)}$
13	$e^{-at}\sin\omega t$	$\dfrac{\omega}{(s+a)^2+\omega^2}$
14	$e^{-at}\cos\omega t$	$\dfrac{s+a}{(s+a)^2+\omega^2}$
15	$\dfrac{1}{a^2}(e^{-at}+at-1)$	$\dfrac{1}{s^2(s+a)}$
16	$\dfrac{\omega_n}{\sqrt{1-\xi^2}}e^{-\xi\omega_n t}\sin\sqrt{1-\xi^2}\omega_n t$	$\dfrac{\omega_n^2}{s^2+2\xi\omega_n s+\omega_n^2}$ $(0<\xi<1)$
17	$\dfrac{-1}{\sqrt{1-\xi^2}}e^{-\xi\omega_n t}\sin(\sqrt{1-\xi^2}\omega_n t-\varphi)$ $\varphi=\arctan\dfrac{\sqrt{1-\xi^2}}{\xi}$	$\dfrac{s}{s^2+2\xi\omega_n s+\omega_n^2}$ $(0<\xi<1)$
18	$1-\dfrac{1}{\sqrt{1-\xi^2}}e^{-\xi\omega_n t}\sin(\sqrt{1-\xi^2}\omega_n t+\varphi)$ $\varphi=\arctan\dfrac{\sqrt{1-\xi^2}}{\xi}$	$\dfrac{\omega_n^2}{s(s^2+2\xi\omega_n s+\omega_n^2)}$ $(0<\xi<1)$

当初始条件为零时，即式中 $f(t)$ 及其各阶导数在 $t=0$ 时的值都为零，则以上各式可以写为

$$\begin{cases} L\left[\int f(t)\,\mathrm{d}t\right] = \dfrac{1}{s}F(s) \\ L\left[\int\int f(t)\,\mathrm{d}t^2\right] = \dfrac{1}{s^2}F(s) \\ \vdots \\ L\left[\underbrace{\int\cdots\int}_{n} f(t)\,\mathrm{d}t^n\right] = \dfrac{1}{s^n}F(s) \end{cases} \qquad (2\text{-}19)$$

4. Shifting theorem

If $F(s) = L[f(t)]$, then

$$L[\mathrm{e}^{-at}f(t)] = F(s+a) \qquad (2\text{-}20)$$

The shifting theorem states that multiplying the primitive function by an exponential function e^{-at} is equivalent to replacing the variable s with the displacement $s+a$ on the corresponding image function $F(s)$ in the complex domain.

5. Delay theorem

If a function $f_1(t)$ is obtained by the function $f(t)$ through moving a period of time τ, that is, $f_1(t) = f(t-\tau)$, and $F(s) = L[f(t)]$, then

$$L[f_1(t)] = L[f(t-\tau)] = \mathrm{e}^{-s\tau}F(s) \qquad (2\text{-}21)$$

The theorem states the Laplace transform of the delay function $f_1(t)$ is equivalent to that the result multiplying the Laplace transform of the primitive function $f(t)$ by an exponential function $\mathrm{e}^{-s\tau}$.

6. Time-proportional theorem

If $F(s) = L[f(t)]$, then

$$L\left[f\left(\dfrac{t}{a}\right)\right] = aF(as) \qquad (2\text{-}22)$$

This theorem states that if the time function $f(t)$ reduces to $f\left(\dfrac{t}{a}\right)$, that is, the primitive function narrows a times in the time plane, its Laplace transform amplifies a times in the complex plane. Conversely, if the primitive function widens, its Laplace transform shrinks, i. e.

$$L[f(at)] = \dfrac{1}{a}F\left(\dfrac{s}{a}\right) \qquad (2\text{-}23)$$

7. Final value theorem

If $F(s) = L[f(t)]$, and $\lim\limits_{t\to\infty} f(t)$ exists, then the steady-state value of the time function $f(t)$ at $t\to\infty$ is

$$\lim_{t\to\infty} f(t) = \lim_{s\to 0} sF(s) \qquad (2\text{-}24)$$

It is convenient to obtain the steady-state value of the dynamic process according to the theorem, but $\lim\limits_{t\to\infty} f(t)$ must exist, which is a restriction on the use of the final value theorem.

8. Initial value theorem

With the initial value theorem, the initial value of the dynamic process can easily be solved according to the image function. The expression of the initial theorem is

$$\begin{cases} L\left[\dfrac{\mathrm{d}f(t)}{\mathrm{d}t^n}\right] = sF(s) \\ L\left[\dfrac{\mathrm{d}^2 f(t)}{\mathrm{d}t^n}\right] = s^2 F(s) \\ \quad\vdots \\ L\left[\dfrac{\mathrm{d}^n f(t)}{\mathrm{d}t^n}\right] = s^n F(s) \end{cases} \tag{2-18}$$

3. 积分定理

如果 $F(s) = L[f(t)]$，则有

$$L\left[\int f(t)\,\mathrm{d}t\right] = \frac{1}{s}F(s) + \frac{1}{s}f^{(-1)}(0)$$

$$L\left[\iint f(t)\,\mathrm{d}t^2\right] = \frac{1}{s^2}F(s) + \frac{1}{s^2}f^{(-1)}(0) + \frac{1}{s}f^{(-2)}(0)$$

$$\vdots$$

$$L\left[\underbrace{\int\cdots\int}_{n} f(t)\,\mathrm{d}t^n\right] = \frac{1}{s^n}F(s) + \frac{1}{s^n}f^{(-1)}(0) + \cdots + \frac{1}{s}f^{(-n)}(0)$$

同样，当式中 $f(t)$ 及其各重积分在 $t=0$ 时的值都为零，则以上各式可以简化为

$$\begin{cases} L\left[\int f(t)\,\mathrm{d}t\right] = \dfrac{1}{s}F(s) \\ L\left[\iint f(t)\,\mathrm{d}t^2\right] = \dfrac{1}{s^2}F(s) \\ \quad\vdots \\ L\left[\underbrace{\int\cdots\int}_{n} f(t)\,\mathrm{d}t^n\right] = \dfrac{1}{s^n}F(s) \end{cases} \tag{2-19}$$

4. 位移定理

如果 $F(s) = L[f(t)]$，则有

$$L[\mathrm{e}^{-at}f(t)] = F(s+a) \tag{2-20}$$

位移定理说明，原函数乘以指数函数 e^{-at}，就等于对应的象函数在复数域中变量由 s 替换为 $s+a$。

5. 延迟定理

如果函数 $f_1(t)$ 为函数 $f(t)$ 在坐标系中右移了一段时间 τ 的结果，即 $f_1(t) = f(t-\tau)$，且 $F(s) = L[f(t)]$，则函数 $f_1(t)$ 的拉氏变换为

$$L[f_1(t)] = L[f(t-\tau)] = \mathrm{e}^{-s\tau}F(s) \tag{2-21}$$

这一定理说明延时函数 $f_1(t)$ 的拉氏变换等于原函数 $f(t)$ 的拉氏变换乘以指数函数 $\mathrm{e}^{-s\tau}$。

6. 时间比例定理

如果 $F(s) = L[f(t)]$，则有

$$L\left[f\left(\frac{t}{a}\right)\right] = aF(as) \tag{2-22}$$

$$\lim_{t \to 0} f(t) = \lim_{s \to \infty} sF(s) \quad (2\text{-}25)$$

9. Convolution theorem

If $F(s) = L[f(t)]$, $G(s) = L[g(t)]$, then

$$L\left[\int_0^t f(t-\tau)g(\tau)\,\mathrm{d}\tau\right] = F(s)G(s) \quad (2\text{-}26)$$

where $\int_0^t f(t-\tau)g(\tau)\,\mathrm{d}\tau = f(t) * g(t)$ is called the convolution of $f(t)$ and $g(t)$.

2.2.4 The inverse Laplace transform

The operation to obtain $F(s)$ from $f(t)$ through integration is called the Laplace transform. Inversely, if the image function $F(s)$ is known, then the operation to obtain the primitive function $f(t)$ corresponding to $F(s)$ is called the inverse Laplace transform. Its formula is

$$f(t) = L^{-1}[F(s)] = \frac{1}{2\pi\mathrm{j}}\int_{\sigma-\mathrm{j}\infty}^{\sigma+\mathrm{j}\infty} F(s)\mathrm{e}^{st}\,\mathrm{d}s \quad (2\text{-}27)$$

Since the above formula is a function of complex variable, it is difficult to calculate directly. In general, the formula is only used as a mathematical definition of the inverse Laplace transform. In practice, the following calculation method is often used: firstly, the image function $F(s)$ is decomposed into the sum of some simple rational fractions, and these fractions are usually typical functions; then the inverse transformation function (i.e. primitive function) can be found out by referring Table 2-1.

Assume that the general expression of $F(s)$ is

$$F(s) = \frac{B(s)}{A(s)} = \frac{b_0 s^m + b_1 s^{m-1} + \cdots + b_{m-1}s + b_m}{s^n + a_1 s^{n-1} + \cdots + a_{n-1}s + a_n} \quad (2\text{-}28)$$

where $a_1, a_2, \cdots, a_n$ and $b_1, b_2, \cdots, b_n$ are real coefficients; m and n are positive integers, and $m < n$.

The equation (2-28) is factorized and divided into the following two cases according to the root of its denominator $A(s)$.

1) When $A(s)$ has different roots, equation (2-28) can be written as

$$F(s) = \frac{B(s)}{A(s)} = \frac{C_1}{s+p_1} + \frac{C_2}{s+p_2} + \cdots + \frac{C_n}{s+p_n} \quad (2\text{-}29)$$

or

$$F(s) = \sum_{i=1}^n \frac{C_i}{s+p_i} \quad (2\text{-}30)$$

where the solution expression of each coefficient is

$$C_i = \lim_{s \to -p_i}(s+p_i) \cdot F(s) \quad (2\text{-}31)$$

or

$$C_i = \frac{B(s)}{A(s)} \cdot (s+p_i)\Big|_{s=-p_i} \quad (2\text{-}32)$$

After all coefficients are determined, according to the inverse Laplace transform, the primitive function can be obtained in the time domain, i.e.

这一定理说明，如果时间函数 $f(t)$ 收缩为 $f\left(\dfrac{t}{a}\right)$，即原函数在时域平面上缩窄 a 倍，则其拉氏变换在复数平面上按 a 倍放大；反之，如果原函数被展宽，则其拉氏变换被缩小，即

$$L[f(at)] = \frac{1}{a}F\left(\frac{s}{a}\right) \tag{2-23}$$

7. 终值定理

如果 $F(s)=L[f(t)]$，且 $\lim\limits_{t\to\infty}f(t)$ 存在，则时间函数 $f(t)$ 在 $t\to\infty$ 时的稳态值为

$$\lim_{t\to\infty}f(t) = \lim_{s\to 0}sF(s) \tag{2-24}$$

根据这一定理来求动态过程的稳态值是很方便的，但即 $\lim\limits_{t\to\infty}f(t)$ 必须存在，这是终值定理使用的限制条件。

8. 初值定理

应用初值定理，可以方便地通过象函数求解动态过程的初始值，其表达式为

$$\lim_{t\to 0}f(t) = \lim_{s\to\infty}sF(s) \tag{2-25}$$

9. 卷积定理

如果 $F(s)=L[f(t)]$，$G(s)=L[g(t)]$，则有

$$L\left[\int_0^t f(t-\tau)g(\tau)\mathrm{d}\tau\right] = F(s)G(s) \tag{2-26}$$

式中，$\int_0^t f(t-\tau)g(\tau)\mathrm{d}\tau = f(t)*g(t)$ 称为 $f(t)$ 和 $g(t)$ 的卷积。

2.2.4 拉普拉斯反变换

由 $f(t)$ 通过积分求解 $F(s)$ 的运算称为拉氏变换，反之，若已知象函数 $F(s)$，求 $F(s)$ 对应的原函数 $f(t)$ 的运算称为拉普拉斯反变换。它的运算公式为

$$f(t) = L^{-1}[F(s)] = \frac{1}{2\pi\mathrm{j}}\int_{\sigma-\mathrm{j}\infty}^{\sigma+\mathrm{j}\infty}F(s)\mathrm{e}^{st}\mathrm{d}s \tag{2-27}$$

由于上式是复变函数，很难直接计算，一般只作为拉氏反变换的数学定义式。而在实际中，常采用下述计算方法：先将象函数 $F(s)$ 分解为一些简单的有理分式之和，这些分式通常都是典型函数；然后由拉氏变换对照表 2-1 查出其反变换函数，即得到了原函数。

设 $F(s)$ 的一般表达式为

$$F(s) = \frac{B(s)}{A(s)} = \frac{b_0 s^m + b_1 s^{m-1} + \cdots + b_{m-1}s + b_m}{s^n + a_1 s^{n-1} + \cdots + a_{n-1}s + a_n} \tag{2-28}$$

式中，a_1，a_2，$\cdots$，a_n 及 b_1，b_2，$\cdots$，b_n 为实系数；m 及 n 为正整数，且 $m<n$。

将式 (2-28) 进行因式分解，根据其分母 $A(s)$ 的根，分为以下两种情况来讨论。

1) $A(s)$ 具有不同的根时，式 (2-28) 可展开为

$$F(s) = \frac{B(s)}{A(s)} = \frac{C_1}{s+p_1} + \frac{C_2}{s+p_2} + \cdots + \frac{C_n}{s+p_n} \tag{2-29}$$

或者

$$f(t) = L^{-1}[F(s)] = L^{-1}\left[\sum_{i=1}^{n} \frac{C_i}{s+p_i}\right] = \sum_{i=1}^{n} C_i e^{-p_i t} \tag{2-33}$$

Example 2-3 Try to figure out the inverse Laplace transform of $F(s) = \dfrac{s+5}{s^2+4s+3}$.

Solution: Expanding $F(s)$ into fractions, we have

$$F(s) = \frac{s+5}{(s+3)(s+1)} = \frac{C_1}{s+3} + \frac{C_2}{s+1}$$

Then according to equation (2-32), the coefficients C_1 and C_2 can be calculated as

$$C_1 = \lim_{s \to -3}(s+3) \cdot F(s) = \lim_{s \to -3}\frac{s+5}{s+1} = -1$$

$$C_2 = \lim_{s \to -1}(s+1) \cdot F(s) = \lim_{s \to -3}\frac{s+5}{s+3} = 2$$

Besides, each coefficient can also be obtained by solving the coefficient equation, i.e.

$$F(s) = \frac{s+5}{(s+3)(s+1)} = \frac{C_1}{s+3} + \frac{C_2}{s+1} = \frac{(C_1+C_2)s + C_1 + 3C_2}{(s+3)(s+1)}$$

The coefficient equations are

$$\begin{cases} C_1 + C_2 = 1 \\ C_1 + 3C_2 = 5 \end{cases}$$

Solving the equations, we have $C_1 = -1$, $C_2 = 2$, which are the same as the previous results.

Thus $F(s)$ can be written as

$$F(s) = \frac{s+5}{(s+3)(s+1)} = -\frac{1}{s+3} + \frac{2}{s+1}$$

Taking the inverse Laplace transform, we have

$$f(t) = 2e^{-t} - e^{-3t}$$

2) When $A(s)$ has repeated roots, equation (2-28) can be written as

$$F(s) = \frac{B(s)}{A(s)} = \frac{C_m}{(s+p_1)^m} + \frac{C_{m-1}}{(s+p_1)^{m-1}} + \cdots + \frac{C_1}{s+p_1} + \frac{C_{m+1}}{s+p_{m+1}} + \cdots + \frac{C_n}{s+p_n} \tag{2-34}$$

where $C_{m+1}, \cdots, C_n$ are the undetermined coefficients of different roots, which can be calculated according to equation (2-31) or (2-32); $C_1, \cdots, C_m$ are the coefficients of repeated roots, which can be calculated as

$$\begin{cases} C_m = \lim_{s \to -p_1}(s+p_1)^m \cdot F(s) \\ C_{m-1} = \lim_{s \to -p_1}\dfrac{d}{ds}[(s+p_1)^m \cdot F(s)] \\ \vdots \\ C_{m-j} = \dfrac{1}{j!}\lim_{s \to -p_1}\dfrac{d^j}{ds^j}[(s+p_1)^m \cdot F(s)] \\ \vdots \\ C_1 = \dfrac{1}{(m-1)!}\lim_{s \to -p_1}\dfrac{d^{(m-1)}}{ds^{(m-1)}}[(s+p_1)^m \cdot F(s)] \end{cases} \tag{2-35}$$

$$F(s) = \sum_{i=1}^{n} \frac{C_i}{s+p_i} \tag{2-30}$$

式中,各系数的求解公式为

$$C_i = \lim_{s \to -p_i} (s+p_i) \cdot F(s) \tag{2-31}$$

或者

$$C_i = \frac{B(s)}{A(s)} \cdot (s+p_i) \Big|_{s=-p_i} \tag{2-32}$$

在所有系数 C_i 都确定后,即可由拉氏反变换得到原函数的时域表达式,即

$$f(t) = L^{-1}[F(s)] = L^{-1}\left[\sum_{i=1}^{n} \frac{C_i}{s+p_i}\right] = \sum_{i=1}^{n} C_i e^{-p_i t} \tag{2-33}$$

例 2-3 求 $F(s) = \dfrac{s+5}{s^2+4s+3}$ 的反拉氏变换。

解:将 $F(s)$ 进行因式分解,可得

$$F(s) = \frac{s+5}{(s+3)(s+1)} = \frac{C_1}{s+3} + \frac{C_2}{s+1}$$

根据式(2-32),计算两个待定系数 C_1 和 C_2,有

$$C_1 = \lim_{s \to -3}(s+3) \cdot F(s) = \lim_{s \to -3}\frac{s+5}{s+1} = -1$$

$$C_2 = \lim_{s \to -1}(s+1) \cdot F(s) = \lim_{s \to -1}\frac{s+5}{s+3} = 2$$

除此之外,还可以用解系数方程的方法,求出各系数,即

$$F(s) = \frac{s+5}{(s+3)(s+1)} = \frac{C_1}{s+3} + \frac{C_2}{s+1} = \frac{(C_1+C_2)s+C_1+3C_2}{(s+3)(s+1)}$$

得到系数方程组为

$$\begin{cases} C_1+C_2 = 1 \\ C_1+3C_2 = 5 \end{cases}$$

解该方程组,得到系数为 $C_1 = -1$,$C_2 = 2$,与前面得到的结果相同。

因此,$F(s)$ 可写为

$$F(s) = \frac{s+5}{(s+3)(s+1)} = -\frac{1}{s+3} + \frac{2}{s+1}$$

进行拉氏反变换,得

$$f(t) = 2e^{-t} - e^{-3t}$$

2)$A(s)$ 有重根时,式(2-28)可展开为

$$F(s) = \frac{B(s)}{A(s)} = \frac{C_m}{(s+p_1)^m} + \frac{C_{m-1}}{(s+p_1)^{m-1}} + \cdots + \frac{C_1}{s+p_1} + \frac{C_{m+1}}{s+p_{m+1}} + \cdots + \frac{C_n}{s+p_n} \tag{2-34}$$

式中,$C_{m+1}, \cdots, C_n$ 为根不相同的项的待定系数,可按照式(2-31)或式(2-32)来计算;$C_1, \cdots, C_m$ 为重根各项的系数,计算方法为

After all coefficients are determined, the primitive function can be obtained in the time domain according to the inverse Laplace transform, i. e.

$$\begin{aligned} f(t) &= L^{-1}[F(s)] \\ &= L^{-1}\left[\frac{C_m}{(s+p_1)^m}+\frac{C_{m-1}}{(s+p_1)^{m-1}}+\cdots+\frac{C_1}{s+p_1}+\frac{C_{m+1}}{s+p_{m+1}}+\cdots+\frac{C_n}{s+p_n}\right] \\ &= \left[\frac{C_m}{(m-1)!}t^{m-1}+\frac{C_{m-1}}{(m-2)!}t^{m-2}+\cdots+C_2t+C_1\right]e^{-p_1t}+\sum_{i=m+1}^{n}C_ie^{-p_it} \end{aligned} \tag{2-36}$$

Example 2-4 Try to figure out the primitive function $f(t)$ of $F(s)=\dfrac{s+2}{s(s+1)^2(s+3)}$.

Solution: Expanding $F(s)$ into fraction, we have

$$F(s)=\frac{C_2}{(s+1)^2}+\frac{C_1}{s+1}+\frac{C_3}{s}+\frac{C_4}{s+3}$$

We should determine the four undetermined coefficients next. According to equation (2-31) and equation (2-35), the coefficients can be calculated as

$$C_3=\lim_{s\to 0}s\cdot F(s)=\lim_{s\to 0}\frac{s+2}{(s+1)^2(s+3)}=\frac{2}{3}$$

$$C_4=\lim_{s\to -3}(s+3)\cdot F(s)=\lim_{s\to -3}\frac{s+2}{s(s+1)^2}=\frac{1}{12}$$

$$C_2=\lim_{s\to -1}(s+1)^2\cdot F(s)=\lim_{s\to -1}\frac{s+2}{s(s+3)}=-\frac{1}{2}$$

$$C_1=\lim_{s\to -1}\frac{d}{ds}[(s+1)^2\cdot F(s)]=\lim_{s\to -1}\frac{d}{ds}\left[\frac{s+2}{s(s+3)}\right]=-\frac{3}{4}$$

Besides, each coefficient can also be obtained by solving the coefficient equations, i. e.

$$\begin{cases} C_1+C_3+C_4=0 \\ C_2+4C_1+5C_3+2C_4=0 \\ 3C_2+3C_1+7C_3+C_4=1 \\ 3C_2=2 \end{cases}$$

The same results are obtained through solving the equations. Substituting the coefficients into $F(s)$, we have

$$F(s)=-\frac{1}{2}\cdot\frac{1}{(s+1)^2}-\frac{3}{4}\cdot\frac{1}{s+1}+\frac{2}{3}\cdot\frac{1}{s}+\frac{1}{12}\cdot\frac{1}{s+3}$$

Then, taking the inverse Laplace transform on the above expression, we have

$$f(t)=-\frac{1}{2}te^{-t}-\frac{3}{4}e^{-t}+\frac{2}{3}+\frac{1}{12}e^{-3t}$$

2.3 The Transfer Function of Control Systems

Due to the complex process of solving differential equations, the analytical method is limited to

$$\begin{cases} C_m = \lim_{s\to -p_1}(s+p_1)^m \cdot F(s) \\ C_{m-1} = \lim_{s\to -p_1}\dfrac{\mathrm{d}}{\mathrm{d}s}\big[(s+p_1)^m \cdot F(s)\big] \\ \quad\vdots \\ C_{m-j} = \dfrac{1}{j!}\lim_{s\to -p_1}\dfrac{\mathrm{d}^j}{\mathrm{d}s^j}\big[(s+p_1)^m \cdot F(s)\big] \\ \quad\vdots \\ C_1 = \dfrac{1}{(m-1)!}\lim_{s\to -p_1}\dfrac{\mathrm{d}^{(m-1)}}{\mathrm{d}s^{(m-1)}}\big[(s+p_1)^m \cdot F(s)\big] \end{cases} \quad (2\text{-}35)$$

在所有的系数都确定后,即可由拉氏反变换得到原函数的时域表达式,即

$$\begin{aligned}f(t) &= L^{-1}[F(s)] \\ &= L^{-1}\left[\dfrac{C_m}{(s+p_1)^m} + \dfrac{C_{m-1}}{(s+p_1)^{m-1}} + \cdots + \dfrac{C_1}{s+p_1} + \dfrac{C_{m+1}}{s+p_{m+1}} + \cdots + \dfrac{C_n}{s+p_n}\right] \\ &= \left[\dfrac{C_m}{(m-1)!}t^{m-1} + \dfrac{C_{m-1}}{(m-2)!}t^{m-2} + \cdots + C_2 t + C_1\right]\mathrm{e}^{-p_1 t} + \sum_{i=m+1}^{n} C_i \mathrm{e}^{-p_i t}\end{aligned}$$

$$(2\text{-}36)$$

下面举一些例子来说明上述公式的应用。

例 2-4 求 $F(s) = \dfrac{s+2}{s(s+1)^2(s+3)}$ 的原函数 $f(t)$。

解: 将 $F(s)$ 进行因式分解,可得

$$F(s) = \dfrac{C_2}{(s+1)^2} + \dfrac{C_1}{s+1} + \dfrac{C_3}{s} + \dfrac{C_4}{s+3}$$

接下来确定四个待定系数,根据式 (2-31) 和式 (2-35),系数的计算为

$$C_3 = \lim_{s\to 0} s \cdot F(s) = \lim_{s\to 0}\dfrac{s+2}{(s+1)^2(s+3)} = \dfrac{2}{3}$$

$$C_4 = \lim_{s\to -3}(s+3)\cdot F(s) = \lim_{s\to -3}\dfrac{s+2}{s(s+1)^2} = \dfrac{1}{12}$$

$$C_2 = \lim_{s\to -1}(s+1)^2 \cdot F(s) = \lim_{s\to -1}\dfrac{s+2}{s(s+3)} = -\dfrac{1}{2}$$

$$C_1 = \lim_{s\to -1}\dfrac{\mathrm{d}}{\mathrm{d}s}\big[(s+1)^2 \cdot F(s)\big] = \lim_{s\to -1}\dfrac{\mathrm{d}}{\mathrm{d}s}\left[\dfrac{s+2}{s(s+3)}\right] = -\dfrac{3}{4}$$

此外,也可以使用解系数方程的方法,得到系数方程组,即

$$\begin{cases} C_1 + C_3 + C_4 = 0 \\ C_2 + 4C_1 + 5C_3 + 2C_4 = 0 \\ 3C_2 + 3C_1 + 7C_3 + C_4 = 1 \\ 3C_2 = 2 \end{cases}$$

解该方程组即可得到与前面相同的结果。将所求系数代入 $F(s)$ 中,得

solving low-order differential equations in most cases, while the numerical method is used to solve high-order differential equations. And even if they can be solved, it is not easy to analyze the influence of the parameters or elements of the system on its dynamic process. Therefore, the transfer function, which is one of the most commonly used mathematical models, is proposed in the study of linear time-invariant systems with single variable. The transfer function is a mathematical model that describes the dynamic relation between the input and output of the system in the complex domain. Meanwhile, the transfer function is an extremely important basic concept in control theory.

2.3.1 Definition of the transfer function

For a linear time-invariant system, when the initial conditions of the input and output are zero, the transfer function of the system is defined as the ratio of the Laplace transform of the output $x_o(t)$ to the Laplace transform of the input $x_i(t)$, i. e.

$$G(s) = \frac{L[x_o(t)]}{L[x_i(t)]} = \frac{X_o(s)}{X_i(s)} \tag{2-37}$$

Assume that the general form of the differential equation for a linear time-invariant system is

$$a_n x_o^{(n)}(t) + a_{n-1} x_o^{(n-1)}(t) + \cdots + a_1 \dot{x}_o(t) + a_0 x_o(t)$$
$$= b_m x_i^{(m)}(t) + b_{m-1} x_i^{(m-1)}(t) + \cdots + b_1 \dot{x}_i(t) + b_0 x_i(t) \quad (n \geq m) \tag{2-38}$$

where $x_o(t)$ is the output of the system; $x_i(t)$ is the input of the system. When the initial conditions $x_o(0), \dot{x}_o(0), \cdots, x_o^{(n-1)}(0)$ and $x_i(0), \dot{x}_i(0), \cdots, x_i^{(m-1)}(0)$ are equal to zero, taking the Laplace transform on the above equation, we have

$$(a_n s^n + a_{n-1} s^{n-1} + \cdots + a_1 s + a_0) X_o(s)$$
$$= (b_m s^m + b_{m-1} s^{m-1} + \cdots + b_1 s + b_0) X_i(s) \tag{2-39}$$

Consequently, we can obtain the general form of the transfer function for a system as

$$G(s) = \frac{X_o(s)}{X_i(s)} = \frac{b_m s^m + b_{m-1} s^{m-1} + \cdots + b_1 s + b_0}{a_n s^n + a_{n-1} s^{n-1} + \cdots + a_1 s + a_0} \quad (n \geq m) \tag{2-40}$$

From the above derivation, if the differential equation of a system is known, the transfer function can be easily obtained. The highest order of s in the denominator of the transfer function represents the order of the system. Therefore, the system described by equation (2-40) is an nth-order system.

2.3.2 Characteristics of the transfer function

The transfer function is a rational function with the complex variable s, and is the bridge between the input and output of the system. It has the following characteristics.

1) The denominator of the transfer function reflects the inherent characteristics of the system, while the numerator of the transfer function reflects the relation between the system and the outside.

2) If the input is already given, the output of the system is completely dependent on the transfer function. Due to $X_o(s) = G(s) X_i(s)$, the output of the system in the time domain can be obtained through the inverse Laplace transform, i. e.

$$F(s) = -\frac{1}{2} \cdot \frac{1}{(s+1)^2} - \frac{3}{4} \cdot \frac{1}{s+1} + \frac{2}{3} \cdot \frac{1}{s} + \frac{1}{12} \cdot \frac{1}{s+3}$$

再对上式进行拉氏反变换，则可得

$$f(t) = -\frac{1}{2}te^{-t} - \frac{3}{4}e^{-t} + \frac{2}{3} + \frac{1}{12}e^{-3t}$$

2.3 控制系统的传递函数

由于微分方程的求解过程比较繁琐，一般情况下，解析法只限于求解低阶微分方程，而数值解法用于求解高阶微分方程。但即使能够求解，也不便于分析系统的参数或环节对其动态过程的影响。因此，人们在对单变量线性定常系统的研究中，提出了传递函数这一最常用的数学模型。传递函数是在复域内描述系统输入、输出之间动态关系的数学模型。传递函数同时也是控制理论中一个极其重要的基本概念。

2.3.1 传递函数的定义

对于线性定常系统，当输入及输出的初始条件为零时，系统的传递函数定义为输出量 $x_o(t)$ 的拉氏变换与输入量 $x_i(t)$ 的拉氏变换之比，即

$$G(s) = \frac{L[x_o(t)]}{L[x_i(t)]} = \frac{X_o(s)}{X_i(s)} \tag{2-37}$$

设线性定常系统微分方程的一般形式为

$$a_n x_o^{(n)}(t) + a_{n-1} x_o^{(n-1)}(t) + \cdots + a_1 \dot{x}_o(t) + a_0 x_o(t)$$
$$= b_m x_i^{(m)}(t) + b_{m-1} x_i^{(m-1)}(t) + \cdots + b_1 \dot{x}_i(t) + b_0 x_i(t) \quad (n \geq m) \tag{2-38}$$

式中，$x_o(t)$ 是系统的输出量；$x_i(t)$ 是系统的输入量。当初始条件 $x_o(0)$，$\dot{x}_o(0)$，$\cdots$，$x_o^{(n-1)}(0)$ 和 $x_i(0)$，$\dot{x}_i(0)$，$\cdots$，$x_i^{(m-1)}(0)$ 均为零时，对上式进行拉氏变换可得

$$(a_n s^n + a_{n-1} s^{n-1} + \cdots + a_1 s + a_0) X_o(s)$$
$$= (b_m s^m + b_{m-1} s^{m-1} + \cdots + b_1 s + b_0) X_i(s) \tag{2-39}$$

故得系统传递函数的一般形式为

$$G(s) = \frac{X_o(s)}{X_i(s)} = \frac{b_m s^m + b_{m-1} s^{m-1} + \cdots + b_1 s + b_0}{a_n s^n + a_{n-1} s^{n-1} + \cdots + a_1 s + a_0} \quad (n \geq m) \tag{2-40}$$

由上述推导可知，若已知系统的微分方程，则可以很容易地求出其传递函数。传递函数分母中 s 的最高阶代表系统的阶次。因此，由式（2-40）描述的系统就是 n 阶系统。

2.3.2 传递函数的特点

传递函数是复变量 s 的有理函数，是系统输入与输出之间的桥梁，具有如下特点。

1) 传递函数的分母反映系统的固有特性，而其分子反映系统与外界之间的关系。

2) 若输入已经给定，则系统的输出完全取决于传递函数。因为 $X_o(s) = G(s)X_i(s)$，所以通过拉氏反变换，便可求得系统在时域内的输出，即

$$x_o(t) = L^{-1}[X_o(s)] = L^{-1}[G(s)X_i(s)] \tag{2-41}$$

$$x_o(t) = L^{-1}[X_o(s)] = L^{-1}[G(s)X_i(s)] \qquad (2\text{-}41)$$

It should be noted that the output is independent on the initial state of the system, because the initial state has been assumed as zero.

3) The order n of s in the denominator of the transfer function must not be less than the order m of s in the numerator. Because the actual system always has inertia so that the output doesn't go ahead of the input, and there must be $n \geq m$.

4) The transfer function can be dimensional or non-dimensional, and its dimension depends on the input and output of the system.

5) The physical structure of the system cannot be described by the transfer function. Different physical systems can have the same transfer function, and such different physical systems are called similar systems; the same physical system can have different forms of transfer functions due to different research purposes.

2.3.3 Forms of the transfer function

In addition to the numerator-denominator polynomial model shown in equation (2-40), the transfer function can also be written in the following forms.

1. Zero-pole model

Factorizing the numerator and denominator of equation (2-40), we have

$$G(s) = \frac{X_o(s)}{X_i(s)} = K_g \frac{(s+z_1)(s+z_2)\cdots(s+z_m)}{(s+p_1)(s+p_2)\cdots(s+p_n)} = K_g \frac{\prod_{i=1}^{m}(s+z_i)}{\prod_{j=1}^{n}(s+p_j)} \quad (n \geq m) \qquad (2\text{-}42)$$

where $K_g = b_m/a_n$ is the transfer coefficient or root-locus gain of a control system; $-z_i$ ($i=1, 2, \cdots, m$) is the zero of a control system; $-p_j$ ($j=1, 2, \cdots, n$) is the pole of a control system.

The model given by equation (2.42) is called the pole-zero gain model. Zeros, poles and gains of a transfer function determine the transient and steady-state performance of the system.

2. Normalization model

Factorizing the numerator and denominator of equation (2-40), we have

$$G(s) = \frac{X_o(s)}{X_i(s)} = K \frac{\prod_{k=1}^{p}(T_k s + 1)\prod_{l=1}^{q}(T_l^2 s^2 + 2\xi_l T_l s + 1)}{s^v \prod_{i=1}^{g}(T_i s + 1)\prod_{j=1}^{h}(T_j^2 s^2 + 2\xi_j T_j s + 1)}$$

$$(p + 2q = m, v + g + 2h = n, n \geq m) \qquad (2\text{-}43)$$

where $K = b_0/a_0$ is the static gain coefficient of a transfer function in a control system, and it determines the magnification of the system's steady-state response; T_i, T_j, T_k, T_l are various time constants of the control system.

The equation (2-43) shows that a complex control system can be composed of several first-order systems and second-order systems, which means that it is convenient to analyze the dynamic characteristics of the complex system.

需要注意，这一输出与系统的初始状态无关，因为此时已设定初始状态为零。

3）传递函数分母中 s 的阶数 n 必不小于分子中 s 的阶数 m。因为实际系统总具有惯性，以使输出不会超前于输入，所以必然有 $n \geq m$。

4）传递函数可以有量纲，也可以无量纲，其量纲取决于系统的输入与输出。

5）传递函数不能描述系统的物理结构。不同的物理系统可以有形式相同的传递函数，这样不同的物理系统称为相似系统；同一个物理系统，由于研究目的的不同，可以有不同形式的传递函数。

2.3.3 传递函数的形式

传递函数除了可以写成式（2-40）所示的分子分母多项式模型式外，还可以写成以下形式。

1. 零极点模型

对式（2-40）中的分子和分母进行因式分解，得

$$G(s) = \frac{X_o(s)}{X_i(s)} = K_g \frac{(s+z_1)(s+z_2)\cdots(s+z_m)}{(s+p_1)(s+p_2)\cdots(s+p_n)} = K_g \frac{\prod_{i=1}^{m}(s+z_i)}{\prod_{j=1}^{n}(s+p_j)} \quad (n \geq m) \quad (2\text{-}42)$$

式中，$K_g = b_m/a_n$ 为控制系统的传递系数或根轨迹增益；$-z_i$（$i=1, 2, \cdots, m$）为控制系统的零点；$-p_j$（$j=1, 2, \cdots, n$）为控制系统的极点。

式（2-42）给出的模型称为零极点增益模型。传递函数的零点、极点和增益决定着系统的瞬态性能和稳态性能。

2. 归一化模型

对式（2-40）中的分子和分母进行因式分解，得

$$G(s) = \frac{X_o(s)}{X_i(s)} = K \frac{\prod_{k=1}^{p}(T_k s+1) \prod_{l=1}^{q}(T_l^2 s^2 + 2\xi_l T_l s + 1)}{s^v \prod_{i=1}^{g}(T_i s+1) \prod_{j=1}^{h}(T_j^2 s^2 + 2\xi_j T_j s + 1)}$$

$$(p+2q = m, v+g+2h = n, n \geq m) \tag{2-43}$$

式中，$K = b_0/a_0$ 为控制系统传递函数的静态增益系数，它决定了系统稳态响应的放大倍数；T_i，T_j，T_k，T_l 为控制系统的各种时间常数。

式（2-43）表明，一个复杂的控制系统可以由若干一阶系统和二阶系统构成，这意味着可以方便地分析复杂系统的动态特性。

2.4 典型环节的传递函数

无论是机械系统还是电气系统，尽管这些系统的物理本质差别很大，但从数学观点来看，可以有完全相同的数学模型，亦即具有相同的动态性能。因此我们可以从数学建模的角

2.4 Transfer Functions of Typical Elements

Whether they are mechanical systems or electrical systems, although the physical essence of these systems varies widely, they could have the identical mathematical model from a mathematical point of view, that is, they could have the same dynamic performance. Therefore, from the perspective of mathematical modeling, we can decompose a complex system into a limited number of typical elements based on mathematical models. Each typical element can be consisted of one or several components. It is convenient to analyze the complex control system by figuring out the transfer functions of these typical elements. There are typical elements commonly used in control systems, including proportion elements, differential elements, integral elements, inertia elements, first-order differential elements, oscillation elements, second-order differential elements, delay elements, etc. These typical elements are discussed as follows.

2.4.1 Proportion element

When the output is proportional to the input, the element whose signal is not distorted or delayed is called the proportion element or amplifying element. Its dynamic equation is

$$x_o(t) = K x_i(t)$$

where $x_o(t)$ is the output; $x_i(t)$ is the input; K is an amplification coefficient or gain of the element. its transfer function is

$$G(s) = \frac{X_o(s)}{X_i(s)} = K \qquad (2\text{-}44)$$

The block diagram of the proportion element is shown in Fig. 2-9.

Example 2-5 Fig. 2-10 shows an operational amplifier, where $u_i(t)$ is the input voltage, $u_o(t)$ is the output voltage, and R_1 and R_2 are the resistors. Figure out the transfer function.

Solution: According to Kirchhoff's law, we have

$$u_o(t) = -\frac{R_2}{R_1} u_i(t)$$

Taking the Laplace transform, we have

$$U_o(s) = -\frac{R_2}{R_1} U_i(s)$$

So, the transfer function can be written as

$$G(s) = \frac{U_o(s)}{U_i(s)} = -\frac{R_2}{R_1} = K$$

It is an inverse proportion element. When $R_1 = R_2$ (i.e. $K = 1$), it is called an inverter.

2.4.2 Differential element

When the output is proportional to the differential of the input, there is

$$x_o(t) = T \dot{x}_i(t)$$

度研究，将一个复杂的系统分解为有限个以数学模型划分的典型环节。每个典型环节可以由一个或若干元件组成。求出这些典型环节的传递函数可为复杂系统的分析提供方便。控制系统中常用的典型环节有：比例环节、微分环节、积分环节、惯性环节、一阶微分环节、振荡环节、二阶微分环节和延时环节等。下面讨论这些典型环节。

2.4.1 比例环节

当输出量与输入量成正比时，信号不失真也不延时的环节称为比例环节或放大环节。其动力学方程为

$$x_o(t) = K x_i(t)$$

式中，$x_o(t)$ 为输出；$x_i(t)$ 为输入；K 为环节的放大系数或增益。其传递函数为

$$G(s) = \frac{X_o(s)}{X_i(s)} = K \qquad (2\text{-}44)$$

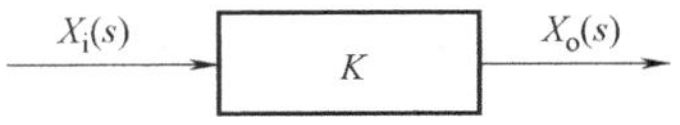

比例环节的框图如图2-9所示。

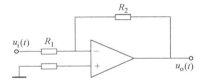

Fig. 2-9　The block diagram of proportion element
（比例环节框图）

例 2-5 如图 2-10 所示的运算放大器中，$u_i(t)$ 为输入电压，$u_o(t)$ 为输出电压，R_1、R_2 为电阻。求其传递函数。

解：根据基尔霍夫定律，得

$$u_o(t) = -\frac{R_2}{R_1} u_i(t)$$

进行拉氏变换，得

$$U_o(s) = -\frac{R_2}{R_1} U_i(s)$$

则传递函数为

$$G(s) = \frac{U_o(s)}{U_i(s)} = -\frac{R_2}{R_1} = K$$

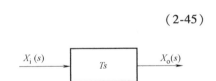

Fig. 2-10　An operational amplifier
（运算放大器）

这是一个反相比例环节。当 $R_1 = R_2$（即 $K = 1$）时，就称为反相器。

2.4.2 微分环节

当输出量正比于输入量的微分时，则有

$$x_o(t) = T \dot{x}_i(t)$$

具有这样方程式的环节称为微分环节。其传递函数为：

$$G(s) = \frac{X_o(s)}{X_i(s)} = Ts \qquad (2\text{-}45)$$

式中，T 为微分环节的时间常数。微分环节的框图如图2-11所示。

Fig. 2-11　The block diagram of differential element（微分环节框图）

微分环节的输出是输入的微分，当输入为单位阶跃函数时，输出应是脉冲函数，这在实际中是不可能的，因此工程上无法制造传递函数为微分环节的元件或装置，

The element with such an equation is called the differential element, and its transfer functions is

$$G(s) = \frac{X_o(s)}{X_i(s)} = Ts \tag{2-45}$$

where T is a time constant of the differential element. The block diagram of the differential element is shown in Fig. 2-11.

The output of the differential element is the derivative of the input. If the input is a unit step function, the output should be a pulse function, but it is impossible in practice. Therefore, it is impossible to design components or devices whose transfer function is a differential element in engineering. Differential elements does not exist alone in the system. Of course, when the inertia of some components is very small, their transfer functions can be approximated as differential elements.

Example 2-6 Fig. 2-12 shows the schematic diagram of a hydraulic damper, in which the spring is rigidly connected to the piston, ignoring the inertial force of the moving part. $x_i(t)$ is the input displacement, $x_o(t)$ is the output displacement, k is the spring stiffness, and c is the viscous damping coefficient. Figure out the transfer function.

Solution: The force balance equation of the piston is

$$kx_o(t) = c[\dot{x}_i(t) - \dot{x}_o(t)]$$

Taking the Laplace transform, we have

$$kX_o(s) = cs[X_i(s) - X_o(s)]$$

So, the transfer function can be written as,

$$G(s) = \frac{X_o(s)}{X_i(s)} = \frac{\frac{c}{k}s}{\frac{c}{k}s + 1} = \frac{Ts}{Ts + 1} \quad (T = c/k)$$

Example 2-7 Fig. 2-13 shows the RC differential circuit. u_i is the input voltage, u_o is the output voltage, i is the current, R is the resistance, and C is the capacitance. Figure out the transfer function.

Solution: According to Kirchhoff's law, the circuit equations are

$$u_i(t) = \frac{1}{C}\int i(t)\,dt + u_o(t)$$

$$u_o(t) = i(t)R$$

Taking the Laplace transform, we have

$$U_i(s) = \frac{1}{RCs}U_o(s) + U_o(s)$$

So, the transfer function can be written as

$$G(s) = \frac{U_o(s)}{U_i(s)} = \frac{RCs}{RCs + 1} = \frac{Ts}{Ts + 1} \quad (T = RC)$$

Above two examples are systems with inertial and differential elements. Only when $|Ts| \ll 1$, $G(s) \approx Ts$ is approximated as a differential element. In fact, the inertia is always involved in the differential elements, and the ideal differential element is only a mathematical assumption.

微分环节在系统中不会单独出现。当然，当有些元件惯性很小时，其传递函数可以近似地看成微分环节。

例 2-6　图 2-12 为液压阻尼器的原理图，其中，弹簧与活塞刚性连接，忽略运动件的惯性力，设 $x_i(t)$ 为输入位移，$x_o(t)$ 为输出位移，k 为弹簧刚度，c 为黏性阻尼系数。求传递函数。

解：活塞的力平衡方程式为

$$kx_o(t) = c[\dot{x}_i(t) - \dot{x}_o(t)]$$

经拉氏变换后有

$$kX_o(s) = cs[X_i(s) - X_o(s)]$$

故传递函数为

$$G(s) = \frac{X_o(s)}{X_i(s)} = \frac{\dfrac{c}{k}s}{\dfrac{c}{k}s + 1} = \frac{Ts}{Ts+1} \quad (T = c/k)$$

例 2-7　图 2-13 所示为 RC 微分电路，其中，u_i 为输入电压，u_o 为输出电压，i 为电流，R 为电阻，C 为电容。求传递函数。

解：根据基尔霍夫定律，得电路方程为

$$u_i(t) = \frac{1}{C}\int i(t)\,dt + u_o(t)$$

$$u_o(t) = i(t)R$$

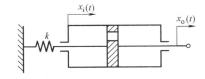

Fig. 2-12　The hydraulic damper
（液压阻尼器）

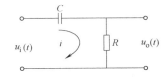

Fig. 2-13　RC differential circuit
（RC 微分电路）

进行拉氏变换，得

$$U_i(s) = \frac{1}{RCs}U_o(s) + U_o(s)$$

则传递函数为

$$G(s) = \frac{U_o(s)}{U_i(s)} = \frac{RCs}{RCs+1} = \frac{Ts}{Ts+1} \quad (T = RC)$$

以上两例都是具有惯性环节和微分环节的系统。仅当 $|Ts| \ll 1$ 时，表达式 $G(s) \approx Ts$ 才近似成为微分环节。实际上，微分环节总是含有惯性环节的，理想的微分环节只是数学上的假设。

微分环节对系统的控制作用如下。

1) 微分环节使输出提前。如对比例环节 K_p 施加一单位斜坡函数 $r(t) = t$ 作为输入，则此环节在时域中的输出 $x_o(t)$ 如图 2-14a 所示；再对此比例环节并联一微分环节 $G_1(s) = K_pTs$，如图 2-14b 所示，则传递函数为

The control effect of the differential element on the system is as follows.

1) The differential element brings forward the output. If a unit ramp function $r(t)=t$, as the input, is applied to the proportion element K_p, the output $x_o(t)$ of the element in the time domain is shown in Fig. 2-14a; then the differential element $G_1(s)=K_p Ts$ is connected with the proportion element in parallel, as is shown in Fig. 2-14b. So, the transfer function is

$$G(s) = \frac{X_o(s)}{R(s)} = K_p(Ts+1)$$

The output increased by the differential element in parallel connection (it is assumed that $K_p = 1$) is

$$x_{o1}(t) = L^{-1}[G_1(s)R(s)] = L^{-1}[TsR(s)] = TL^{-1}[sR(s)] = TL^{-1}[1/s] = T \cdot 1(t)$$

It makes the original output moves vertically up by T to obtain a new output. As shown in Fig. 2-14a, the output of the system is increased by T at each moment, i.e. the original output moves to the left by T. In other words, the new output at time t_1 reaches what the original output reaches at time t_2.

The output of the differential element is the derivative $T\dot{x}_i(t)$ of the input, which reflects the changing trend of the input. Thus, we can predict the changing trend of the system input through the differential element, and it is possible to apply the compensation in advance in order to improve the performance of the system.

2) The differential element can increase the damping of the system. As is shown in Fig. 2-15a, the transfer function of the system is

$$G_1(s) = \frac{\dfrac{K_p K}{s(Ts+1)}}{1+\dfrac{K_p K}{s(Ts+1)}} = \frac{K_p K}{Ts^2+s+K_p K}$$

For the system, the proportion element K_p is connected with the differential element $K_p T_d s$ in parallel, as is shown in Fig. 2-15b, and its transfer function is

$$G_2(s) = \frac{\dfrac{K_p K(T_d s+1)}{s(Ts+1)}}{1+\dfrac{K_p K(T_d s+1)}{s(Ts+1)}} = \frac{K_p K(T_d s+1)}{Ts^2+(1+K_p K T_d)s+K_p K}$$

Comparing the above two equations, $G_1(s)$ and $G_2(s)$ are both transfer functions of the second-order system. The coefficient of the second term in the denominator is related to the damping. The coefficient of s is 1 in $G_1(s)$ while the coefficient $1+K_p K T_d > 1$ in $G_2(s)$. Therefore, after introducing the differential element, the damping of the system increases.

3) The differential element has the function of enhancing noise. Through the differential element, we can predict the input, we can also predict the noise (i.e., disturbance). It should be noted that enhancing the sensitivity of the noise increases the error caused by the disturbance.

2.4.3 Integral element

When the output is proportional to the integral of the input, there is

$$G(s) = \frac{X_o(s)}{R(s)} = K_p(Ts+1)$$

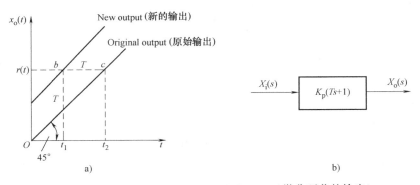

Fig. 2-14 The output of the differential element （微分环节的输出）

并联微分环节所增加的输出可表示为（假设 $K_p = 1$）为

$$x_{o1}(t) = L^{-1}[G_1(s)R(s)] = L^{-1}[TsR(s)] = TL^{-1}[sR(s)] = TL^{-1}[1/s] = T \cdot 1(t)$$

它使原输出竖直向上平移了 T，得到新输出。如图 2-14a 所示，系统在每一时刻的输出都增加了 T，即原输出向左平移 T。换言之，在 t_2 时刻的原输出在 t_1 时刻就已达到。

微分环节的输出是输入的导数 $T\dot{x}_i(t)$，它反映了输入的变化趋势，因此可以通过微分环节预测输入的变化情况，并有可能对系统提前施加校正作用以提高系统的性能。

2）微分环节可增加系统的阻尼。如图 2-15a 所示，系统的传递函数为

$$G_1(s) = \frac{\dfrac{K_p K}{s(Ts+1)}}{1 + \dfrac{K_p K}{s(Ts+1)}} = \frac{K_p K}{Ts^2 + s + K_p K}$$

对系统的比例环节 K_p 并联微分环节 $K_p T_d s$，如图 2-15b 所示，其传递函数为

$$G_2(s) = \frac{\dfrac{K_p K(T_d s+1)}{s(Ts+1)}}{1 + \dfrac{K_p K(T_d s+1)}{s(Ts+1)}} = \frac{K_p K(T_d s+1)}{Ts^2 + (1+K_p K T_d)s + K_p K}$$

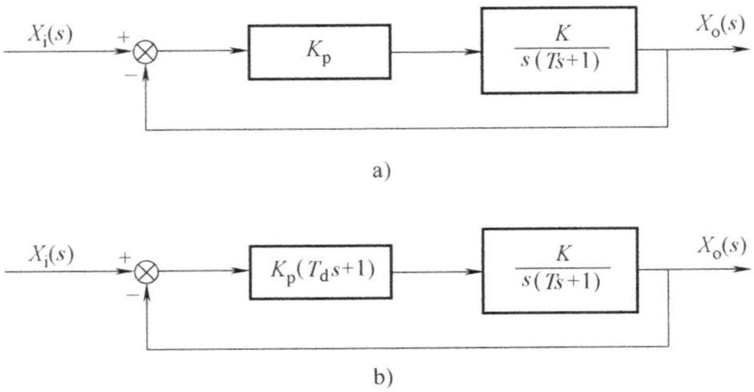

Fig. 2-15 The output of differential element （微分环节的输出）

$$x_o(t) = \frac{1}{T}\int x_i(t)\,dt$$

The element with such an equation is called the integral element, and its transfer function is

$$G(s) = \frac{X_o(s)}{X_i(s)} = \frac{1}{Ts} \qquad (2\text{-}46)$$

where T is a time constant of the integral element. The block diagram of the integral element is shown in Fig. 2-16.

Example 2-8 Fig. 2-17 shows the transmission mechanism of the rack and pinion. The rotation speed $n(t)$ of the pinion is set as the input invariant, and the displacement $x(t)$ of the rock is set as the output variable. Figure out the transfer function.

Solution: According to the speed relation between the rack and pinion, we have

$$x(t) = \int \pi D n(t)\,dt$$

where D is the pitch diameter of the pinion.

Taking the Laplace transform, the transfer function can be written as

$$G(s) = \frac{X(s)}{N(s)} = \frac{\pi D}{s} = \frac{K}{s} \qquad (K = \pi D)$$

2.4.4 Inertial element

When the kinetic equation is a first-order differential equation, there is

$$T\dot{x}_o + x_o = K x_i$$

The element with such an equation is called the inertial element. Its transfer function is

$$G(s) = \frac{X_o(s)}{X_i(s)} = \frac{K}{Ts+1} \qquad (2\text{-}47)$$

where K is an amplification coefficient; T is a time constant. The block diagram of the inertial element is shown in Fig. 2-18.

In such components, there are always energy-storage components and energy-consuming components so that the output cannot reproduce immediately for the input with the mutant form, and the output always lags behind the input.

Example 2-9 Fig. 2-19 shows a spring-damper system. $x_i(t)$ is the input displacement, $x_o(t)$ is the output displacement, k is the spring stiffness, and c is the damping coefficient. Figure out the transfer function.

Solution: According to Newton's law, we have

$$k[x_i(t) - x_o(t)] = c\dot{x}_o(t)$$

Taking the Laplace transform, we have

$$csX_o(s) + kX_o(s) = kX_i(s)$$

So, the transfer function can be written as

$$G(s) = \frac{X_o(s)}{X_i(s)} = \frac{1}{\frac{c}{k}s+1} = \frac{1}{Ts+1} \qquad (T = c/k)$$

比较上述两式可知，$G_1(s)$ 与 $G_2(s)$ 均为二阶系统的传递函数，其分母中第二项的系数与阻尼有关，$G_1(s)$ 中 s 的系数为 1，$G_2(s)$ 中 s 的系数 $1+K_p KT_d>1$。所以，引入微分环节后，系统的阻尼增加了。

3）微分环节具有强化噪声的作用。可以通过微分环节对输入进行预测，也可以对噪声（即干扰）进行预测。需要注意的是，提高噪声的灵敏度会增加干扰引起的误差。

2.4.3 积分环节

当输出量正比于输入量的积分时，有

$$x_o(t) = \frac{1}{T}\int x_i(t)\,\mathrm{d}t$$

具有这样方程式的环节称为积分环节，其传递函数为

$$G(s) = \frac{X_o(s)}{X_i(s)} = \frac{1}{Ts} \tag{2-46}$$

式中，T 为积分环节的时间常数。积分环节的框图如图 2-16 所示。

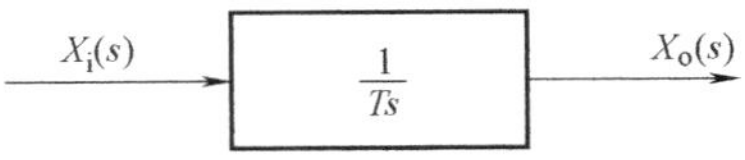

例 2-8 如图 2-17 所示为齿轮齿条传动机构，取齿轮的转速 $n(t)$ 为输入量，齿条的位移 $x(t)$ 为输出量。求传递函数。

解：由齿轮齿条间的转速关系，有

$$x(t) = \int \pi D n(t)\,\mathrm{d}t$$

式中，D 为齿轮节圆直径。

取拉氏变换后，得传递函数为

$$G(s) = \frac{X(s)}{N(s)} = \frac{\pi D}{s} = \frac{K}{s} \quad (K = \pi D)$$

Fig. 2-16 The block diagram of integral element（积分环节框图）

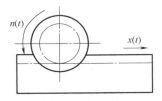

Fig. 2-17 The transmission mechanism of the rack and pinion（齿轮齿条传动机构）

2.4.4 惯性环节

当动力学方程为一阶微分方程时，有

$$T\dot{x}_o + x_o = Kx_i$$

具有这样方程的环节称为惯性环节。其传递函数为

$$G(s) = \frac{X_o(s)}{X_i(s)} = \frac{K}{Ts+1} \tag{2-47}$$

式中，K 为放大系数；T 为时间常数。惯性环节的框图如图 2-18 所示。

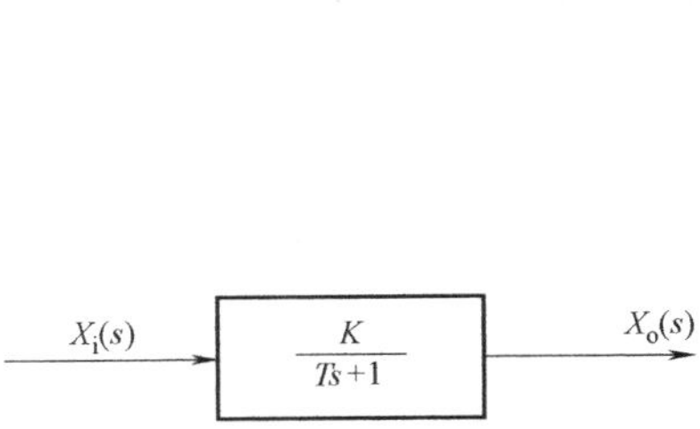

Fig. 2-18 The block diagram of an inertial element（惯性环节框图）

The system is the inertia element. The reason is that there exists the elastic energy-storage component k and the resistive energy-consuming component c.

2.4.5 First-order differential element

The motion equation of the first-order differential element can be described as
$$x_o(t) = K[T\dot{x}_i(t) + x_i(t)]$$
Its transfer function is
$$G(s) = \frac{X_o(s)}{X_i(s)} = K(Ts+1) \tag{2-48}$$
where K is an amplification coefficient; T is a time constant. The block diagram of the first-order differential element is shown in Fig. 2-20.

Example 2-10 Fig. 2-21 shows the RC circuit. u_i is the input voltage, i is the output current, R is the resistance, and C is the capacitance. Figure out the transfer function.

Solution: According to Kirchhoff's law, we have
$$u_i(t) = i_2(t)R$$
$$u_i(t) = \frac{1}{C}\int i_1(t)\,dt$$
$$i(t) = i_1(t) + i_2(t)$$
Taking the Laplace transform, the transfer function is
$$G(s) = \frac{I(s)}{U_i(s)} = Cs + \frac{1}{R} = \frac{1}{R}(RCs+1) = K(Ts+1)$$
where $K = 1/R$; $T = RC$.

Like the differential element, the first-order differential element does not exist alone in the system. It is often combined with other typical elements to describe the motion characteristics of the component or system.

2.4.6 Second-order oscillation element

If the relation between input and output can be expressed as
$$T^2 \ddot{x}_o(t) + 2\xi T \dot{x}_o(t) + x_o(t) = x_i(t)$$
the element with such second order differential equation is called the second-order oscillation element. Its transfer function is
$$G(s) = \frac{X_o(s)}{X_i(s)} = \frac{1}{T^2s^2 + 2\xi Ts + 1} \tag{2-49}$$
where T is a time constant; ξ is a damping ratio. Sometimes, the transfer function is also written as
$$G(s) = \frac{\omega_n^2}{s^2 + 2\xi\omega_n s + \omega_n^2} \tag{2-50}$$
where ω_n is an undamped natural frequency of the system, and $\omega_n = 1/T$.

The block diagram of the second-order oscillation element is shown in Fig. 2-22.

在这类元件中,总含有储能元件和耗能元件,以致对于突变形式的输入来说,输出不能立即复现,输出总落后于输入。

例 2-9 图 2-19 所示为弹簧-阻尼系统,设 $x_i(t)$ 为输入位移,$x_o(t)$ 为输出位移,k 为弹簧刚度,c 为阻尼系数。求传递函数。

解:根据牛顿定律,有

$$k[x_i(t)-x_o(t)] = c\dot{x}_o(t)$$

进行拉氏变换并整理,有

$$csX_o(s) + kX_o(s) = kX_i(s)$$

故传递函数为

$$G(s) = \frac{X_o(s)}{X_i(s)} = \frac{1}{\frac{c}{k}s+1} = \frac{1}{Ts+1} \quad (T = c/k)$$

本系统为惯性环节,原因是含有弹性储能元件 k 和阻性耗能元件 c。

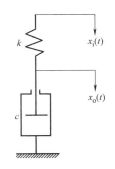

Fig. 2-19　A spring-damper system(弹簧-阻尼系统)

2.4.5　一阶微分环节

一阶微分环节的运动方程可描述为

$$x_o(t) = K[T\dot{x}_i(t) + x_i(t)]$$

其传递函数为

$$G(s) = \frac{X_o(s)}{X_i(s)} = K(Ts+1) \qquad (2-48)$$

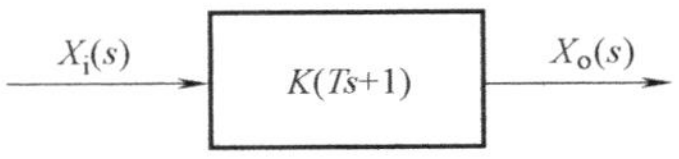

式中,K 为放大系数;T 为时间常数。一阶微分环节的框图如图 2-20 所示。

Fig. 2-20　The block diagram of first-order differential element(一阶微分环节框图)

例 2-10 图 2-21 所示为 RC 电路,u_i 为输入电压,i 为输出电流,R 为电阻,C 为电容。求传递函数。

解:根据基尔霍夫定律,有

$$u_i(t) = i_2(t)R$$

$$u_i(t) = \frac{1}{C}\int i_1(t)\,\mathrm{d}t$$

$$i(t) = i_1(t) + i_2(t)$$

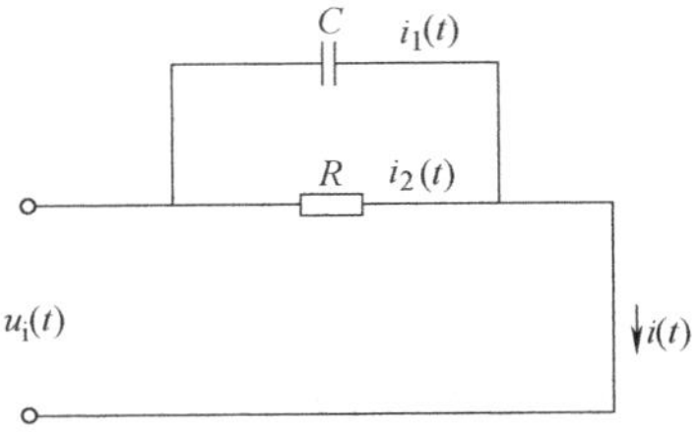

进行拉氏变换并整理,得传递函数为

$$G(s) = \frac{I(s)}{U_i(s)} = Cs + \frac{1}{R} = \frac{1}{R}(RCs+1) = K(Ts+1)$$

式中,$K = 1/R$;$T = RC$。

Fig. 2-21　The RC circuit(RC 电路)

与微分环节一样,一阶微分环节在系统中也不会单独出现,它往往与其他典型环节组合在一起来描述元件或系统的运动特性。

2.4.6　二阶振荡环节

如果输入和输出之间的关系可表达为

Example 2-11 Fig. 2-23 shows a rotational inertia system, rotational i. e. a inertia-damper-spring system for a motion. J is the rotational inertia of the rotor, c is the viscous damping coefficient, k is the spring torsional stiffness. The torque M that is externally applied on the system is set as the input, and the rotor angle θ is set as the output. Figure out the transfer function.

Solution: According to theoretical mechanics, the system's dynamics equation is

$$J\ddot{\theta} + c\dot{\theta} + k\theta = M$$

Taking the Laplace transform, the transfer function can be written as

$$G(s) = \frac{\theta(s)}{M(s)} = \frac{1}{Js^2 + cs + k} = \frac{K}{s^2 + 2\xi\omega_n s + \omega_n^2}$$

where, $K = \frac{1}{J}$; $\xi = \frac{c}{2\sqrt{kJ}}$; $\omega_n = \sqrt{\frac{k}{J}}$.

Example 2-12 Fig. 2-24 shows the RLC circuit. L is the inductance, R is the resistance, and C is the capacitance. $u_i(t)$ is the input voltage, and $u_o(t)$ is the output voltage. Figure out the transfer function.

Solution: According to Kirchhoff's law, we have

$$u_i(t) = L\dot{i}_L(t) + u_o(t)$$

$$u_o(t) = i_R(t)R = \frac{1}{C}\int i_C(t)\,dt$$

$$i_L(t) = i_R(t) + i_C(t)$$

Taking the Laplace transform, we have

$$LCs^2 U_o(s) + \frac{L}{R}sU_o(s) + U_o(s) = U_i(s)$$

So, the transfer function of the system is

$$G(s) = \frac{U_o(s)}{U_i(s)} = \frac{1}{LCs^2 + \frac{L}{R}s + 1} = \frac{\omega_n^2}{s^2 + 2\xi\omega_n s + \omega_n^2}$$

where $\omega_n = \sqrt{\frac{1}{LC}}$; $\xi = \frac{1}{2R}\sqrt{\frac{L}{C}}$.

2.4.7 Second-order differential element

The motion equation of the second-order differential element is

$$x_o(t) = K[T^2\ddot{x}_i(t) + 2\xi T\dot{x}_i(t) + x_i(t)]$$

Its transfer function is

$$G(s) = \frac{X_o(s)}{X_i(s)} = K(T^2 s^2 + 2\xi Ts + 1) \tag{2-51}$$

The characteristics of the element are determined by K, T and ξ. It should be noted that only when the differential equation has complex roots, it is called second-order differential element. If it has real roots only, the element can be considered as two first-order differential elements in series.

$$T^2\ddot{x}_o(t)+2\xi T\dot{x}_o(t)+x_o(t)=x_i(t)$$

则具有这样二阶微分方程的环节称为二阶振荡环节。其传递函数为

$$G(s)=\frac{X_o(s)}{X_i(s)}=\frac{1}{T^2s^2+2\xi Ts+1} \qquad (2\text{-}49)$$

式中，T 为环节的时间常数；ξ 为阻尼比。有时，又将传递函数写成

$$G(s)=\frac{\omega_n^2}{s^2+2\xi\omega_n s+\omega_n^2} \qquad (2\text{-}50)$$

式中，ω_n 为系统无阻尼固有频率，且 $\omega_n=1/T$。

二阶振荡环节的框图如图 2-22 所示。

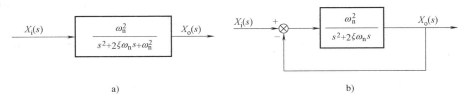

Fig. 2-22　The block diagram of the second-order oscillation element（二阶振荡环节框图）

例 2-11　图 2-23 所示为转动惯量系统，即做旋转运动的惯量-阻尼-弹簧系统，转子转动惯量为 J，黏性阻尼系数为 c，弹簧扭转刚度为 k。在系统外部施加一扭矩 $M(t)$ 作为输入，以转子转角 $\theta(t)$ 作为输出。求系统的传递函数。

解：根据理论力学知识，系统动力学方程为

$$J\ddot{\theta}(t)+c\dot{\theta}(t)+k\theta(t)=M(t)$$

进行拉氏变换并整理，得传递函数为

$$G(s)=\frac{\theta(s)}{M(s)}=\frac{1}{Js^2+cs+k}=\frac{K}{s^2+2\xi\omega_n s+\omega_n^2}$$

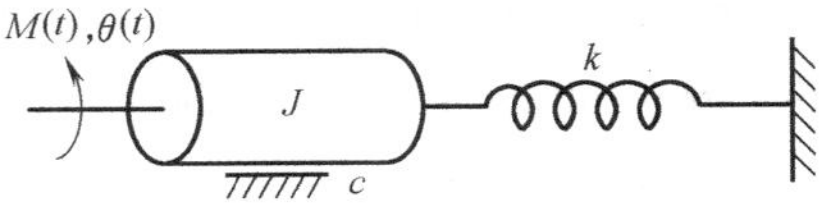

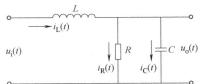

Fig. 2-23　A rotational inertia system
（转动惯量系统）

式中，$K=\dfrac{1}{J}$；$\xi=\dfrac{c}{2\sqrt{kJ}}$；$\omega_n=\sqrt{\dfrac{k}{J}}$。

例 2-12　图 2-24 所示为 RLC 电路，L 为电感，R 为电阻，C 为电容，$u_i(t)$ 为输入电压，$u_o(t)$ 为输出电压。求传递函数。

解：根据基尔霍夫定律，有

$$u_i(t)=L\dot{i}_L(t)+u_o(t)$$
$$u_o(t)=i_R(t)R=\frac{1}{C}\int i_C(t)\mathrm{d}t$$
$$i_L(t)=i_R(t)+i_C(t)$$

Fig. 2-24　The RLC circuit（RLC 电路）

进行拉氏变换并整理，得

$$LCs^2U_o(s)+\frac{L}{R}sU_o(s)+U_o(s)=U_i(s)$$

2.4.8 Delay element

Delay elements are units that cause a time-shift τ in the input signal without affecting the signal characteristics. A system with a delay element is called a delay system. The delay element generally coexists with other elements.

The relation between the output and input of the delay element is
$$x_o(t) = x_i(t-\tau)$$
where τ is the delay time (or time-shift). The transfer function of the delay element is
$$G(s) = \frac{L[x_o(t)]}{L[x_i(t)]} = \frac{L[x_i(t-\tau)]}{L[x_i(t)]} = \frac{X_i(s)e^{-\tau s}}{X_i(s)} = e^{-\tau s} \quad (2\text{-}52)$$

The block diagram of the delay element is shown in Fig. 2-25. The delay element is different from the inertia element. The output of the inertia element delays for a period of time to approach the required output, and it generates the output since the beginning of the input. From the beginning of the input to the time τ, the delay element doesn't generates the output. After τ, the output is exactly equal to the input, and there is no other hysteresis process. In short, the output is the same as the input in the delay element, but it is delayed by a time interval τ.

Example 2-13 Fig. 2-26 shows the schematic diagram of strip thickness measurement during rolling. When the strip is rolled out at point A, a thickness deviation Δh_1 (h is the desired thickness) is generated, but the thickness deviation is detected by the thickness gauge until it reaches point B. Take Δh_1 as the input signal $x_i(t)$ and Δh_2 (i.e. the strip thickness deviation detected by the thickness gauge) as the output signal $x_o(t)$. Figure out its transfer function.

Solution: Assume that the distance from the thickness gauge to the frame is L, and the speed of the strip rolling is v, then the detection delay time is $\tau = L/v$. Then, the relation between the output signal of the thickness gauge and the input signal is
$$x_o(t) = x_i(t-\tau)$$
So, its transfer function is
$$G(s) = \frac{X_o(s)}{X_i(s)} = e^{-\tau s}$$

The detection element is an example of describing the motion characteristics by a delay element. However, the pure delay element is rare in control systems, and the delay element often appears together with other elements.

2.5 Function Block Diagrams of Control Systems

A system can be composed of a number of elements that are represented by blocks, and the blocks are connected with corresponding variables and signal lines to form a block diagram of the system. For a system, the mathematical model of each element, the relation among the variables and the direction the signal flows can be specifically and visually represented by the system block diagram. In fact, the block diagram is a graphical representation of the mathematical model of the sys-

故系统的传递函数为

$$G(s) = \frac{U_o(s)}{U_i(s)} = \frac{1}{LCs^2 + \frac{L}{R}s + 1} = \frac{\omega_n^2}{s^2 + 2\xi\omega_n s + \omega_n^2}$$

式中，$\omega_n = \sqrt{\frac{1}{LC}}$；$\xi = \frac{1}{2R}\sqrt{\frac{L}{C}}$。

2.4.7 二阶微分环节

二阶微分环节的运动方程为

$$x_o(t) = K[T^2 \ddot{x}_i(t) + 2\xi T \dot{x}_i(t) + x_i(t)]$$

其传递函数为

$$G(s) = \frac{X_o(s)}{X_i(s)} = K(T^2 s^2 + 2\xi Ts + 1) \tag{2-51}$$

该环节的特性由 K、T 和 ξ 所决定。需要注意，只有当微分方程具有复根时，才称其为二阶微分环节。如果只具有实根，则可以认为这个环节是两个一阶微分环节串联而成的。

2.4.8 延时环节

延时环节是在不影响信号特性的情况下使输入信号中产生移位时间 τ 的环节。具有延时环节的系统称为延时系统。延时环节一般与其他环节共存，而不单独存在。

延时环节的输出 $x_o(t)$ 与输入 $x_i(t)$ 之间的关系为

$$x_o(t) = x_i(t - \tau)$$

式中，τ 为延迟时间或移位时间。延时环节的传递函数为

$$G(s) = \frac{L[x_o(t)]}{L[x_i(t)]} = \frac{L[x_i(t-\tau)]}{L[x_i(t)]} = \frac{X_i(s)e^{-\tau s}}{X_i(s)} = e^{-\tau s} \tag{2-52}$$

延时环节的框图如图 2-25 所示。延时环节与惯性环节不同，惯性环节的输出会延迟一段时间才接近所要求的输出量，但它从输入开始时刻起就已存在输出。延时环节在输入开始之初的时间 τ 内并无输出，在 τ 后，输出就完全等于输入，且不再有其他滞后过程。简言之，延时环节的输出与输入相同，只是输出延迟了一段时间间隔 τ。

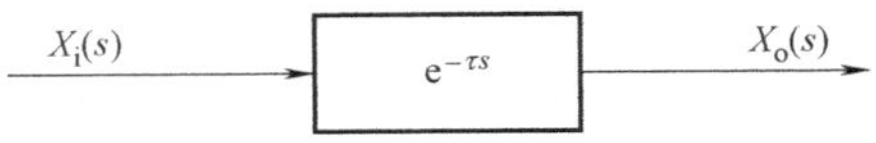

Fig. 2-25 The block diagram of delay element（延时环节框图）

例 2-13 图 2-26 所示为轧钢时的带钢厚度检测示意图。带钢在 A 点轧出时，产生厚度偏差 Δh_1（h 为要求的理想厚度），但是，这一厚度偏差在到达 B 点时才为测厚仪所检测到。取 Δh_1 为输入信号 $x_i(t)$，取 Δh_2（即测厚仪检测到的带钢厚度偏差）为输出信号 $x_o(t)$。求其传递函数。

解：设测厚仪到机架的距离为 L，带钢轧制速度为 v，则检测延迟时间为 $\tau = L/v$。故测厚仪输出信号与输入信号之间的关系为

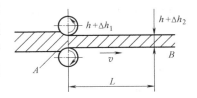

Fig. 2-26 The schematic diagram of strip thickness measurement（带钢厚度检测示意图）

tem. It provides information about the dynamic performance of the system. It can reveal and evaluate the impact of each element on the system as well. According to the block diagram, the transfer function of the system can be obtained by the operational transformation. Therefore, it is very convenient to describe, analyze, and calculate the system by using the block diagrams, and it is widely used.

The block diagram includes the following structural elements.

(1) Signal line   The signal line is a straight line with an arrow. The arrow indicates the direction of signal transmission. The signal variables can be marked on the line, as is shown in Fig. 2-27a.

(2) Function block   The function block is used to represent the transfer function of the element, as is shown in Fig. 2-27b. The arrow pointing to the block represents the input signal. The arrow leaving the block represents the output signal, and the transfer function between the input and output is represented within the block.

(3) Summing point   The summing point, also called the comparator, is the graphical representation of algebraic operation among signals, as is shown in Fig. 2-27c. At the summing point, the output signal is equal to the algebraic sum of all the input signals, and the symbol "+" or "−" next to the arrow of each input signal indicates the sign of the input signal in the algebraic operation. There could be multiple inputs at the summing point (their dimension must be the same), but the output is unique.

(4) Tie point   The tie point indicates the transmission of the same signal in different directions, and each signal connected to the tie point is the same, as is shown in Fig. 2-27d.

2.5.1 Basic connection modes of control systems

In order to figure out the transfer function of the entire system, it is necessary to study the connection among the various elements in the system. The following section will describe how to calculate the transfer function of the system according to the connection modes among the elements.

1. Series connection

The characteristic of the series connection is that the output of the previous element is the input of the next element. Fig. 2-28 shows the connection of two series elements, and its transfer function is

$$G(s) = \frac{X_o(s)}{X_i(s)} = \frac{X_o(s)}{X_1(s)} \cdot \frac{X_1(s)}{X_i(s)} = G_2(s) \cdot G_1(s)$$

In general, assuming that a system consists of n elements connected in series, we have

$$G(s) = G_1(s) G_2(s) \cdots G_n(s) = \prod_{i=1}^{n} G_i(s) \tag{2-53}$$

In other words, the transfer function of the system is the product of the transfer functions of the series elements.

2. Parallel connection

The characteristic of the parallel connection is that the input of all the elements are identical, and the final output is the algebraic sum of the output of each element. Fig. 2-29 shows the connec-

$$x_o(t) = x_i(t-\tau)$$

因而其传递函数为

$$G(s) = \frac{X_o(s)}{X_i(s)} = e^{-\tau s}$$

该检测环节是用延时环节描述运动特性的一个实例。但在控制系统中,单纯的延时环节是很少的,延时环节往往与其他环节一起出现。

2.5 控制系统的框图模型

一个系统可由用方框表示的若干环节组成,方框之间用相应的变量及信号线联系起来,就构成系统的框图。针对一个系统,各环节的数学模型、各变量之间的相互关系以及信号流向可以用系统框图具体而直观地表示出来。事实上,框图是系统数学模型的一种图解表示方法,它提供了关于系统动态性能的有关信息,并且可以揭示和评价每个组成环节对系统的影响。根据框图,可通过运算变换求得系统的传递函数。故用框图来描述、分析、计算系统较为方便,因而被广泛应用。

框图包括以下几种结构要素。

(1) 信号线 信号线是带有箭头的直线,箭头表示信号传递的方向,线上可以标记传递的信号变量,如图2-27a所示。

(2) 函数方框 函数方框用来表示环节的传递函数,如图2-27b所示。指向方框的箭头表示输入信号,离开方框的箭头表示输出信号,输入、输出之间的传递函数在方框中表示。

(3) 相加点 相加点又称比较器,是信号之间代数运算的图解表示,如图2-27c所示。在相加点处,输出信号等于各输入信号的代数和,每个输入信号箭头旁的"+"或"-"表示该输入信号在代数运算中的符号。相加点处可以有多个输入(其量纲必须相同),但输出是唯一的。

(4) 引出点 引出点表示同一信号向不同方向的传递,与引出点相连的各个信号是相同的,如图2-27d所示。

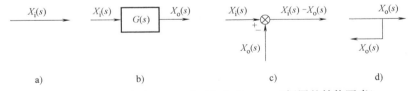

Fig. 2-27 Structural elements of a block diagram (框图的结构要素)

2.5.1 控制系统的基本连接方式

为了求得整个系统的传递函数,需要研究系统中各环节间的联系,本小节介绍怎样根据环节之间的连接来计算系统的传递函数。

1. 串联连接

串联连接的特点是,前一个环节的输出量是后一个环节的输入量。图2-28所示为两个环节串联连接,其传递函数为

tion of two parallel elements, and its transfer function of the system is

$$G(s) = \frac{X_o(s)}{X_i(s)} = \frac{X_1(s) + X_2(s)}{X_i(s)} = \frac{X_1(s)}{X_i(s)} + \frac{X_2(s)}{X_i(s)} = G_1(s) + G_2(s)$$

In general, assuming that a system consists of n elements connected in parallel, we have

$$G(s) = G_1(s) + G_2(s) + \cdots + G_n(s) = \sum_{i=1}^{n} G_i(s) \qquad (2\text{-}54)$$

In other words, the transfer function of the system is the sum of the transfer functions of the parallel elements.

3. Feedback connection

Fig. 2-30 shows the feedback connection between the two elements. In fact, it is the most basic function block diagram of a closed-loop system.

In the Fig. 2-30, $G(s)$ is called the forward path transfer function, which is the ratio of the output $X_o(s)$ to the deviation $E(s)$, i.e.

$$G(s) = \frac{X_o(s)}{E(s)} \qquad (2\text{-}55)$$

$H(s)$ is called the feedback loop transfer function, i.e.

$$H(s) = \frac{B(s)}{X_o(s)} \qquad (2\text{-}56)$$

The product of the forward path transfer function $G(s)$ and the feedback loop transfer function $H(s)$ is defined as the open-loop transfer function $G_K(s)$ of the system, which is the ratio of the feedback signal $B(s)$ to the deviation $E(s)$, i.e.

$$G_K(s) = \frac{B(s)}{E(s)} = G(s)H(s) \qquad (2\text{-}57)$$

Due to the same dimension regarding $B(s)$ and $E(s)$ at the summing point, the open-loop transfer function is dimensionless.

The ratio of the output signal $X_o(s)$ to the input signal $X_i(s)$ is defined as the closed-loop transfer function $\Phi(s)$ of the system, i.e.

$$\Phi(s) = \frac{X_o(s)}{X_i(s)} \qquad (2\text{-}58)$$

According to Fig. 2-30, we have

$$E(s) = X_i(s) \mp B(s) = X_i(s) \mp X_o(s)H(s)$$
$$X_o(s) = G(s)E(s) = G(s)[X_i(s) \mp X_o(s)H(s)]$$
$$= G(s)X_i(s) \mp G(s)X_o(s)H(s)$$

Therefore, the closed-loop transfer function can be obtained as

$$\Phi(s) = \frac{X_o(s)}{X_i(s)} = \frac{G(s)}{1 \pm G(s)H(s)} = \frac{G(s)}{1 \pm G_K(s)} \qquad (2\text{-}59)$$

where, "+" corresponds to the negative feedback; "-" corresponds to the positive feedback. The positive feedback increases the deviation signal by enhancing the input part, while the negative feedback reduces the deviation signal by weakening the input part. Therefore, the negative feedback

$$G(s) = \frac{X_o(s)}{X_i(s)} = \frac{X_o(s)}{X_1(s)} \cdot \frac{X_1(s)}{X_i(s)} = G_2(s) \cdot G_1(s)$$

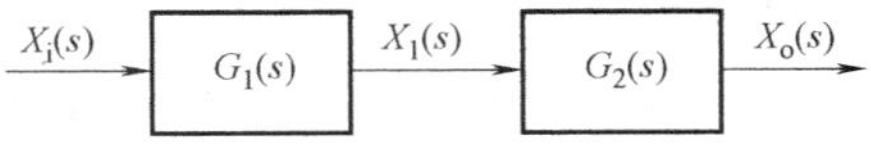

一般地，设一个系统由 n 个环节串联而成，则有

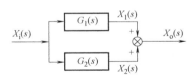

Fig. 2-28 Series connection（串联连接）

$$G(s) = G_1(s) G_2(s) \cdots G_n(s) = \prod_{i=1}^{n} G_i(s) \qquad (2\text{-}53)$$

换言之，系统的传递函数是各串联环节传递函数之积。

2. 并联连接

并联连接的特点是，所有环节的输入量是共同的，最终的输出量为各环节输出量的代数和。图 2-29 所示为两个环节并联连接，其系统传递函数为

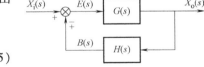

Fig. 2-29 Parallel connection
（并联连接）

$$G(s) = \frac{X_o(s)}{X_i(s)} = \frac{X_1(s) + X_2(s)}{X_i(s)} = \frac{X_1(s)}{X_i(s)} + \frac{X_2(s)}{X_i(s)} = G_1(s) + G_2(s)$$

一般地，设系统由 n 个环节并联而成，则有

$$G(s) = G_1(s) + G_2(s) + \cdots + G_n(s) = \sum_{i=1}^{n} G_i(s) \qquad (2\text{-}54)$$

换言之，系统的传递函数是各并联环节传递函数之和。

3. 反馈连接

图 2-30 所示是两个环节之间的反馈连接，实际上它是闭环系统最基本的函数框图。

图 2-30 中，$G(s)$ 称为前向通道传递函数，它是输出 $X_o(s)$ 与偏差 $E(s)$ 之比，即

$$G(s) = \frac{X_o(s)}{E(s)} \qquad (2\text{-}55)$$

Fig. 2-30 Feedback connection
（反馈连接）

$H(s)$ 称为反馈回路传递函数，即

$$H(s) = \frac{B(s)}{X_o(s)} \qquad (2\text{-}56)$$

前向通道传递函数 $G(s)$ 与反馈回路传递函数 $H(s)$ 之乘积定义为系统的开环传递函数 $G_K(s)$，它是反馈信号 $B(s)$ 与偏差 $E(s)$ 之比，即

$$G_K(s) = \frac{B(s)}{E(s)} = G(s) H(s) \qquad (2\text{-}57)$$

由于 $B(s)$ 与 $E(s)$ 在相加点的量纲相同，因此，开环传递函数无量纲。

输出信号 $X_o(s)$ 与输入信号 $X_i(s)$ 之比，定义为系统的闭环传递函数 $\Phi(s)$，即

$$\Phi(s) = \frac{X_o(s)}{X_i(s)} \qquad (2\text{-}58)$$

由图 2-30 可知

$$E(s) = X_i(s) \mp B(s) = X_i(s) \mp X_o(s) H(s)$$

$$X_o(s) = G(s) E(s) = G(s) [X_i(s) \mp X_o(s) H(s)]$$

$$= G(s) X_i(s) \mp G(s) X_o(s) H(s)$$

connection is mainly adopted in a control system.

The negative feedback closed-loop system as is shown in Fig. 2-31 is called the unit negative feedback system when $H(s)=1$. According to equation (2-59), the closed-loop transfer function of the unit negative feedback system is

$$\Phi(s) = \frac{G(s)}{1+G(s)}$$

The closed-loop characteristic equation of the system is

$$D(s) = 1 \pm G_K(s) = 0 \tag{2-60}$$

For closed-loop systems, this section introduces the concept of open-loop transfer function $G_K(s)$ and closed-loop transfer function $\Phi(s)$. For open-loop systems, the transfer function of the system cannot be called open-loop transfer function, while it should be called transfer function of open-loop system.

2.5.2 The closed-loop control system under disturbances

Fig. 2-32 shows the closed-loop control system under disturbance. The disturbance signal is also an input to the system. For example, the load of a machine, the error of a mechanical transmission system, the change of environment, the electrical noise of a system, etc. can all affect the output of the system in the form of input. For a linear system, the output of each input can be calculated separately, and each corresponding output can be superimposed to obtain the total output of the system.

Assuming that $X(s) \neq 0$, $N(s) = 0$, the output $Y_1(s)$ of the system is

$$Y_1(s) = \frac{G_1(s)G_2(s)}{1+G_1(s)G_2(s)H(s)} X(s) \tag{2-61}$$

Assuming that $X(s) = 0$, $N(s) \neq 0$, the output $Y_2(s)$ of the system is

$$Y_2(s) = \frac{G_2(s)}{1+G_1(s)G_2(s)H(s)} N(s) \tag{2-62}$$

If the input and disturbance are applied to the system simultaneously, the output $Y(s)$ of the system is obtained by superimposing the above two outputs, i.e.

$$Y(s) = Y_1(s) + Y_2(s) = \frac{G_2(s)}{1+G_1(s)G_2(s)H(s)} [G_1(s)X(s) + N(s)] \tag{2-63}$$

If $|G_1(s)H(s)| \gg 1$, $|G_K| = |G_1(s)G_2(s)H(s)| \gg 1$, according to equation (2-62), the output $Y_2(s)$ caused by the disturbance is

$$Y_2(s) = \frac{G_2(s)}{1+G_1(s)G_2(s)H(s)} N(s) \approx \frac{G_2(s)}{G_1(s)G_2(s)H(s)} N(s) \approx \frac{1}{G_1(s)H(s)} N(s) \approx \delta N(s)$$

Since $|G_1(s)H(s)| \gg 1$, δ is a minimum value. The output variable $Y_2(s)$ caused by the disturbance approaches to zero, and the disturbance is effectively suppressed. If there is no feedback loop in the system, i.e. $H(s) = 0$, the system becomes an open-loop system. Then, the output $Y_2(s) = G_2(s)N(s)$ cannot be eliminated, and all errors are generated by the output. Therefore, it can be concluded that the closed-loop control system has good anti-disturbance performance.

由此可得闭环传递函数为

$$\Phi(s) = \frac{X_o(s)}{X_i(s)} = \frac{G(s)}{1 \pm G(s)H(s)} = \frac{G(s)}{1 \pm G_K(s)} \tag{2-59}$$

式中,"+"对应负反馈;"-"对应正反馈。正反馈是通过增强输入部分使偏差信号增大,而负反馈是通过削弱输入部分使偏差信号减小。因此,在控制系统中,主要采用负反馈连接。

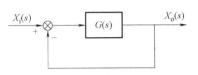

如图 2-31 所示的负反馈闭环系统,当 $H(s)=1$ 时,称为单位负反馈系统。由式(2-59)可得,单位负反馈系统的闭环传递函数为

Fig. 2-31 Unit negative feedback system
(单位负反馈系统)

$$\Phi(s) = \frac{G(s)}{1 + G(s)}$$

系统的闭环特征方程为

$$D(s) = 1 \pm G_K(s) = 0 \tag{2-60}$$

针对闭环系统,本节引入了开环传递函数 $G_K(s)$ 及闭环传递函数 $\Phi(s)$ 的概念。对于开环系统,系统的传递函数不能称为开环传递函数,而应称为开环系统传递函数。

2.5.2 扰动作用下的闭环控制系统

图 2-32 所示为在扰动作用下的闭环控制系统。扰动信号也是系统的一种输入量。例如,机器的负载、机械传动系统的误差、环境的变化、系统中的电气噪声等都能以输入的形式影响系统的输出量。对于线性系统,可以单独计算每个输入量作用下的输出量,再将各个相应的输出量叠加得到系统的总输出量。

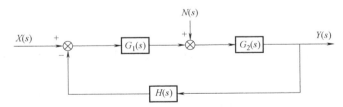

Fig. 2-32 The closed-loop control system under the disturbance
(扰动作用下的闭环控制系统)

设 $X(s) \neq 0$,$N(s) = 0$,系统的输出 $Y_1(s)$ 为

$$Y_1(s) = \frac{G_1(s)G_2(s)}{1 + G_1(s)G_2(s)H(s)} X(s) \tag{2-61}$$

设 $X(s) = 0$,$N(s) \neq 0$,系统的输出 $Y_2(s)$ 为

$$Y_2(s) = \frac{G_2(s)}{1 + G_1(s)G_2(s)H(s)} N(s) \tag{2-62}$$

若输入和扰动同时施加于该系统,则系统的输出 $Y(s)$ 由上述两个输出叠加得到,即

2.5.3 The drawing of block diagrams

In order to draw the block diagram of a control system, the following steps must be applied.

1) Write the motion equation for each component of the system.

2) Take the Laplace transform on each equation and transform it into an input-output relation under zero initial conditions.

3) Represent each input-output relation by using a block diagram unit.

4) Connect the same signals of block diagram units, draw the input signal of the system on the left side and the output signal of the system on the right side to form a complete block diagram of the control system.

Example 2-14 Draw the block diagram of a passive network shown in Fig. 2-33.

Solution: The differential equations of the passive network are established as

$$\begin{cases} u_i(t) = i_2(t)R_1 + u_o(t) \\ i_2(t)R_1 = \dfrac{1}{C}\int i_1(t)\,dt \\ i(t) = i_1(t) + i_2(t) \\ u_o(t) = i(t)R_2 \end{cases}$$

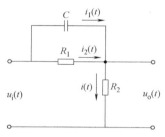

Fig. 2-33 The RC passive network （RC 无源网络）

Taking the Laplace transform on the above equations under zero initial conditions, we have

$$\begin{cases} U_i(s) = I_2(s)R_1 + U_o(s) \\ I_2(s)R_1 = \dfrac{1}{Cs}I_1(s) \\ I(s) = I_1(s) + I_2(s) \\ U_o(s) = I(s)R_2 \end{cases}$$

According to the above equations, the block diagram of each element is depicted in Fig. 2-34.

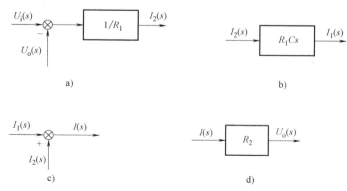

Fig. 2-34 The block diagram of each part （各环节框图）

Finally, an entire block diagram of the system, which can be obtained by combining the block

$$Y(s) = Y_1(s) + Y_2(s) = \frac{G_2(s)}{1+G_1(s)G_2(s)H(s)}[G_1(s)X(s)+N(s)] \qquad (2\text{-}63)$$

若 $|G_1(s)H(s)| \gg 1$，$|G_K| = |G_1(s)G_2(s)H(s)| \gg 1$，则由式（2-62）可知，扰动引起的输出 $Y_2(s)$ 为

$$Y_2(s) = \frac{G_2(s)}{1+G_1(s)G_2(s)H(s)}N(s) \approx \frac{G_2(s)}{G_1(s)G_2(s)H(s)}N(s) \approx \frac{1}{G_1(s)H(s)}N(s) \approx \delta N(s)$$

由于 $|G_1(s)H(s)| \gg 1$，因此 δ 为极小值，则由扰动引起的输出量 $Y_2(s)$ 趋近于零，有效地抑制了干扰。而如果系统没有反馈回路，即 $H(s)=0$，则系统成为一开环系统，此时输出 $Y_2(s) = G_2(s)N(s)$ 无法被消除，全部误差由该输出产生。由此可见，闭环控制系统具有良好的抗干扰性能。

2.5.3 框图的绘制

绘制控制系统的框图的流程如下。

1）列写系统各组成部分的运动方程。
2）在零初始条件下，对各方程进行拉氏变换，并整理成输入输出关系式。
3）采用框图单元表示每一个输入输出关系式。
4）将各框图单元相同的信号连接起来，并将系统的输入画在左侧，输出画在右侧，构成控制系统完整的框图。

例 2-14　绘制图 2-33 所示无源网络的框图。

解：先列写该网络的微分方程，得

$$\begin{cases} u_i(t) = i_2(t)R_1 + u_o(t) \\ i_2(t)R_1 = \dfrac{1}{C}\int i_1(t)\,\mathrm{d}t \\ i(t) = i_1(t) + i_2(t) \\ u_o(t) = i(t)R_2 \end{cases}$$

对上述各式在零初始条件下分别进行拉氏变换，有

$$\begin{cases} U_i(s) = I_2(s)R_1 + U_o(s) \\ I_2(s)R_1 = \dfrac{1}{Cs}I_1(s) \\ I(s) = I_1(s) + I_2(s) \\ U_o(s) = I(s)R_2 \end{cases}$$

根据上述各式绘制各环节框图如图 2-34 所示。

最后，将上面各环节的框图组合，得到系统的总框图如图 2-35 所示。

2.5.4 框图的简化

实际的自动控制系统通常用多回路的函数框图表示，其框图甚为复杂。为便于分析与计

diagrams of the above elements, is shown in Fig. 2-35.

2.5.4 The simplification of block diagrams

For the actual automatic control system, it is usually represented by the function block diagram of multi-loop, which is very complex. In order to facilitate its analysis and calculation, the complex block diagram needs to be simplified. For this purpose, two solving methods are introduced below.

1. The equivalent transformation of function block diagrams

The operation and transformation of the block diagrams should be carried out according to the equivalent principle, which means the input and output are unchanged, and the mathematical relation between the input and output is unchanged. Obviously, the essence of the transformation is equivalent to the elimination of the equations describing the system, and the relation between the input and output of the system is obtained. Table 2-2 lists some algebraic rules for the transformation of typical function block diagrams.

Example 2-15 Simplify the block diagram shown in Fig. 2-36.

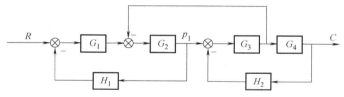

Fig. 2-36 The block diagram of a system （系统框图）

Solution: According to Fig. 2-36, it can be seen that there are no serial blocks, no parallel blocks, and no feedback loop to be simplified. As is shown in Fig. 2-37a, move summing point s_2 ahead of its left block and exchang its position with summing point s_1; then move tie point p_2 behind its right block, and exchang the positions of p_2 and p_3.

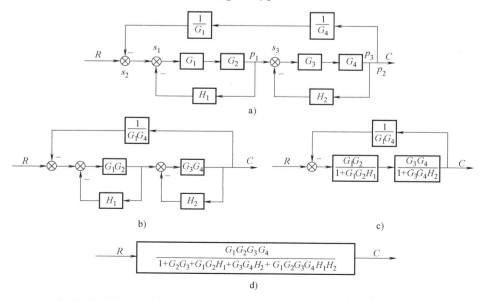

Fig. 2-37 The simplification process of the block diagram （框图简化过程）

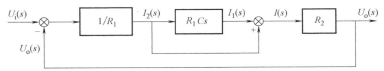

Fig. 2-35　The block diagram of the system for the passive network

（无源网络系统框图）

算，需要对复杂框图进行简化，为此，下文将介绍两种求解方法。

1. 框图的等效变换

框图的运算和变换应按等效原则进行，即变换前后，输入量和输出量不变，输入输出之间的数学关系不变。显然，变换的实质相当于对描述系统的方程组进行消元，求出系统输入输出的关系式。表 2-2 列举了一些典型函数框图变换的代数法则。

例 2-15　对图 2-36 所示的框图进行简化。

解：由图 2-36 可知，图中没有可以简化的串联方框、并联方框，也没有可以简化的反馈回路。因此，如图 2-37a 所示，首先将相加点 s_2 移动到它左侧方框的前面，并与相加点 s_1 交换位置；然后，将分支点 p_2 移动到它右侧方框的后面，并交换 p_2 和 p_3 的位置。

可以看出，由 $G_1(s)$ 和 $G_2(s)$ 组成的部分以及由 $G_3(s)$ 和 $G_4(s)$ 组成的部分均确认为串联连接，并可以简化为如图 2-37b 所示的等效方框。可以看出有两个简单的负反馈回路，并可以简化成如图 2-37c 所示的等效方框。接着，我们可以把给定的框图简化成如图 2-37d 所示单一的方框。最后，获得系统传递函数为

$$\Phi(s) = \frac{C(s)}{R(s)} = \frac{G_1 G_2 G_3 G_4}{1 + G_2 G_3 + G_1 G_2 H_1 + G_3 G_4 H_2 + G_1 G_2 G_3 G_4 H_1 H_2}$$

2. 梅逊增益公式

对于连接关系比较复杂的系统框图，用上述简化方法求取系统的传递函数有时非常困难，还容易出错。此时可利用梅逊增益公式，直接由框图求取系统的传递函数，而不用对框图进行简化。

梅逊增益公式可表示为

$$G(s) = \frac{X_o(s)}{X_i(s)} = \frac{1}{\Delta} \sum_k P_k \cdot \Delta_k \tag{2-64}$$

式中，$G(s)$ 为系统的传递函数；P_k 为第 k 条前向通道的传递函数；Δ 为框图特征式，Δ_k 为框图特征式关于第 k 条前向通道的余子式，即为去掉与第 k 条前向通道相接触的回路后的特征式。

框图特征式可表示为

$$\Delta = 1 - \sum L_a + \sum L_b L_c - \sum L_d L_e L_f + \cdots$$

式中，$\sum L_a$ 为所有不同回路的传递函数增益之和；$\sum L_b L_c$ 为所有互不接触的两个回路的传递函数增益乘积的和；$\sum L_d L_e L_f$ 为所有均互不接触的三个回路的传递函数增益乘积的和。

例 2-16　利用梅逊增益公式，求图 2-38 所示系统的传递函数。

It can be seen that the part formed by $G_1(s)$, $G_2(s)$ and the part formed by $G_3(s)$, $G_4(s)$ are all identified as series connection, and can be simplified to the equivalent blocks as is shown in Fig. 2-37b; then two negative feedback loops can be identified and simplified to the equivalent blocks as is shown in Fig. 2-37c; next we can simplify the given block diagram to a simple block as is shown in Fig. 2-37d. Finally, the transfer function of the system is obtained as

$$\Phi(s) = \frac{C(s)}{R(s)} = \frac{G_1 G_2 G_3 G_4}{1 + G_2 G_3 + G_1 G_2 H_1 + G_3 G_4 H_2 + G_1 G_2 G_3 G_4 H_1 H_2}$$

2. Mason's gain formula

For a system block diagram with a complicated connection relation, it is sometimes difficult and error-prone to obtain the transfer function of the system by applying the above-described simplification method. Under this circumstances, the transfer function of the system can be directly obtained from the block diagram by using the Mason's gain formula without simplifying the block diagram.

The Mason's gain formula can be expressed as

$$G(s) = \frac{X_o(s)}{X_i(s)} = \frac{1}{\Delta} \sum_k P_k \cdot \Delta_k \qquad (2\text{-}64)$$

where $G(s)$ is the transfer function of the system; P_k is the transfer function of the kth forward path; Δ is the determinant of the block diagram, Δ_k is the cofactor of the block diagram determinant for the kth forward path, i.e., the determinant with the loops touching the kth forward path removed.

The determinant of the block diagram can be described as

$$\Delta = 1 - \sum L_a + \sum L_b L_c - \sum L_d L_e L_f + \cdots$$

where $\sum L_a$ is the sum of the transfer function gains of all different loops; $\sum L_b L_c$ is the sum of the transfer function gain products of all combination of two non-touching loops; $\sum L_d L_e L_f$ is the sum of the transfer function gain products of all combination of three non-touching loops.

Example 2-16 By using Mason's gain formula, figure out the transfer function of the system shown in Fig. 2-38.

Solution: There are two forward paths and five feedback loops in the system, and the loops touch each other.

$$P_1 = G_1 G_2 G_3 \quad P_2 = G_1 G_4 \quad L_1 = -G_1 G_2 H_1 \quad L_2 = -G_2 G_3 H_2$$
$$L_3 = -G_1 G_2 G_3 H_3 \quad L_4 = -G_1 G_4 H_3 \quad L_5 = -G_4 H_2$$
$$\Delta = 1 + G_1 G_2 H_1 + G_2 G_3 H_2 + G_1 G_2 G_3 H_3 + G_1 G_4 H_3 + G_4 H_2$$

Since five loops are in touch with the two forward paths, we have

$$\Delta_1 = 1 \quad \Delta_2 = 1$$

According to Mason's gain formula, the transfer function of the system is obtained as

$$G(s) = \frac{C(s)}{R(s)} = \frac{G_1 G_2 G_3 + G_1 G_4}{1 + G_1 G_2 H_1 + G_2 G_3 H_2 + G_1 G_2 G_3 H_3 + G_1 G_4 H_3 + G_4 H_2}$$

Example 2-17 By using Mason's gain formula, figure out the transfer function of the system shown in Fig. 2-39.

Solution: There is one forward path and four feedback loops in the system, and there is a pair of loops (L_1 and L_2) that are non-touching, so we have

Table 2-2 Algebraic rules for the transformation of block diagrams（框图变换的代数法则）

Number(序号)	Original block diagrams(原框图)	Equivalent block diagrams(等效框图)
1		
2		
3		
4		
5		
6		
7		
8		
9		
10		
11		
12		
13		

$$P_1 = G_1G_2G_3G_4G_5G_6 \quad L_1 = -G_2G_3H_2 \quad L_2 = G_4G_5H_3$$
$$L_3 = -G_1G_2G_3G_4G_5G_6H_1 \quad L_4 = -G_3G_4H_4$$
$$\Delta = 1 + G_2G_3H_2 - G_4G_5H_3 + G_1G_2G_3G_4G_5G_6H_1 + G_3G_4H_4 - G_2G_3G_4G_5H_2H_3$$
$$\Delta_1 = 1$$

According to Mason formula, the transfer function of the system is obtained as

$$G(s) = \frac{C(s)}{R(s)} = \frac{G_1G_2G_3G_4G_5G_6}{1 + G_2G_3H_2 - G_4G_5H_3 + G_1G_2G_3G_4G_5G_6H_1 + G_3G_4H_4 - G_2G_3G_4G_5H_2H_3}$$

2.6 Mathematical Models of Typical Control Systems

A complex physical system is often composed of various components, which can be electrical, mechanical, hydraulic, pneumatic, optical, thermodynamic, etc. According to the physical principle, the analytical method can be used to find the differential equation of the system, as long as the input and output of the system are determined, and then the transfer function of the system can be deduced. Discussion about some examples of typical systems is as follows.

2.6.1 Mechanical systems

1. Translational systems

A translational motion is a type of motion that takes place along a straight line. As is shown in Fig. 2-40, the basic elements of a translational system generally are mass, damper and spring. The common variables used to describe the dynamic performance of the translational system are force, displacement, velocity and acceleration.

(1) Mass Fig. 2-40a shows a mass M subjected to an applied force $f_M(t)$, and its output is a displacement $x(t)$. According to Newton's second law, we have

$$f_M(t) = Ma = M\frac{d^2x(t)}{dt^2} \quad (2\text{-}65)$$

where a represents the acceleration. It is pointed out that the mass is rigid.

(2) Damper Fig. 2-40b shows a damper subjected to an applied force $f_B(t)$, and its output is a displacement $x(t)$. A mathematical model of the damper is

$$f_B(t) = Bv = B\frac{dx(t)}{dt} \quad (2\text{-}66)$$

where v represents the velocity; B is the coefficient of viscous friction.

(3) Spring Fig. 2-40c shows a spring subjected to an applied force $f_K(t)$, and its output is a displacement $x(t)$. According to Hooke's law, we have

$$f_K(t) = Kx(t) \quad (2\text{-}67)$$

where, K represents the spring coefficient.

Example 2-18 Fig. 2-41 shows a simple spring-mass-damper mechanical system, where an external force $f(t)$ is applied and the motion is indicated by the displacement $x(t)$. Figure out the transfer function of the system.

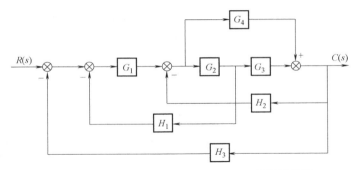

Fig. 2-38 The block diagram of a system（系统框图）

解：本系统有两条前向通道，五个反馈回路，且回路均互相接触。所以

$$P_1 = G_1G_2G_3 \quad P_2 = G_1G_4$$

$$L_1 = -G_1G_2H_1 \quad L_2 = -G_2G_3H_2 \quad L_3 = -G_1G_2G_3H_3 \quad L_4 = -G_1G_4H_3 \quad L_5 = -G_4H_2$$

$$\Delta = 1 + G_1G_2H_1 + G_2G_3H_2 + G_1G_2G_3H_3 + G_1G_4H_3 + G_4H_2$$

又因为五个回路均与两条前向通道接触，所以

$$\Delta_1 = 1 \quad \Delta_2 = 1$$

则根据梅逊增益公式可求得该系统的传递函数为：

$$G(s) = \frac{C(s)}{R(s)} = \frac{G_1G_2G_3 + G_1G_4}{1 + G_1G_2H_1 + G_2G_3H_2 + G_1G_2G_3H_3 + G_1G_4H_3 + G_4H_2}$$

例 2-17 利用梅逊增益公式，求图 2-39 所示系统的传递函数。

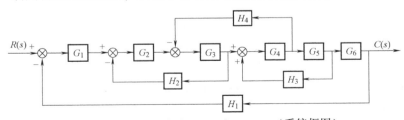

Fig. 2-39 The block diagram of a system（系统框图）

解：本系统有一条前向通道，四个反馈回路 L_1、L_2、L_3、L_4，有一对回路（L_1 与 L_2）互不接触，则有

$$P_1 = G_1G_2G_3G_4G_5G_6 \quad L_1 = -G_2G_3H_2 \quad L_2 = G_4G_5H_3 \quad L_3 = -G_1G_2G_3G_4G_5G_6H_1 \quad L_4 = -G_3G_4H_4$$

$$\Delta = 1 + G_2G_3H_2 - G_4G_5H_3 + G_1G_2G_3G_4G_5G_6H_1 + G_3G_4H_4 - G_2G_3G_4G_5H_2H_3 \quad \Delta_1 = 1$$

则根据梅逊增益公式，可求得该系统的传递函数为

$$G(s) = \frac{C(s)}{R(s)} = \frac{G_1G_2G_3G_4G_5G_6}{1 + G_2G_3H_2 - G_4G_5H_3 + G_1G_2G_3G_4G_5G_6H_1 + G_3G_4H_4 - G_2G_3G_4G_5H_2H_3}$$

2.6 典型控制系统的数学模型

一个复杂物理系统常常由多种元件组成，可以是电气元件、机械元件、液压元件、气动元件、光学元件、热力学元件等。根据物理原理，只要确定这些系统输入和输出，就可以应用解析法求出系统的微分方程，从而推导出系统的传递函数。下面对一些典型系统的例子进

Solution: Taking mass M as the research object, according to Newton's second law, we have

$$f(t) - Kx(t) - B\frac{dx(t)}{dt} = M\frac{d^2x(t)}{dt^2}$$

Arranging the above equation, we have

$$M\frac{d^2x(t)}{dt^2} + B\frac{dx(t)}{dt} + Kx(t) = f(t)$$

Taking the Laplace transform on the above equation under the zero initial conditions, we have

$$(Ms^2 + Bs + K)X(s) = F(s)$$

So, the transfer function of the system is

$$\frac{X(s)}{F(s)} = \frac{1}{Ms^2 + Bs + K}$$

2. Rotational systems

A rotational motion of a subject describes the motion around a fixed axis. The variables used to describe the dynamic performance of the rotational system are torque, angular displacement, rotational velocity and rotational acceleration. The linear rotational systems are analogues to the linear translational systems. As is shown in Fig. 2-42, the basic elements of a rotational system generally are moment of inertia element, viscous friction element and torsion spring.

(1) **Moment of inertia element** Fig. 2-42a shows the Moment of inertia element, and its mathematical model is defined as

$$T(t) = J\frac{d^2\theta(t)}{dt^2} \tag{2-68}$$

where $T(t)$ represents the applied torque; J is the moment of inertia; $\theta(t)$ is the angle of rotation.

(2) **Viscous friction element** Fig. 2-42b shows the viscous friction element, and its mathematical model is defined as

$$T(t) = B\frac{d\theta(t)}{dt} \tag{2-69}$$

where B is the damping coefficient.

(3) **Torsion spring** Fig. 2-42c shows the torsion spring element, and its mathematical model is defined as

$$T(t) = K\theta(t) \tag{2-70}$$

where K represents the torsion spring coefficient.

Example 2-19 The torsional vibration control of the long shaft in a machine is the problem that should be solved in mechanical engineering. According to the flywheel and mass distribution on the shaft, the shaft can be equivalent to a simplified mode of several concentrated mass wheels and torsion springs in connection, as is shown in Fig. 2-43, where J_1, J_2 are the equivalent moments of inertia of the shaft, K_1, K_2 are the stiffness of the torsion spring, and B is the angular friction coefficient of the shaft. The rotation angle $\theta_i(t)$ at one end of the shaft is selected as the input, and the rotation angle $\theta_o(t)$ at the other end is the output. Establish a mathematical model of the system.

行讨论。

2.6.1 机械系统

1. 平移系统

平移运动是一种沿直线进行的运动。如图 2-40 所示，平移系统的基本元件一般包括质量、阻尼器和弹簧。通常用来描述平移系统动态性能的变量有力、位移、速度和加速度。

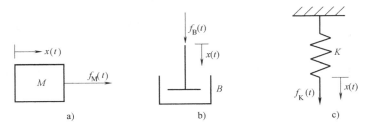

Fig. 2-40　The basic elements of a translational system（平移系统的基本元件）
a) Mass（质量）　b) Damper（阻尼器）　c) Spring（弹簧）

（1）质量　图 2-40a 所示质量 M 受到作用力 $f_M(t)$ 的作用，其输出为位移 $x(t)$。根据牛顿第二定律，有

$$f_M(t) = Ma = M\frac{d^2 x(t)}{dt^2} \tag{2-65}$$

式中，a 表示加速度。需要指出，这里的质量是刚性的。

（2）阻尼器　图 2-40b 所示阻尼器受到作用力 $f_B(t)$ 的作用，其输出为位移 $x(t)$。阻尼器的数学模型为

$$f_B(t) = Bv = B\frac{dx(t)}{dt} \tag{2-66}$$

式中，v 表示速度；B 是黏性摩擦系数。

（3）弹簧　图 2-40c 所示弹簧受到作用力 $f_K(t)$ 的作用，其输出为位移 $x(t)$。根据胡克定律，有

$$f_K(t) = Kx(t) \tag{2-67}$$

式中，K 是弹簧系数。

例 2-18　图 2-41 所示是一个简单的弹簧-质量-阻尼器机械系统，系统受到外力 $f(t)$ 的作用，其运动用位移 $x(t)$ 表示，试建立系统的数学模型。

解：以质量 M 为研究对象，根据牛顿第二定律，有

$$f(t) - Kx(t) - B\frac{dx(t)}{dt} = M\frac{d^2 x(t)}{dt^2}$$

整理上式，有

$$M\frac{d^2 x(t)}{dt^2} + B\frac{dx(t)}{dt} + Kx(t) = f(t)$$

在零初始条件下进行拉氏变换，得

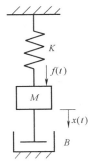

Fig. 2-41　The model of a translational system（平移运动模型）

Solution: Assuming that the intermediate variable of J_1 and J_2 is $\theta_m(t)$, according to the kinetic principle, we have

$$\begin{cases} T_1(t) = K_1[\theta_i(t) - \theta_m(t)] \\ T_1(t) - T_2(t) = J_1 \dfrac{d^2\theta_m(t)}{dt^2} \\ T_2(t) = K_2[\theta_m(t) - \theta_o(t)] \\ T_2(t) - B\dfrac{d\theta_o(t)}{dt} = J_2\dfrac{d^2\theta_o(t)}{dt^2} \end{cases}$$

Taking the Laplace transform on the above equations under the zero initial conditions, we have

$$\begin{cases} T_1(s) = K_1[\theta_i(s) - \theta_m(s)] \\ T_1(s) - T_2(s) = J_1 s^2 \theta_m(s) \\ T_2(s) = K_2[\theta_m(s) - \theta_o(s)] \\ T_2(s) - Bs\theta_o(s) = J_2 s^2 \theta_o(s) \end{cases}$$

Each of the above equations can be understood as an element. Draw a block diagram of each element and combine them into the entire block diagram of the system, as is shown in Fig. 2-44.

Therefore, the transfer function of the system is obtained by simplifying the block diagram of the system as

$$G(s) = \dfrac{K_1 K_2}{J_1 J_2 s^4 + J_1 B s^3 + [(K_1+K_2)J_2 + K_2 J_1]s^2 + (K_1+K_2)Bs + K_1 K_2}$$

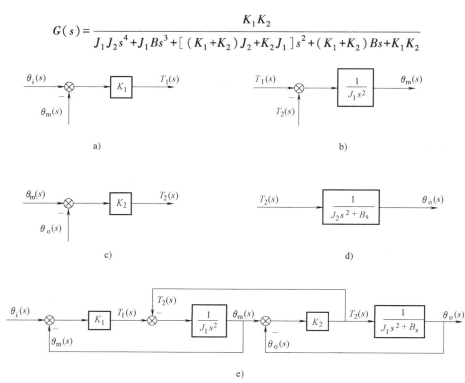

Fig. 2-44 The block diagrams of each part for the system and its entire block diagram
(系统各部分的框图及总的框图)

$$(Ms^2+Bs+K)X(s)=F(s)$$

因此，系统的传递函数为

$$\frac{X(s)}{F(s)}=\frac{1}{Ms^2+Bs+K}$$

2. 旋转系统

一个物体的旋转运动可描述为围绕固定轴的运动。用来描述旋转系统动态性能的变量有扭矩、角位移、转速和旋转加速度。线性转动系统类似于线性平移系统。如图 2-42 所示，转动系统的基本元件一般包括转动惯量元件、黏性摩擦元件和扭簧。

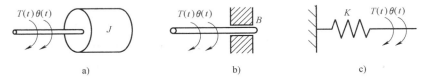

Fig. 2-42 The basic elements of a rotational system（旋转系统的基本元件）
a) Moment of inertia element（转动惯量元件） b) Viscous friction element（黏性摩擦元件）
c) Torsion spring（扭簧）

（1）转动惯量元件 图 2-42a 所示为转动惯量元件，其数学模型可以定义为

$$T(t)=J\frac{\mathrm{d}^2\theta(t)}{\mathrm{d}t^2} \tag{2-68}$$

式中，$T(t)$ 为外加的力矩；J 为转动惯量；$\theta(t)$ 为旋转的角度。

（2）黏性摩擦元件 图 2-42b 所示为黏性摩擦元件，其数学模型可以定义为

$$T(t)=B\frac{\mathrm{d}\theta(t)}{\mathrm{d}t} \tag{2-69}$$

式中，B 是阻尼系数。

（3）扭簧 图 2-42c 所示为扭簧，其数学模型可以定义为

$$T(t)=K\theta(t) \tag{2-70}$$

式中，K 是扭簧系数。

例 2-19 机器中长轴的扭转振动控制是机械工程中应解决的问题。根据轴上的飞轮和质量分布情况，可以将轴等效为几个集中质量轮和扭簧相连接的简化模型，如图 2-43 所示。其中，轴的等效转动惯量为 J_1、J_2，扭簧的刚度为 K_1、K_2，轴的角摩擦系数为 B。选择轴一端的转角 $\theta_i(t)$ 为输入，另一端的转角 $\theta_o(t)$ 为输出。建立系统的数学模型。

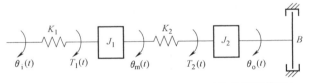

Fig. 2-43 The torsion model of an axis（轴的扭转模型）

解：设 J_1 和 J_2 的中间变量为 $\theta_m(t)$，根据动力学原理，可得

2.6.2 Electrical systems

As is shown in Fig. 2-45, the elements in an electrical system that we should consider are resistor, capacitor, and inductor. They are analogous to the elements of the mechanical system. Here, we directly establish the relation between the voltage and current for each element in the electrical system.

(1) Resistor Fig. 2-45a shows the element of resistor, and according to Ohm's law, we have

$$U_R(t) = Ri(t) \tag{2-71}$$

where $U_R(t)$ is the voltage of the resistor; $i(t)$ is the input current; R is the resistance.

(2) Capacitor Fig. 2-45b shows the element of capacitor, and the relation between the voltage $U_C(t)$ of the capacitor and the input current $i(t)$ can be expressed as

$$U_C(t) = \frac{1}{C}\int i(t)\,\mathrm{d}t \tag{2-72}$$

where C is the capacitance coefficient.

(3) Inductor Fig. 2-45c shows the element of inductor, and according to Faraday's law, its voltage $U_L(t)$ is defined as

$$U_L(t) = L\frac{\mathrm{d}i(t)}{\mathrm{d}t} \tag{2-73}$$

where L is the inductance coefficient.

Example 2-20 Fig. 2-46 shows the passive filter network where two RC circuits are connected in series. The voltage of the network at one end is selected as the input, and the voltage of the network at the other end is selected as the output. Establish a mathematical model of the system.

Solution: Assume that the intermediate variable is $U_m(s)$ in the series element. According to Kirchhoff's law, we have

$$\begin{cases} u_i(t) = R_1 i_1(t) + u_m(t) \\ u_m(t) = R_2 i_2(t) + u_o(t) \\ u_m(t) = \frac{1}{C_1}\int [i_1(t) - i_2(t)]\,\mathrm{d}t \\ u_o(t) = \frac{1}{C_2}\int i_2(t)\,\mathrm{d}t \end{cases}$$

Taking the Laplace transform on the above equations under the zero initial conditions, we have

$$\begin{cases} U_i(s) = R_1 I_1(s) + U_m(s) \\ U_m(s) = R_2 I_2(s) + U_o(s) \\ U_m(s) = \frac{1}{C_1 s}[I_1(s) - I_2(s)] \\ U_o(s) = \frac{1}{C_2 s} I_2(s) \end{cases}$$

$$\begin{cases} T_1(t) = K_1[\theta_i(t) - \theta_m(t)] \\ T_1(t) - T_2(t) = J_1 \dfrac{d^2\theta_m(t)}{dt^2} \\ T_2(t) = K_2[\theta_m(t) - \theta_o(t)] \\ T_2(t) - B\dfrac{d\theta_o(t)}{dt} = J_2 \dfrac{d^2\theta_o(t)}{dt^2} \end{cases}$$

在零初始条件下，对上式进行拉氏变换，得

$$\begin{cases} T_1(s) = K_1[\theta_i(s) - \theta_m(s)] \\ T_1(s) - T_2(s) = J_1 s^2 \theta_m(s) \\ T_2(s) = K_2[\theta_m(s) - \theta_o(s)] \\ T_2(s) - Bs\theta_o(s) = J_2 s^2 \theta_o(s) \end{cases}$$

以上方程组中每个式子可以理解为一个环节。画出每个环节的框图并组合成系统框图，如图 2-44 所示。

简化系统框图，可得系统的传递函数为

$$G(s) = \dfrac{K_1 K_2}{J_1 J_2 s^4 + J_1 B s^3 + [(K_1 + K_2)J_2 + K_2 J_1]s^2 + (K_1 + K_2)Bs + K_1 K_2}$$

2.6.2 电气系统

如图 2-45 所示，电气系统中需要考虑的元件有电阻、电容和电感。它们类似于机械系统的元件。在这里，我们直接给出电气系统中每个元件的电压和电流之间的关系。

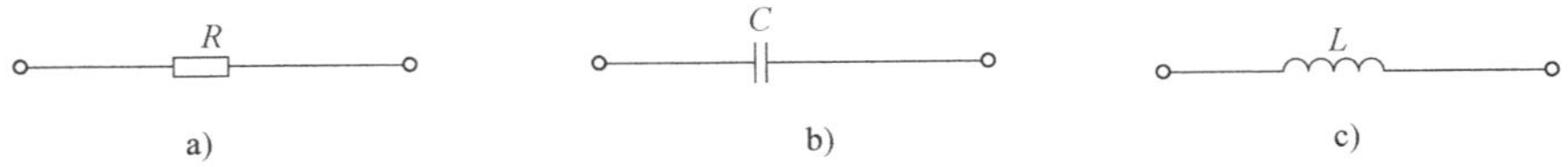

Fig. 2-45　The elements of an electrical system（电气系统的元件）
a）Resistor（电阻）　b）Capacitor（电容器）　c）Inductor（电感）

（1）电阻　图 2-45a 所示为电阻，根据欧姆定律，有

$$U_R(t) = Ri(t) \tag{2-71}$$

式中，$U_R(t)$ 为电阻的电压；$i(t)$ 为输入电流；R 为电阻值。

（2）电容　图 2-45b 所示为电容，电容电压 $U_C(t)$ 与输入电流 $i(t)$ 之间的关系可以表示为

$$U_C(t) = \dfrac{1}{C} \int i(t) \, dt \tag{2-72}$$

式中，C 是电容系数。

（3）电感　图 2-45c 所示为电感，根据法拉第定律，其电压 $U_L(t)$ 可以定义为

$$U_L(t) = L \dfrac{di(t)}{dt} \tag{2-73}$$

式中，L 是电感系数。

Draw a block diagram of each element and combine them into the entire block diagram of the system, as is shown in Fig. 2-47.

Therefore, by simplifying the block diagram of the system, the transfer function of the system is as

$$G(s) = \frac{1}{R_1 R_2 s^2 + (R_1 C_1 + R_2 C_2 + R_1 C_2)s + 1}$$

例 2-20 图 2-46 所示为两个 RC 电路串联组成的无源滤波网络，选择一端的电压为输入，另一端的电压为输出，建立系统的数学模型。

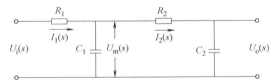

Fig. 2-46　The passive filter network（无源滤波网络）

解：假设在串联环节中设置中间变量为 $U_m(s)$，根据基尔霍夫定律，可得

$$\begin{cases} u_i(t) = R_1 i_1(t) + u_m(t) \\ u_m(t) = \dfrac{1}{C_1}\int [i_1(t) - i_2(t)] dt \end{cases} \quad \begin{matrix} u_m(t) = R_2 i_2(t) + u_o(t) \\ u_o(t) = \dfrac{1}{C_2}\int i_2(t) dt \end{matrix}$$

在零初始条件下，对上式进行拉氏变换得：

$$\begin{cases} U_i(s) = R_1 I_1(s) + U_m(s) \\ U_m(s) = R_2 I_2(s) + U_o(s) \\ U_m(s) = \dfrac{1}{C_1 s}[I_1(s) - I_2(s)] \\ U_o(s) = \dfrac{1}{C_2 s} I_2(s) \end{cases}$$

根据上面式子分别画出框图并组合成系统框图，如图 2-47 所示。

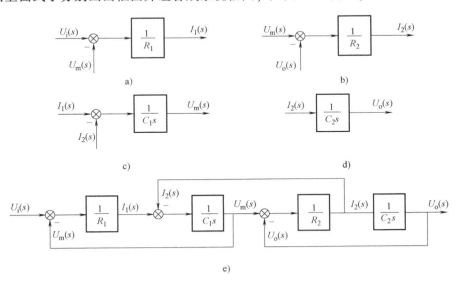

Fig. 2-47　The block diagrams of each part for the network and its entire block diagram
（网络各部分的框图及总的框图）

简化系统的框图，可得系统的传递函数为

$$G(s) = \dfrac{1}{R_1 R_2 s^2 + (R_1 C_1 + R_2 C_2 + R_1 C_2)s + 1}$$

Chapter 3   Time-Domain Analysis of Control Systems

第3章　控制系统的时域分析

The previous chapter describes how to build a mathematical model of a control system. It is an important step to analyze and design the system by using mathematical models. This chapter focuses on the characteristics analysis of the control system with time domain analysis.

The so-called time domain analysis method is applying a given input signal to the system and evaluating the system performance by studying the time response of the control system. Since the output of the system is generally a time function, the response is a time domain response. The method can directly analyze the system in the time domain, which is intuitive, accurate and clear, especially for first-order and second-order systems.

3.1 Transient Response of Control Systems

3.1.1 Typical input signals

To obtain the time domain response of a system, a given input signal needs to be applied to the system. In actual systems, although there are various input signals, they can be divided into deterministic signals and non-deterministic signals. A deterministic signal means that the relation between its variable and an independent variable can be described by a certain function. For example, in order to study the dynamic characteristics of a machine tool, a force $F = A\sin\omega t$ generated by an electromagnetic exciter is applied to the machine tool. The force is a deterministic signal. A non-deterministic signal means that the relation between its variable and an independent variable cannot be described by a certain function. That is, the relation between its variable and independent variable is random, and it is only subject to some statistical laws. For example, when a workpiece is processed on a lathe, the cutting force is a non-deterministic signal. Due to random factors such as the unevenness of the workpiece material and the change of the actual tool angle, it is impossible to express the changing law of the cutting force with a time function. As the input of the system is diverse, it is necessary to give some typical input signals when analyzing and designing the system, and then evaluate the time response of the system to the typical input signals. Although the input signal is rarely a typical input signal in a real-world environment, there is a relation between the time response of the system to a typical input signal and the time response of the system to any signal. Since

$$\frac{X_{01}(s)}{X_{i1}(s)} = G(s) = \frac{X_{02}(s)}{X_{i2}(s)} \tag{3-1}$$

it can be seen that the response of the system to any other input signal can be known as long as the response of the system to a typical input signal is obtained. The selection of typical input signals should meet the following requirements.

1) The selected input signal should reflect most of the actual conditions of the system at work.

2) The form of the selected input signal should be as simple as possible so that it can be expressed and analyzed by a mathematical model.

3) The input signals that enable the system to operate under the worst situation should be selected as typical input signals.

第3章 控制系统的时域分析

上一章讲述了如何建立控制系统的数学模型,用数学模型分析和设计系统是一个重要步骤。本章主要介绍用时域分析法对控制系统的特性进行分析。

所谓时域分析法,就是对系统施加一个给定输入信号,然后通过研究控制系统的时间响应来评价系统的性能。由于系统的输出量一般是时间的函数,故这种响应为时间响应。此方法可直接在时域中对系统进行分析,具有直观、准确、清楚的特点,尤其适用于一阶和二阶系统。

3.1 控制系统的瞬态响应

3.1.1 典型输入信号

欲求系统时间响应,则必须对系统施加一个给定输入信号。在实际系统中,输入信号虽然多种多样,但可分为确定性信号和非确定性信号。确定性信号是指其变量和自变量之间的关系能够用确定性函数描述的信号。例如,为了研究机床的动态特性,用电磁激振器给机床施加一个作用力 $F=A\sin\omega t$,这个作用力就是一个确定性信号。非确定性信号是其变量和自变量之间的关系不能用某一个确定的函数描述的信号,也就是说,它的变量和自变量之间的关系是随机的,只服从于某些统计规律。例如,车床加工工件时,切削力就是非确定性信号,由于工件材料的不均匀性和刀具实际角度的变化等随机因素的影响,切削力的变化规律是无法用确定的时间函数表示的。由于系统的输入具有多样性,因此在分析和设计系统时,需要给定一些典型的输入信号,然后评价系统对这些典型输入信号的时间响应。尽管在现实环境中,输入信号很少是典型输入信号,但系统对典型输入信号的时间响应和系统对任意信号的时间响应之间存在一定的关系。由于

$$\frac{X_{01}(s)}{X_{i1}(s)} = G(s) = \frac{X_{02}(s)}{X_{i2}(s)} \tag{3-1}$$

因此只要得出系统对典型输入信号的响应,就能知道系统对其他任意输入信号的响应。典型输入信号的选取应满足以下几个要求。

1) 选取的输入信号应反映系统工作时的大部分实际情况。
2) 所选输入信号的形式应尽可能简单,便于用数学模型表达分析及处理。
3) 应选取那些能使系统工作在最糟糕情况下的信号作为典型输入信号。

如果系统在典型输入信号下的性能满足要求,则可以断言系统在实际输入信号下的性能也令人满意。常用的典型输入信号介绍如下。

1. 单位阶跃信号

阶跃信号的输入变量有一个突然的定量变化,如图 3-1a 所示。其幅值高度等于 1 个单位时称为单位阶跃信号。有时它又称为位置信号,其数学表达式为

$$x_i(t) = 1(t) = \begin{cases} 1 & (t \geqslant 0) \\ 0 & (t < 0) \end{cases} \tag{3-2}$$

其拉氏变换式为

$$L[1(t)] = 1/s \tag{3-3}$$

阶跃信号是评价系统动态性能时常用的一种典型输入信号。实际工作环境中电源的突然

If the system performance can be satisfied under typical input signals, it can be asserted that the system performance under the actual input signal is also satisfactory. The typical input signals commonly used are described as follows.

1. Unit step signal

The input variable of step signal has a sudden quantitative change, as is shown in Fig. 3-1a. When its amplitude is equal to one unit, it is called the unit step signal. Sometimes, it is also called the position signal, and its mathematical expression is

$$x_i(t) = 1(t) = \begin{cases} 1 & (t \geq 0) \\ 0 & (t < 0) \end{cases} \tag{3-2}$$

Its Laplace transform is

$$L[1(t)] = 1/s \tag{3-3}$$

The step signal is a typical input signal that is often used to evaluate the dynamic performance of the system. In the actual working environment, it can be regarded as the step signal such as the sudden turning on and off of the power supply, the sudden change of the load, the conversion of the switch, etc.

2. Unit ramp signal

The input variable of ramp signal varies with the same velocity, as is shown in Fig. 3-1b. The signal whose slope is equal to one is called the unit ramp signal. Sometimes, it is also called the unit speed signal, and its mathematical expression is

$$x_i(t) = r(t) = \begin{cases} t & (t \geq 0) \\ 0 & (t < 0) \end{cases} \tag{3-4}$$

Its Laplace transform is

$$L[r(t)] = 1/s^2 \tag{3-5}$$

In practical work, the signal when the CNC machine tool is machining inclined plane and the signal when the large ship lock is rising and falling at a constant speed can be simulated by the ramp signal.

3. Unit parabolic signal

The input variable of parabolic signal varies with the same acceleration, as is shown in Fig. 3-1c. So the unit parabolic signal is also called the unit acceleration signal. Its mathematical expression is

$$x_i(t) = a(t) = \begin{cases} \dfrac{1}{2}t^2 & (t \geq 0) \\ 0 & (t < 0) \end{cases} \tag{3-6}$$

Its Laplace transform is

$$L[a(t)] = \frac{1}{s^3} \tag{3-7}$$

In actual working environment, this kind of signal is often used in a servo system to analyze its steady-state accuracy. For example, the feed instruction signal of the servo system to move with equal acceleration can be simulated by the parabolic signal.

接通与断开、负载的突变、开关的转换等均可视为阶跃信号。

2. 单位斜坡信号

斜坡信号的输入变量是等速度变化的,如图 3-1b 所示。其斜率等于 1 时称为单位斜坡信号,有时它又称为单位速度信号,其数学表达式为

$$x_i(t) = r(t) = \begin{cases} t & (t \geqslant 0) \\ 0 & (t < 0) \end{cases} \tag{3-4}$$

其拉氏变换为

$$L[r(t)] = 1/s^2 \tag{3-5}$$

实际工作中,数控机床加工斜面时的信号、大型船闸匀速升降时的信号等均可用斜坡信号模拟。

3. 单位抛物线信号

抛物线信号的输入变量是等加速度变化的,如图 3-1c 所示。故单位抛物线信号也称为单位加速度信号,其数学表达式为

$$x_i(t) = a(t) = \begin{cases} \dfrac{1}{2}t^2 & (t \geqslant 0) \\ 0 & (t < 0) \end{cases} \tag{3-6}$$

其拉氏变换为

$$L[a(t)] = \frac{1}{s^3} \tag{3-7}$$

在实际工作环境中,这类信号经常用于分析随动系统的稳态精度。例如,随动系统做等加速度移动的进给指令信号可用抛物线信号模拟。

4. 单位脉冲信号

单位脉冲信号的数学表达式为

$$x_i(t) = \delta(t) = \begin{cases} \dfrac{1}{h} & (0 \leqslant t \leqslant h, h \to 0) \\ 0 & (t < 0, t > h) \end{cases} \tag{3-8}$$

此处,定义脉冲面积

$$\int_{-\infty}^{+\infty} \delta(t) \, \mathrm{d}t = 1$$

单位脉冲信号如图 3-1d 所示,其脉冲高度为无穷大,持续时间为无穷小,脉冲面积为 1,因此单位脉冲信号的强度为 1。

应该指出,符合这种数学定义的理想脉冲函数在工程中是不可能出现的。为尽量接近单位脉冲信号,通常用宽度为 h 而高度为 $1/h$ 的信号作为单位脉冲信号。实际工作环境中的时间很短的冲击力、天线上的阵风扰动等可用脉冲信号模拟。单位脉冲信号的拉氏变换为

$$L[\delta(t)] = 1 \tag{3-9}$$

以上所述的单位脉冲信号、单位阶跃信号、单位斜坡信号以及单位抛物线信号之间的关系为

4. Unit pulse signal

The mathematical expression of the unit pulse signal is

$$x_i(t) = \delta(t) = \begin{cases} \dfrac{1}{h} & (0 \leqslant t \leqslant h,\ h \to 0) \\ 0 & (t < 0,\ t > h) \end{cases} \quad (3\text{-}8)$$

Here, the pulse area is defined as

$$\int_{-\infty}^{+\infty} \delta(t)\,\mathrm{d}t = 1$$

The unit pulse signal can be depicted as is shown in Fig. 3-1d. The pulse height is infinite. The duration is infinitesimal and the pulse area is 1. So the intensity of the unit pulse signal is 1.

It should be noted that an ideal pulse function that conforms to the mathematical definition is unlikely to occur in engineering. In order to be as close as possible to the unit pulse signal, a signal with a width h and a height of $1/h$ is usually used as a unit pulse signal. In actual working environment, short-time impact forces, and gust disturbances on the antenna can be simulated by pulse signals. The Laplace transform of unit pulse signal is

$$L[\delta(t)] = 1 \quad (3\text{-}9)$$

The relationship among the unit pulse signal, the unit step signal, the unit ramp signal and the unit parabolic signal described above is

$$\begin{cases} \delta(t) = \dfrac{\mathrm{d}}{\mathrm{d}t}[1(t)] \\ 1(t) = \dfrac{\mathrm{d}}{\mathrm{d}t}[r(t)] = \dfrac{\mathrm{d}}{\mathrm{d}t}[t \cdot 1(t)] \\ r(t) = \dfrac{\mathrm{d}}{\mathrm{d}t}[a(t)] \end{cases} \quad (3\text{-}10)$$

Therefore, according to the characteristics of the linear time-invariant system, if these signals are input to the same system separately, the relationship among their outputs is

$$\begin{cases} x_{o\delta}(t) = \dfrac{\mathrm{d}}{\mathrm{d}t}[x_{o1}(t)] \\ x_{o1}(t) = \dfrac{\mathrm{d}}{\mathrm{d}t}[x_{or}(t)] \\ x_{or}(t) = \dfrac{\mathrm{d}}{\mathrm{d}t}[x_{oa}(t)] \end{cases} \quad (3\text{-}11)$$

5. Unit sine signal

A signal as is shown in Fig. 3-1e is a unit sine signal. Its mathematical expression is

$$x_i(t) = \sin\omega t \quad (3\text{-}12)$$

Its Laplace transform is

$$L[x_i(t)] = \dfrac{\omega}{s^2 + \omega^2} \quad (3\text{-}13)$$

In actual working environment, it can be considered as the sine signal such as power supply fluctuations, mechanical vibrations, component noise jamming, etc. The sine signal is an input sig-

$$\begin{cases} \delta(t) = \dfrac{\mathrm{d}}{\mathrm{d}t}[1(t)] \\ 1(t) = \dfrac{\mathrm{d}}{\mathrm{d}t}[r(t)] = \dfrac{\mathrm{d}}{\mathrm{d}t}[t \cdot 1(t)] \\ r(t) = \dfrac{\mathrm{d}}{\mathrm{d}t}[a(t)] \end{cases} \quad (3\text{-}10)$$

因此，根据线性定常系统的特性，若分别将这些信号输入同一系统，则它们的输出量之间的关系为

$$\begin{cases} x_{o\delta}(t) = \dfrac{\mathrm{d}}{\mathrm{d}t}[x_{o1}(t)] \\ x_{o1}(t) = \dfrac{\mathrm{d}}{\mathrm{d}t}[x_{or}(t)] \\ x_{or}(t) = \dfrac{\mathrm{d}}{\mathrm{d}t}[x_{oa}(t)] \end{cases} \quad (3\text{-}11)$$

由以上关系可知，对于同一线性定常系统，对输入信号导数的响应等于系统对该输入信号响应的导数；对输入信号积分的响应等于系统对该输入信号响应的积分；积分常数由零初始条件确定。

5. 单位正弦信号

图 3-1e 所示信号即为单位正弦信号，其数学表达式为

$$x_i(t) = \sin\omega t \quad (3\text{-}12)$$

其拉氏变换为

$$L[x_i(t)] = \dfrac{\omega}{s^2 + \omega^2} \quad (3\text{-}13)$$

在实际工作环境中，电源的波动、机械振动、元件的噪声干扰等均可视为正弦信号。正弦信号是系统或元件做动态性能实验时广泛采用的输入信号。

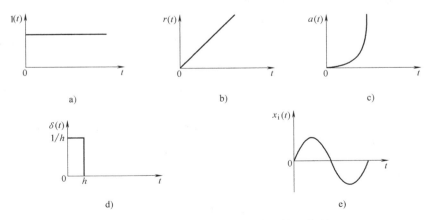

Fig. 3-1　Typical input signals（典型输入信号）
a) Step function（阶跃函数）　b) Ramp function（斜坡函数）　c) Parabolic function（抛物线函数）
d) Pulse function（脉冲函数）　e) Sine function（正弦函数）

nal that is widely used in systems or components for dynamic performance experiments.

Which signal is selected as the typical input signal depends on the specific working conditions of different systems. In fact, several signals are often used to test the same system. Taking the NC feed servo system as an example, during the development phase, the sine signal is often used to evaluate the frequency response of the system to improve the system design; during the debugging phase, the step response with load is performed to evaluate the smoothness of the transition process, and the time response with constant speed is performed to evaluate its steady state error.

3.1.2 Time response

In order to clearly understand the time response of a system, a RLC circuit network shown in Fig. 3-2 is taken as an example to analyze the output signal response of the system under the unit step signal input. In Fig. 3-2, $u_i(t)$ is the input voltage of the system, and $u_o(t)$ is the output voltage of the system. According to Kirchhoff's law, the mathematical model of the system is established as

$$\begin{cases} u_i(t) = L\dfrac{di(t)}{dt} + i(t)R + \dfrac{1}{C}\int i(t)dt \\ u_o(t) = \dfrac{1}{C}\int i(t)dt \end{cases} \tag{3-14}$$

Taking the Laplace transform on the above equations and then eliminating the intermediate variable $I(s)$ under zero initial conditions, the transfer function can be obtained as

$$G(s) = \frac{U_o(s)}{U_i(s)} = \frac{1}{LCs^2 + RCs + 1} = \frac{\omega_n^2}{s^2 + 2\xi\omega_n s + \omega_n^2} \tag{3-15}$$

where $\omega_n = \sqrt{\dfrac{1}{LC}}$ is the undamped natural frequency of the system; $\xi = \dfrac{1}{2}R\sqrt{\dfrac{C}{L}}$ is the damping ratio of the system.

The characteristic equation of the system is

$$s^2 + 2\xi\omega_n s + \omega_n^2 = 0 \tag{3-16}$$

In the case of the underdamping state ($0 < \xi < 1$), the characteristic root (or system pole) is

$$s_{1,2} = -\xi\omega_n \pm j\omega_d$$

where $\omega_d = \omega_n\sqrt{1-\xi^2}$ is called the damped natural frequency. When the input voltage is a step signal, the Laplace transform of the output response is

$$U_o(s) = \frac{\omega_n^2}{s(s^2 + 2\xi\omega_n s + \omega_n^2)} = \frac{1}{s} - \frac{s+\xi\omega_n}{(s+\xi\omega_n)^2 + \omega_d^2} - \frac{\omega_d}{(s+\xi\omega_n)^2 + \omega_d^2}\frac{\xi\omega_n}{\omega_d} \tag{3-17}$$

Taking the inverse Laplace transform, we have

$$u_o(t) = 1 - \frac{e^{-\xi\omega_n t}}{\sqrt{1-\xi^2}}\sin(\omega_d t + \beta) \quad (t \geq 0) \tag{3-18}$$

where $\beta = \arctan\dfrac{\sqrt{1-\xi^2}}{\xi}$.

选择哪一种信号作为典型输入信号，应视不同系统的具体工作状况而定。在实际中，往往采用几种信号测试同一个系统。以数控进给随动系统为例，在研制阶段，常用正弦信号来评价系统的频率响应，以改进系统的设计；在调试阶段，则带上负载观察阶跃响应以了解过渡过程的平稳性，并观察恒速的时间响应以了解其稳态误差。

3.1.2 时间响应

为明确地了解系统的时间响应，这里以如图 3-2 所示的 RLC 电路网络为例，分析系统在单位阶跃信号输入作用下的输出信号响应情况。图 3-2 中，$u_i(t)$ 为系统的输入电压，$u_o(t)$ 为系统的输出电压。根据基尔霍夫定律，建立的系统数学模型为

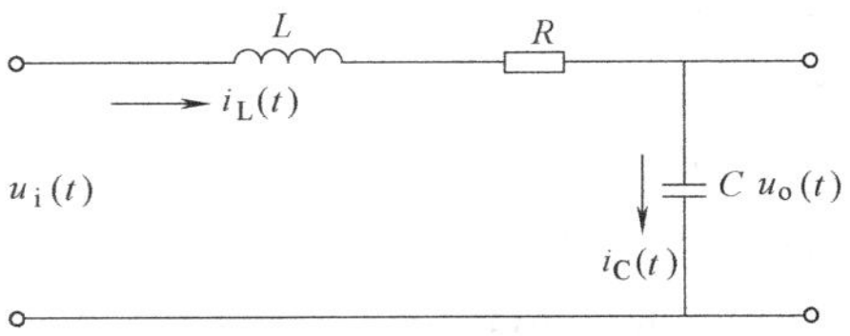

$$\begin{cases} u_i(t) = L\dfrac{\mathrm{d}i(t)}{\mathrm{d}t} + i(t)R + \dfrac{1}{C}\int i(t)\mathrm{d}t \\ u_o(t) = \dfrac{1}{C}\int i(t)\mathrm{d}t \end{cases} \tag{3-14}$$

Fig. 3-2　RLC circuit network
（RLC 电路网络）

在零初始条件下，进行拉氏变换并消去中间变量 $I(s)$，得传递函数为

$$G(s) = \dfrac{U_o(s)}{U_i(s)} = \dfrac{1}{LCs^2 + RCs + 1} = \dfrac{\omega_n^2}{s^2 + 2\xi\omega_n s + \omega_n^2} \tag{3-15}$$

式中，$\omega_n = \sqrt{\dfrac{1}{LC}}$ 为系统的无阻尼固有频率；$\xi = \dfrac{1}{2}R\sqrt{\dfrac{C}{L}}$ 为系统的阻尼比。

系统的特征方程为

$$s^2 + 2\xi\omega_n s + \omega_n^2 = 0 \tag{3-16}$$

在欠阻尼（$0 < \xi < 1$）情况下，其特征根（系统极点）为

$$s_{1,2} = -\xi\omega_n \pm \mathrm{j}\omega_d$$

式中，$\omega_d = \omega_n\sqrt{1-\xi^2}$ 称为有阻尼固有频率。当输入电压为阶跃信号时，输出响应的拉氏变换为

$$U_o(s) = \dfrac{\omega_n^2}{s(s^2 + 2\xi\omega_n s + \omega_n^2)} = \dfrac{1}{s} - \dfrac{s+\xi\omega_n}{(s+\xi\omega_n)^2 + \omega_d^2} - \dfrac{\omega_d}{(s+\xi\omega_n)^2 + \omega_d^2}\dfrac{\xi\omega_n}{\omega_d} \tag{3-17}$$

取拉氏反变换，得

$$u_o(t) = 1 - \dfrac{\mathrm{e}^{-\xi\omega_n t}}{\sqrt{1-\xi^2}}\sin(\omega_d t + \beta) \quad (t \geq 0) \tag{3-18}$$

式中，$\beta = \arctan\dfrac{\sqrt{1-\xi^2}}{\xi}$。

式（3-18）就是 RLC 电路网络在输入单位阶跃电压信号时系统输出的时间响应（全响应），系统输出的全响应一般由瞬态响应和稳态响应两部分组成。一个系统在输入信号作用下，总会由某一初始状态变成另一种新的状态，由于实际系统中总会有一些储能和耗能元件，使得输出量不能立即跟随输入量的变化，因而输出在达到稳态之前就必然会有一个过渡过程，这个过渡过程称为瞬态响应。在瞬态响应中，系统的动态性能都会充分表现出来，如

Equation (3-18) is the time response (full response) of the system output when the unit step voltage signal is fed to the RLC circuit network. The full response of a system output is generally composed of a transient response and a steady state response. A system will always change from an initial state to another state with the input signal. As there are always some energy-storage and energy-consuming components in the actual system, the output of the system cannot immediately follow the change of input. To this end, there must be a transient process before the system reaching the steady-state output. And the transient process is called a transient response. In the transient response, the dynamic performance of the system will be fully demonstrated, such as whether the response is rapid, whether there is oscillation, and whether the oscillation process is intense. The response depends on the system structure or the transient component of the real part of the pole, corresponding to the second term in the right part of equation (3-18). It represents the motion caused by the initial condition of the linear system without a control signal. The motion is customarily called the free motion, or free response, or zero input response, Which characterizes the inherent characteristics of the system. When the transition process ends, the system reaches a steady state, and the output state of the system when $t \rightarrow \infty$ is called a steady-state response. The response depends on the steady-state component of the input signal, corresponding to the first term in the right part of equation (3-18). The steady state response characterizes the extent to which the system output ultimately reproduces the input. The value of the steady-state response is called the steady-state value (or the final value of the output), i.e. $x_o(\infty)$. Therefore, the full response of the system represents the motion of the linear system with the control signal, which is called the forced motion or forced response.

3.1.3 Time-domain performance indicators of control systems

According to the analysis of the control system requirements in Chapter 1, it can be known that the stability of the control system is the prerequisite for its normal operation. The accuracy shows how static precision of the system is after a new equilibrium is established. Obviously, there is an intermediate process in which the external action is applied to the system and then a new equilibrium state occurs. The process is called the transient response process. The performance of the transient response process can usually be measured by corresponding indicators, such as the rapidity, static accuracy, relative stability, etc. These indicators are called time domain performance indicators.

Time-domain performance indicators are usually defined in terms of the damping oscillation process under the input of the unit step signal. The time-domain performance indicators of the system under other typical inputs can be directly or indirectly related to the above mentioned indicators. The transient response of the actual control system is often addressed as a damped oscillation process before reaching the steady state. To illustrate the transient response characteristics of the control system under the unit step input signal, the following performance indicators are usually used as is shown in Fig. 3-3.

1. Delay time t_d

Delay time t_d is defined as the time required for the response curve to rise from zero to 50% of

响应是否迅速、是否有振荡、振荡过程是否激烈。该部分可通过取决于系统结构或极点实部的瞬态分量来描述，对应式（3-18）右侧第二项，表示线性系统在没有控制信号作用时由初始条件引起的运动，习惯上称为自由运动自由响应或零输入响应，它表征系统的固有特性。过渡过程一结束系统就达到了稳态，时间 $t \to \infty$ 时系统的输出状态称为稳态响应。该部分可通过取决于输入控制信号的稳态分量来描述，对应式（3-18）右侧第一项。稳态响应表征系统输出量最终复现输入量的程度。稳态响应值称为稳态值（输出终值），即 $x_o(\infty)$。因此，系统的全响应表示线性系统在控制信号作用下的运动，也称为强迫运动或强迫响应。

3.1.3 控制系统的时域性能指标

由第一章对控制系统要求的分析可知，控制系统的稳定性是其正常工作的前提条件。而准确性则说明系统在建立新的平衡状态以后，其静态精度如何。显然，存在一个系统由受到外作用开始到新的平衡状态出现的中间过程，即瞬态响应过程。瞬态响应过程的性能，如快速性、静态精度、系统的相对稳定性等，通常可用相应的指标来衡量。这些指标称为时域性能指标。

时域性能指标通常以系统在单位阶跃输入作用下的衰减振荡过程来定义。系统在其他典型输入作用下定义的时域性能指标，均可以直接或间接与上述指标建立关系。实际控制系统的瞬态响应在达到稳态以前，常常表现为阻尼振荡过程，为了说明控制系统在单位阶跃输入信号下的瞬态响应特性，通常采用下列一些性能指标，如图 3-3 所示。

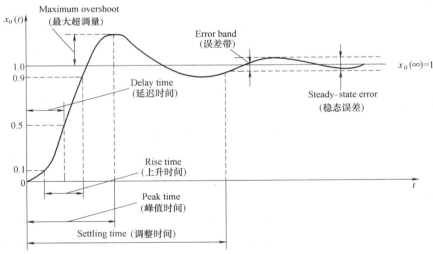

Fig. 3-3 Typical unit step response of a control system
（控制系统的典型单位阶跃响应）

1. 延迟时间 t_d

延迟时间 t_d 是指响应曲线从零上升到稳态值的 50% 所需要的时间，由定义得

$$x_o(t_d) = x_o(\infty) \times 50\% \tag{3-19}$$

2. 上升时间 t_r

上升时间 t_r 是指在瞬态过程中，响应曲线从零时刻到首次达到稳态值所需的时间（对于过阻尼的情况，则一般指响应曲线从稳态值的 10% 上升到稳态值的 90% 所需的时间），即

the steady state value, which can be written as
$$x_o(t_d) = x_o(\infty) \times 50\% \tag{3-19}$$

2. Rise time t_r

In the transient process, the time required for the response curve to rise from zero to the first steady state value is called rise time (for the overdamping situation, the time is to rise from 10% to 90% of the steady state value), i.e.
$$x_o(t_r) = x_o(\infty) \tag{3-20}$$

3. Peak time t_p

The time required for the response curve to rise from zero to the first peak value is called peak time, i.e.
$$x_o(t_p) = \max\{x_o(t)\} \tag{3-21}$$

4. Maximum overshoot M_p

Under the input of the unit step signal, the difference between the maximum peak value of the response curve and the steady state value is called maximum overshoot. The value of the maximum overshoot is also used to measure the relative stability of the system. Usually it is expressed as a percentage, i.e.
$$M_p = \frac{x_o(t_p) - x_o(\infty)}{x_o(\infty)} \times 100\% \tag{3-22}$$

5. Settling time t_s

Settling time t_s is the transient process time (from $t=0$) that the deviation between the response curve and its corresponding steady-state value reaches the allowable range Δ (generally $\Delta = \pm 5\%$ or $\Delta = \pm 2\%$). It is expressed by the following inequality
$$|x_o(t) - x_o(\infty)| \leq \Delta \cdot x_o(\infty) \quad (t \geq t_s) \tag{3-23}$$

6. Number of Oscillation N

If the curve $x_o(t)$ intersects with the horizontal line of the steady state value $x_o(\infty)$ during the transition time $0 \leq t \leq t_s$, then half the number of the intersection (i.e the number of "crossovers") is defined as the number of oscillations N.

7. Steady-state error ε_s

Steady-state error ε_s is the difference between the expected output of the system and the steady-state value of the actual output. For a unit feedback system, the actual steady-state output is required to accurately follow the change of a given input without error. Thus, the input is the expected output, and the steady-state error is equal to the difference between the given value of the input and the steady state value of the actual output. , i.e.
$$\varepsilon_s = \lim_{t \to \infty} [x_i(t) - x_o(t)] \tag{3-24}$$

In a control system, the delay time t_d, rise time t_r and peak time t_p can be used to evaluate the response speed of the system. The smaller these time values of the response curve are, the faster the response speed of the system is. The maximum overshoot M_p and the oscillation number N are used to evaluate the degree of the system damping. Increasing the damping can reduce the response

$$x_o(t_r) = x_o(\infty) \tag{3-20}$$

3. 峰值时间 t_p

峰值时间 t_p 是指响应曲线从零时刻到达第一个峰值所需的时间,即

$$x_o(t_p) = \max\{x_o(t)\} \tag{3-21}$$

4. 最大超调量 M_p

最大超调量 M_p 是指在单位阶跃输入下,响应曲线的最大峰值与稳态值的差。最大超调量也用来度量系统的相对稳定性,通常用百分数表示,即

$$M_p = \frac{x_o(t_p) - x_o(\infty)}{x_o(\infty)} \times 100\% \tag{3-22}$$

5. 调节时间 t_s

调节时间 t_s 是指响应曲线与其对应于稳态值之间的偏差达到容许范围 Δ(一般取 $\Delta = \pm 5\%$ 或 $\Delta = \pm 2\%$)所经历的瞬态过程时间(从 $t=0$ 开始计时),由下列不等式表示

$$|x_o(t) - x_o(\infty)| \leq \Delta \cdot x_o(\infty) \quad (t \geq t_s) \tag{3-23}$$

6. 振荡次数 N

若曲线 $x_o(t)$ 与稳态值 $x_o(\infty)$ 所在水平线在过渡过程时间 $0 \leq t \leq t_s$ 内相交,把相交次数(即穿越次数)的一半定义为振荡次数 N。

7. 稳态误差 ε_s

稳态误差 ε_s 是系统的期望输出与实际输出的稳态值之差。对于单位反馈系统,要求实际的稳态输出量无误差且准确地跟随给定输入量的变化。故输入量就是期望输出量,稳态误差也就等于给定输入量与实际输出稳态值之差,即

$$\varepsilon_s = \lim_{t \to \infty}[x_i(t) - x_o(t)] \tag{3-24}$$

在控制系统中,延迟时间 t_d、上升时间 t_r 及峰值时间 t_p 都可用来评价系统的响应速度,响应曲线的这些时间值越小,说明系统的响应速度越快;最大超调量 M_p 和振荡次数 N 用于评价系统的阻尼程度,增大阻尼可使响应超调量减小,并可以减弱系统的振荡。换言之,系统的动态平稳性变好,但响应速度减慢;调节时间 t_s 是一个综合性指标,可同时反映响应速度和阻尼程度,它取决于系统的阻尼和无阻尼固有频率,该时间值越小,则系统的相对稳定性越好,且系统的动态过程结束得越迅速;稳态误差 ε_s 的大小反映控制系统的稳态性能,稳态误差 ε_s 越大,系统的稳态性越差,反之相反。

3.2 一阶系统的数学模型和时间响应

3.2.1 一阶系统的数学模型

用一阶微分方程描述的系统称为一阶系统。它的典型形式是一阶惯性环节,其动态微分方程为

$$T\dot{x}_o(t) + x_o(t) = x_i(t) \tag{3-25}$$

系统的传递函数为

overshoot, and can weaken the oscillation performance of the system. In other words, the dynamic stability of the system becomes good, but the response speed becomes slow. The settling time is a comprehensive index, which can reflect the response speed and damping degree simultaneously. It depends on the damping and undamped natural frequency of the system. If the settling time value is small, the relative stability of the system is better and the dynamic process of the system can end quickly. The steady-state error reflects the steady-state performance of the control system. The greater the steady-state error is, the worse the steady state of the system is, and vice versa.

3.2 Mathematical Models and Time Response of First-Order Systems

3.2.1 Mathematical models of first-order systems

A system described by a first-order differential equation is called a first-order system. Its typical form is a first-order inertia element, and its dynamic differential equation is

$$T\dot{x}_o(t) + x_o(t) = x_i(t) \tag{3-25}$$

The transfer function of the system is

$$G(s) = \frac{X_o(s)}{X_i(s)} = \frac{1}{Ts+1} \tag{3-26}$$

Taking calculation, the pole of the system is $-1/T$. It lies in the left half of the s complex plane.

The function block diagram of the first-order system is shown in Fig. 3-4, and the open-loop transfer function of the unit feedback system is

$$G_K(s) = \frac{K}{s} \tag{3-27}$$

Comparing equation (3-26) and equation (3-27), we have the relation between the time constant T of the first-order system and the open-loop gain K as

$$T = \frac{1}{K}$$

The time constant T is a characteristic parameter of a first-order system, and has the dimension of time unit "second". For different systems, T consists of different physical quantities, and it shows the inherent characteristics of the first-order system itself independent of the external action. For example, in Chapter 2, the spring-damping system shown in Fig. 2-19 has a time constant $T = c/k$ (c is the damping coefficient of the damper and k is the stiffness of the spring).

3.2.2 Time response of first-order systems

1. The unit step response of a first-order system

When the unit step signal $1(t)$ is applied to a first-order system, the Laplace transform of the unit step response of the system is

$$X_o(s) = G(s)X_i(s) = \frac{1}{Ts+1} \cdot \frac{1}{s}$$

第3章 控制系统的时域分析

$$G(s) = \frac{X_o(s)}{X_i(s)} = \frac{1}{Ts+1} \qquad (3\text{-}26)$$

可求得系统的极点为 $-\frac{1}{T}$,它位于 s 复平面的左半平面。

一阶系统的函数框图如图 3-4 所示,其单位反馈系统的开环传递函数为

$$G_K(s) = \frac{K}{s} \qquad (3\text{-}27)$$

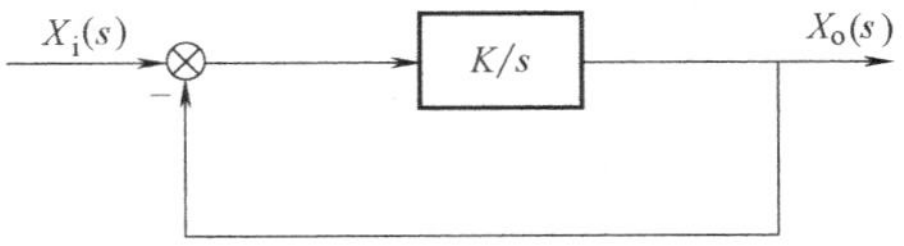

由式 (3-26) 和式 (3-27) 相比较可知,一阶系统的时间常数 T 与开环增益 K 之间的关系为

$$T = \frac{1}{K}$$

Fig. 3-4 Block diagram of the first-order system
(一阶系统的框图)

时间常数 T 是一阶系统的特征参数,具有时间单位"秒"的量纲。对于不同的系统,T 由不同的物理量组成,它显示一阶系统本身与外界作用无关的固有特性。例如第 2 章中,图 2-19 所示弹簧-阻尼系统,其时间常数 $T=c/k$(c 是阻尼器的阻尼系数,k 是弹簧的刚度)。

3.2.2 一阶系统的时间响应

1. 一阶系统的单位阶跃响应

当单位阶跃信号 $1(t)$ 作用于一阶系统时,系统的单位阶跃响应的拉氏变换为

$$X_o(s) = G(s)X_i(s) = \frac{1}{Ts+1} \cdot \frac{1}{s}$$

进行拉氏反变换,可得到单位阶跃输入的时间响应(称为单位阶跃响应)为

$$x_o(t) = L^{-1}[X_o(s)] = L^{-1}\left(\frac{1}{Ts+1} \cdot \frac{1}{s}\right) = 1 - e^{-\frac{t}{T}} \quad (t \geqslant 0) \qquad (3\text{-}28)$$

式中,右边第一项是单位阶跃响应的稳态分量,取决于输入信号 $1(t)$,它等于单位阶跃信号的幅值;第二项是瞬态分量,它与系统的极点 $-\frac{1}{T}$ 有关,当 $t \to \infty$ 时,瞬态分量趋于零,曲线 $x_o(t)$ 随时间的变化如图 3-5a 所示,它是一条按指数规律单调上升的曲线。

一阶系统单位阶跃响应的特性描述如下。

1) 由式 (3-28) 可知,响应曲线单调上升且无振荡地收敛于稳态值 $x_o(\infty) = 1$,因此,一阶系统是稳定系统。

2) 一阶系统的时间常数 T 是输出达到稳态值的 63.2% 所需的时间,即

$$x_o(T) = 1 - e^{-1} = 0.632$$

而响应曲线在 $t=0$ 点的切线斜率正好为时间常数 T 的倒数,即

$$\left.\frac{dx_o(t)}{dt}\right|_{t=0} = \left.\frac{1}{T}e^{-\frac{t}{T}}\right|_{t=0} = \frac{1}{T}$$

由上式及图 3-5 分析可知,时间常数 T 反映了系统响应速度的快慢,时间常数 T 越小,$x_o(t)$ 上升速度越快,达到稳态值所用的时间越短,系统的惯性越小。反之,T 越大,系统响应越缓慢,惯性越大,如图 3-5b 所示。所以 T 的大小反映了一阶系统惯性的大小。

3) 一阶系统的调整时间 t_s 是系统从开始响应到进入稳态所经过的时间(或称为过渡过

Taking inverse Laplace transform, the time response of the unit step input (called the unit step response) is obtained as

$$x_o(t) = L^{-1}[X_o(s)] = L^{-1}\left(\frac{1}{Ts+1}\frac{1}{s}\right) = 1 - e^{-\frac{t}{T}} \quad (t \geq 0) \quad (3\text{-}28)$$

where the first term on the right is the steady-state component of the unit step response, which is equal to the amplitude of the unit step signal, depending on the input signal $1(t)$; The second term is the transient component, which is related to the pole $-1/T$ of the system. When $t \to \infty$, the transient component approaches to zero. The curve $x_o(t)$ that changes with time is shown in Fig. 3-5a, and it is a curve that rises monotonically in terms of the exponential law.

The characteristics of the unit step response for the first-order system is described as follows.

1) According to equation (3-28), the response curve monotonously rises and converges to the steady-state value $x_o(\infty) = 1$ without oscillation. Hence, the first-order system is a stable system.

2) The time constant T of the first-order system is the time required for the output to reach 63.2% of the steady-state value, i.e.

$$x_o(T) = 1 - e^{-1} = 0.632$$

The tangential slope of the response curve at $t = 0$ is exactly the reciprocal of the time constant T, i.e.

$$\left.\frac{dx_o(t)}{dt}\right|_{t=0} = \left.\frac{1}{T}e^{-\frac{t}{T}}\right|_{t=0} = \frac{1}{T}$$

Taking analysis of the above equation and Fig. 3-5, We can know that the time constant T reflects the speed of the system response. The smaller the time constant T is, the faster the rising speed of $x_o(t)$ is, the shorter the time to reach the steady state value is, and the smaller the inertia of the system is. Otherwise, the larger the time constant T is, the slower the response of the system is, and the greater the inertia is, as is shown in Fig. 3-5b. Therefore, the size of T reflects the magnitude of the inertia of the first-order system.

3) The settling time t_s of the first-order system is the time from the start of the response to the steady state (or called transition time). In theory, when the system reaches the steady state after the transition ends, the time t is required to be infinite, i.e. $t \to \infty$. As for $t \to \infty$, there should be a requirement for a specific quantized value in engineering, i.e. the specific requirement about what value should the output reach when the transient process ends. It is related to the accuracy required by the system. According to the definition of the control system performance index in the previous section, if the system transient process is considered over when the output value reaches 98% of the steady state value, according to equation (3-28), the response value can be obtained as $x_o(4T) = 0.98$, when $t = 4T$. So the settling time is $t_s = 4T$ (error range $\Delta = \pm 2\%$). If the system transient process is considered over when the output value reaches 95% of the steady state value, the response value can be obtained as $x_o(3T) = 0.95$, when $t = 3T$. So the settling time is $t_s = 3T$ (error range $\Delta = \pm 5\%$). It should be noted that the settling time only reflects the characteristics of the system and is independent of the input and output. It is a comprehensive performance indicator for eval-

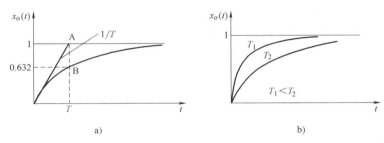

Fig. 3-5 Unit step response curve of a first-order system（一阶系统的单位阶跃响应曲线）

程时间）。从理论上讲，系统结束瞬态进入稳态，要求时间 t 趋向无穷大，即 $t \to \infty$。而在工程应用中，应对 $t \to \infty$ 有一个具体量化值的要求，即输出量达到什么值就算瞬态过程结束的具体要求。这与系统要求的精度有关。按上一节对控制系统性能指标的分析可知，若认为系统在输出值达到稳态值的98%时瞬态过程结束，由式（3-28）可以求出当 $t = 4T$ 时，响应值 $x_o(4T) = 0.98$，因此调整时间 $t_s = 4T$（误差范围 $\Delta = \pm 2\%$）；若认为系统在输出值达到稳态值的95%时瞬态过程结束，可以求出当 $t = 3T$ 时，响应值 $x_o(3T) = 0.95$，因此调整时间 $t_s = 3T$（误差范围 $\Delta = \pm 5\%$）。应当指出调整时间只反映系统的特性，与输入、输出无关。它是评价系统响应快慢和阻尼程度的综合性能指标。

2. 一阶系统的单位脉冲响应

将式（3-28）代入式（3-11），得一阶系统的单位脉冲响应为

$$x_{o\delta}(t) = \frac{d}{dt}[x_o(t)] = \frac{1}{T}e^{-\frac{t}{T}} \quad (t \geq 0) \tag{3-29}$$

一阶系统的单位脉冲响应曲线是单调下降的指数曲线，如图 3-6 所示。而且，函数 $x_{o\delta}(t)$ 只有瞬态项 $\frac{1}{T}e^{-\frac{t}{T}}$，其稳态项为 0。

3. 一阶系统的单位斜坡响应

将式（3-28）代入式（3-11），并在零初始条件下进行积分运算，可得一阶系统的单位斜坡响应为

$$x_{or}(t) = t - T + Te^{-\frac{t}{T}} \quad (t \geq 0) \tag{3-30}$$

进而可得一阶系统的单位斜坡响应曲线，如图 3-7 所示。

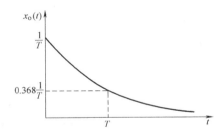

Fig. 3-6 Unit pulse response curve of a first-order system
（一阶系统的单位脉冲响应）

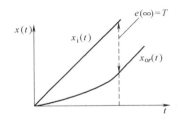

Fig. 3-7 Unit ramp response curve of a first-order system
（一阶系统的单位斜坡响应曲线）

uating the response speed and damping of the system.

2. The unit pulse response of a first-order system

Substituting equation (3-28) into equation (3-11), the unit impulse response of the first-order system is

$$x_{o\delta}(t) = \frac{d}{dt}[x_o(t)] = \frac{1}{T}e^{-\frac{t}{T}} \quad (t \geq 0) \tag{3-29}$$

The unit impulse response curve of a first-order system is a monotonically decreasing exponential curve, as is shown in Fig. 3-6. Moreover, function $x_{o\delta}(t)$ has only a transient term $\frac{1}{T}e^{-\frac{t}{T}}$, and its steady-state term is 0.

3. The unit ramp response of a first-order system

Substituting equation (3-28) into equation (3-11) and taking an integral operation under zero initial conditions, the unit slope response of the first-order system can be obtained as.

$$x_{or}(t) = t - T + Te^{-\frac{t}{T}} \quad (t \geq 0) \tag{3-30}$$

Consequently, the unit slope response curve of the first-order system can be drawn as is shown in Fig. 3-7.

When $t \to \infty$, $\varepsilon_s = e(\infty) = T$. Hence, when the input is a ramp signal, the steady-state error of the first-order system is T. It shows that the first-order system has a steady-state error when tracking a unit ramp signal. Obviously, the smaller the time constant T is, the smaller the steady-state error of the system is.

Example 3-1 If the transfer function of a system is $G(s) = \dfrac{a}{\tau s + a + 1}$, what is the steady-state value of the unit step response function of the system? What is the settling time of the transition process?

Solution: (1) Calculate the steady-state value of the unit step response

1) According to the final value theorem of the Laplace transform, we have

$$x_o(\infty) = \lim_{s \to 0} sX_o(s) = \lim_{s \to 0} \frac{1}{s} \frac{a}{\tau s + a + 1} = \frac{a}{a+1}$$

2) Converting the transfer function to the standard form of the first-order system, we have

$$G(s) = \frac{a}{\tau s + a + 1} = \frac{\dfrac{a}{a+1}}{\dfrac{\tau}{a+1}s + 1} = \frac{a}{a+1} \cdot \frac{1}{\dfrac{\tau}{a+1}s + 1}$$

3) From the standard form of the system transfer function, it can be seen that the system is essentially formed by a proportional element and a standard inertia element in series. So the steady-state value of the system is as

$$x_o(\infty) = \frac{a}{a+1}$$

当 $t\to\infty$ 时，$\varepsilon_s = e(\infty) = T$。故当输入为斜坡信号时，一阶系统的稳态误差为 T。这说明一阶系统在跟踪单位斜坡信号时，存在稳态误差。显然，时间常数 T 越小，则该系统的稳态误差越小。

例 3-1 某一系统的传递函数为 $G(s) = \dfrac{a}{\tau s + a + 1}$，则该系统单位阶跃响应函数的稳态值是多少？过渡过程的调整时间 t_s 为多少？

解：（1）求单位阶跃响应的稳态值

1）根据拉普拉斯变换的终值定理，有

$$x_o(\infty) = \lim_{s\to 0} sX_o(s) = \lim_{s\to 0} s \frac{1}{s} \frac{a}{\tau s + a + 1} = \frac{a}{a+1}$$

2）把传递函数化为一阶系统的标准形式，有

$$G(s) = \frac{a}{\tau s + a + 1} = \frac{\dfrac{a}{a+1}}{\dfrac{\tau}{a+1}s + 1} = \frac{a}{a+1} \cdot \frac{1}{\dfrac{\tau}{a+1}s + 1}$$

3）从传递函数的标准形式可以看出该系统实质是一个比例环节和一个标准惯性环节串联而成的，所以系统的稳态值为

$$x_o(\infty) = \frac{a}{a+1}$$

（2）求过渡过程的调整时间　该系统的时间常数为 $T = \dfrac{\tau}{a+1}$，所以

$$t_s = \begin{cases} \dfrac{4\tau}{a+1} & (\Delta = 2\%) \\ \dfrac{3\tau}{a+1} & (\Delta = 5\%) \end{cases}$$

例 3-2 系统的框图如图 3-8 所示，已知传递函数 $G(s) = \dfrac{10}{0.2s+1}$。欲采用负反馈的办法将调节时间 t_s 减小为原来的 0.1 倍，并保证总的放大倍数不变，试确定 K_h 和 K_o 的值。

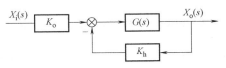

Fig. 3-8　Block diagram of example 3-2（例 3-2 框图）

解：1）求出整个系统的传递函数，有

$$G'(s) = \frac{K_o G(s)}{1 + K_h G(s)} = \frac{10K_o}{0.2s + 1 + 10K_h}$$

2）将传递函数化为标准形式，有

$$G'(s) = \frac{\dfrac{10K_o}{1+10K_h}}{\dfrac{0.2}{1+10K_h}s + 1}$$

3）将 $G'(s)$ 和标准形式进行比较，可得

(2) Calculate the settling time of the transition process Since the time constant of the system is $T=\dfrac{\tau}{a+1}$, we have

$$t_s = \begin{cases} \dfrac{4\tau}{a+1} & (\Delta=2\%) \\ \dfrac{3\tau}{a+1} & (\Delta=5\%) \end{cases}$$

Example 3-2 The structural block diagram of a system is shown in Fig. 3-8. The transfer function is $G(s)=\dfrac{10}{0.2s+1}$. A negative feedback is to be adopted to reduce the settling time t_s to 0.1 times of the original time and the total magnification times is kept unchanged. Try to figure out values of K_h and K_o.

Solution: 1) Calculating the transfer function of the entire system, we have

$$G'(s) = \dfrac{K_o G(s)}{1+K_h G(s)} = \dfrac{10K_o}{0.2s+1+10K_h}$$

2) Converting the transfer function to the standard form, we have

$$G'(s) = \dfrac{\dfrac{10K_o}{1+10K_h}}{\dfrac{0.2}{1+10K_h}s+1}$$

3) Comparing $G'(s)$ with the standard form, we have

$$\begin{cases} \dfrac{10K_o}{1+10K_h}=10 \\ \dfrac{0.2}{1+10K_h}=0.02 \end{cases} \Rightarrow \begin{cases} K_h=0.9 \\ K_o=10 \end{cases}$$

Example 3-3 Two inertia elements with different time constants are connected in series, and the system transfer function is $G(s)=\dfrac{X_o(s)}{X_i(s)}=\dfrac{1}{10s+1}\dfrac{1}{s+1}$. Figure out the unit step response of the system.

Solution: According to the system transfer function, we have $T_1=10$ and $T_2=1$. Two poles of the system transfer function are marked on the s-plane (representing the pole by "×"), and the pole distribution of the system is obtained as is shown in Fig. 3-9. It can be seen that $s_1=-1/T_1=-0.1$, $s_2=-1/T_2=-1$, and s_1 is closer to the imaginary axis on the complex plane, while it is the pole of the element with the larger time constant.

Inputting the unit step signal to the system, i.e. $X_i(s)=\dfrac{1}{s}$, the output of the system is

$$X_o(s)=X_i(s)G(s)=\dfrac{1}{10s+1}\dfrac{1}{s+1}\dfrac{1}{s}$$

The above equation can be transformed into

$$\begin{cases} \dfrac{10K_o}{1+10K_h} = 10 \\ \dfrac{0.2}{1+10K_h} = 0.02 \end{cases} \Rightarrow \begin{cases} K_h = 0.9 \\ K_o = 10 \end{cases}$$

例 3-3 两个时间常数不同的惯性环节串联在一起,系统传递函数为 $G(s) = \dfrac{X_o(s)}{X_i(s)} = \dfrac{1}{10s+1}\dfrac{1}{s+1}$。求该系统的单位阶跃响应。

解:根据系统传递函数,可得 $T_1 = 10$,$T_2 = 1$。把系统传递函数的两个极点标在 s 平面上(用 "×" 表示极点),得到该系统的极点分布如图 3-9 所示,其中 $s_1 = -\dfrac{1}{T_1} = -0.1$,$s_2 = -\dfrac{1}{T_2} = -1$。可以看出在复平面上,$s_1$ 更靠近虚轴,而 s_1 正是时间常数较大环节的极点。

给系统输入单位阶跃信号,即 $X_i(s) = \dfrac{1}{s}$,则输出为

$$X_o(s) = X_i(s)G(s) = \dfrac{1}{10s+1}\dfrac{1}{s+1}\dfrac{1}{s}$$

上式可化为

$$X_o(s) = \dfrac{A}{10s+1} + \dfrac{B}{s+1} + \dfrac{C}{s}$$

式中,A、B、C 为待定系数,可采用系数比较法求出,得

$$A = -\dfrac{1}{0.09}, \quad B = \dfrac{1}{9}, \quad C = 1$$

则有

$$X_o(s) = -\dfrac{1}{0.09}\dfrac{1}{10s+1} + \dfrac{1}{9}\dfrac{1}{s+1} + \dfrac{1}{s}$$

取拉氏反变换,得

$$x_o(t) = 1 - 1.11\mathrm{e}^{-\frac{t}{10}} + 0.11\mathrm{e}^{-t} = x_{o1}(t) + x_{o2}(t) + x_{o3}(t) \tag{3-31}$$

式中,$x_{o1}(t) = 1$;$x_{o2}(t) = -1.11\mathrm{e}^{-\frac{t}{10}}$;$x_{o3}(t) = 0.11\mathrm{e}^{-t}$。它们对应的响应曲线如图 3-10 所示。从响应曲线可以看出,整个系统的瞬态响应取决于时间常数较大的环节,时间常数小的环节对系统的瞬态响应影响很小。从极点分布来看,靠近虚轴的极点在瞬态响应中起主导作用,

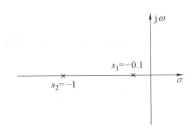

Fig. 3-9 Poles distribution
(极点分布图)

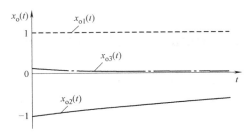

Fig. 3-10 Response curve
(响应曲线)

$$X_o(s) = \frac{A}{10s+1} + \frac{B}{s+1} + \frac{C}{s}$$

where A, B and C are undetermined coefficients, and they can be obtained by employing the coefficient comparison method. So we have

$$A = -\frac{1}{0.09},\ B = \frac{1}{9},\ C = 1$$

Then we have

$$X_o(s) = -\frac{1}{0.09}\frac{1}{10s+1} + \frac{1}{9}\frac{1}{s+1} + \frac{1}{s}$$

Taking the inverse Laplace transform, we have

$$x_o(t) = 1 - 1.11e^{-\frac{t}{10}} + 0.11e^{-t} = x_{o1}(t) + x_{o2}(t) + x_{o3}(t) \tag{3-31}$$

where $x_{o1}(t) = 1$; $x_{o2}(t) = -1.11e^{-\frac{t}{10}}$; $x_{o3}(t) = 0.11e^{-t}$. The corresponding response curves are shown in Fig. 3-10. It can be seen from the response curves that the transient response of the whole system depends on the element with the larger time constant, and the element with the smallest time constant has little influence on the transient response of the system. From the pole distribution, the pole near the imaginary axis plays a leading role in the transient response, and the pole far from the imaginary axis has little effect. When the ratio of the two poles' distance to the imaginary axis exceeds five times, the effect of the furthest pole from the imaginary axis can be approximately negligible in the transient response. The conclusion is useful for discussing high-order systems. If there are two poles satisfying the above conditions and there is no zero around the nearest pole or pair near the imaginary axis, a high-level system with multiple poles can be approximatively simplified to a low-order system like a first-order or a second-order system.

Ignoring the furthest pole from the imaginary axis, we have

$$X_o(s) = \frac{1}{10s+1}\frac{1}{s}$$

Taking the inverse Laplace transform, we have

$$x_o(t) = 1 - e^{-\frac{t}{10}} \tag{3-32}$$

For instance, when $t = 60$, $x_o(60) = 0.9772$ according to equation (3-31) while $x_o(60) = 0.9975$ according to equation (3-32). It can be seen that the difference between two equations' results is very small.

3.3 Mathematical Models and Time Response of Second-Order Systems

The response characteristics of a second-order system is often viewed as a benchmark when analyzing or designing a system. Although not all systems are second-order systems in practice, it is possible for third-order or higher-order systems to be approximated with a second-order system, or their responses can be expressed as a synthesis of first-order and second-order system responses. Therefore, the response of the second-order system is discussed in detail here.

距离虚轴较远的极点影响很小。当两极点到虚轴距离的比值超过五倍时，远离虚轴的极点在瞬态响应中的作用可近似地忽略不计。这个结论对于讨论高阶系统很有用，如果两个极点满足上述条件，并且靠近虚轴最近的一个或一对极点周围没有零点，则可以把多个极点的高阶系统近似地简化成低阶系统，如一阶系统或二阶系统。

忽略与虚轴距离最远的极点，有

$$X_o(s) = \frac{1}{10s+1} \frac{1}{s}$$

取拉氏反变换，得

$$x_o(t) = 1 - e^{-\frac{t}{10}} \tag{3-32}$$

例如当 $t=60$ 时，根据式（3-31）计算得 $x_o(60)=0.9772$，根据式（3-32）计算得 $x_o(60)=0.9975$，可见两个方程的计算结果相差很小。

3.3 二阶系统的数学模型和时间响应

在分析或设计系统时，二阶系统的响应特性常被视为一种基准。虽然在实际中并非每一个系统都是二阶系统，但是对于三阶或更高阶系统，有可能用二阶系统近似，或者其响应可以表示为一、二阶系统响应的合成。因此，这里将对二阶系统的响应进行重点讨论。

3.3.1 二阶系统的数学模型

用二阶微分方程描述的系统称为二阶系统。从物理意义上讲，二阶系统至少包含两个储能元件，能量有可能在两个元件之间交换，引起系统的往复振荡。所以，典型的二阶系统就是二阶振荡环节。其动态微分方程及传递函数分别为

$$\frac{d^2 x_o(t)}{dt^2} + 2\xi\omega_n \frac{dx_o(t)}{dt} + \omega_n^2 x_o(t) = \omega_n^2 x_i(t) \tag{3-33}$$

$$G(s) = \frac{X_o(s)}{X_i(s)} = \frac{\omega_n^2}{s^2 + 2\xi\omega_n s + \omega_n^2} \tag{3-34}$$

式中，ω_n 为系统无阻尼振荡的固有频率；ξ 为系统的阻尼比。不同系统的 ω_n 和 ξ 值取决于各系统的元件参数。ω_n 和 ξ 是二阶系统的特征参数，它们表明二阶系统本身与外界无关的特性。

典型单位负反馈二阶系统框图如图 3-11 所示，其开环传递函数

$$G_K(s) = \frac{\omega_n^2}{s(s+2\xi\omega_n)} \tag{3-35}$$

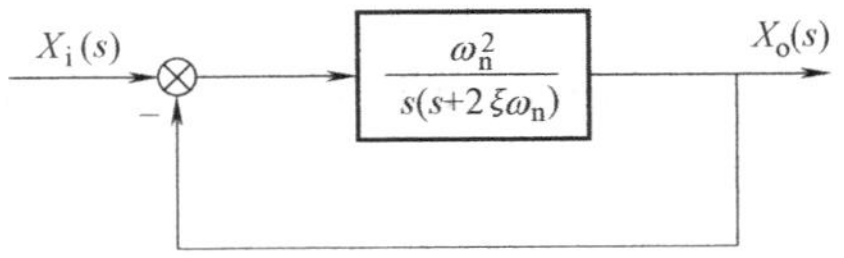

Fig. 3-11 Block diagram of a second-order system with unit negative feedback
（单位负反馈二阶系统框图）

由二阶系统的微分方程或传递函数可得系统的特征方程为

$$s^2 + 2\xi\omega_n s + \omega_n^2 = 0$$

则方程的两个特征根为

$$s_{1,2} = -\xi\omega_n \pm \omega_n\sqrt{\xi^2 - 1} \tag{3-36}$$

3.3.1 Mathematical models of second-order systems

A system described by second-order differential equations is called a second-order system. Physically, a second-order system contains at least two energy-storage components. The energy may be exchanged between the two components, leading to the reciprocating oscillation of the system. Therefore, a typical second-order system is a second-order oscillation element. Its dynamic differential equation and transfer function are respectively

$$\frac{d^2 x_o(t)}{dt^2} + 2\xi\omega_n \frac{dx_o(t)}{dt} + \omega_n^2 x_o(t) = \omega_n^2 x_i(t) \tag{3-33}$$

$$G(s) = \frac{X_o(s)}{X_i(s)} = \frac{\omega_n^2}{s^2 + 2\xi\omega_n s + \omega_n^2} \tag{3-34}$$

where ω_n is the natural frequency of the undamped oscillation of the system; ξ is the damping ratio of the system. The values of ω_n and ξ in different systems depends on the component parameters of each system. ω_n and ξ are the characteristic parameters of the second-order system, which indicate the characteristics of the second-order system itself independent of the outside world.

The block diagram of a typical second-order system with the unit negative feedback is shown in Fig. 3-11. The open-loop transfer function is

$$G_K(s) = \frac{\omega_n^2}{s(s + 2\xi\omega_n)} \tag{3-35}$$

The characteristic equation of the system is obtained by the differential equation or transfer function of the second-order system, i.e.

$$s^2 + 2\xi\omega_n s + \omega_n^2 = 0$$

Two characteristic roots of the equation are

$$s_{1,2} = -\xi\omega_n \pm \omega_n \sqrt{\xi^2 - 1} \tag{3-36}$$

The characteristic roots of the equation are the poles of the transfer function. And the characteristic roots of the second-order system are also different as the damping ratios are different.

1) When $\xi = 0$, it is called the zero damping state. The system has a pair of pure virtual roots, i.e.

$$s_{1,2} = \pm j\omega_n \tag{3-37}$$

The pole distribution is shown in Fig. 3-12a. The system is called an undamped system.

2) When $0 < \xi < 1$. it is called the underdamping state. The equation has a pair of conjugate complex roots with negative real parts, i.e.

$$s_{1,2} = -\xi\omega_n \pm j\omega_n \sqrt{1 - \xi^2} \tag{3-38}$$

The pole distribution is shown in Fig. 3-13a. The system is called an underdamping system.

3) When $\xi = 1$, it is called the critical damping state. The system has a pair of equal negative real roots, i.e.

$$s_{1,2} = -\xi\omega_n \tag{3-39}$$

The pole distribution is shown in Fig. 3-14a.

方程的特征根就是传递函数的极点，并且随着阻尼比 ξ 取值不同，二阶系统的特征根也不同。

1) 当 $\xi=0$ 时，称为零阻尼状态。系统有一对纯虚根，即

$$s_{1,2} = \pm j\omega_n \tag{3-37}$$

极点分布如图 3-12a 所示。这时的系统称为无阻尼系统。

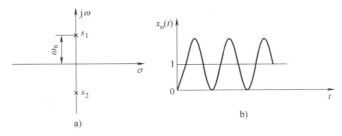

Fig. 3-12　Pole distribution and step response curve of a second-order system with $\xi=0$
（$\xi=0$ 时二阶系统的极点分布及阶跃响应曲线）

2) 当 $0<\xi<1$ 时，称为欠阻尼状态，方程有一对实部为负数的共轭复根，即

$$s_{1,2} = -\xi\omega_n \pm j\omega_n\sqrt{1-\xi^2} \tag{3-38}$$

极点分布如图 3-13a 所示。这时的系统称为欠阻尼系统。

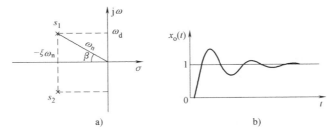

Fig. 3-13　Pole distribution and step response curve of a second-order system with $0<\xi<1$
（$0<\xi<1$ 时二阶系统的极点分布及阶跃响应曲线）

3) 当 $\xi=1$，称为临界阻尼状态。系统有一对相等的负实根，即

$$s_{1,2} = -\xi\omega_n \tag{3-39}$$

极点分布如图 3-14a 所示。这时的系统称为临界阻尼系统。

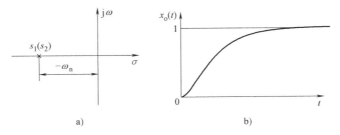

Fig. 3-14　Pole distribution and step response curve of a second-order system with $\xi=1$
（$\xi=1$ 时二阶系统的极点分布及阶跃响应曲线）

The system is called a critical damping system.

4) When $\xi > 1$, it is called the overdamping state, the equation has two negative real roots, i.e.

$$s_{1,2} = -\xi\omega_n \pm \omega_n\sqrt{\xi^2-1} \tag{3-40}$$

The pole distribution is shown in Fig. 3-15a. The system is called an overdamping system.

5) When $\xi < 0$, it is called the negative damping state. The roots of the equation are

$$s_{1,2} = -\xi\omega_n \pm \omega_n\sqrt{1-\xi^2} \tag{3-41}$$

The pole distribution is shown in Fig. 3-16a and Fig. 3-16c. The poles of the system are in the right half of the s-plane when the system is in the negative damping state.

There are many examples for second-order systems, such as the aforementioned RCL network, boosters with inertial loads, mass-spring-damping mechanical systems, etc.

3.3.2 Time response of second-order systems

1. The unit step response of a second-order system

The unit step response for different damping ratios of second-order systems is discussed as follows. The Laplace transform of the unit step input signal is $X_i(s) = \dfrac{1}{s}$, and the Laplace transform of the unit step response for the second-order system is

$$X_o(s) = G(s)X_i(s) = \frac{1}{s}\frac{\omega_n^2}{s^2+2\xi\omega_n s+\omega_n^2} \tag{3-42}$$

1) When $\xi = 0$, we have

$$X_o(s) = \frac{1}{s}\frac{\omega_n^2}{s^2+\omega_n^2} = \frac{1}{s} - \frac{s}{s^2+\omega_n^2}$$

Taking the inverse Laplace transform of the system, we have

$$x_o(t) = 1 - \cos\omega_n t \quad (t \geqslant 0) \tag{3-43}$$

The response curve is shown in Fig. 3-12b. It can be observed that the response curve of the zero damping system is an equal amplitude oscillation curve, and its oscillation frequency is ω_n.

2) When $0 < \xi < 1$, we have.

$$X_o(s) = \frac{\omega_n^2}{s(s^2+2\xi\omega_n s+\omega_n^2)} = \frac{1}{s} - \frac{s+\xi\omega_n}{(s+\xi\omega_n)^2+\omega_d^2} - \frac{\omega_d}{(s+\xi\omega_n)^2+\omega_d^2}\frac{\xi\omega_n}{\omega_d}$$

where ω_d is a damping natural frequency, and $\omega_d = \omega_n\sqrt{1-\xi^2}$. Taking the inverse Laplace transform, we have

$$x_o(t) = 1 - \frac{e^{-\xi\omega_n t}}{\sqrt{1-\xi^2}}\sin(\omega_d t+\beta) \quad (t \geqslant 0) \tag{3-44}$$

The response curve is shown in Fig. 3-13b. It can be seen that the response curve of the underdamping system is a damped oscillation curve. In equation (3-44), $\beta = \arctan(\sqrt{1-\xi^2}/\xi)$ or $\beta = \arccos\xi$, as is shown in Fig. 3-13a. It can be seen that the closer the root is to the negative real ax-

4) 当 $\xi>1$ 时,称为过阻尼状态,方程有两个不等的负实根,即

$$s_{1,2} = -\xi\omega_n \pm \omega_n \sqrt{\xi^2 - 1} \tag{3-40}$$

极点分布如图 3-15a 所示。这时的系统称为过阻尼系统。

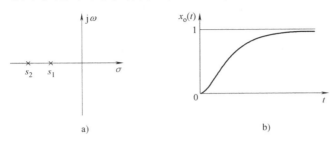

Fig. 3-15　Pole distribution and step response curve of a second-order system with $\xi>1$
($\xi>1$ 时二阶系统的极点分布及阶跃响应曲线)

5) 当 $\xi<0$ 时,为负阻尼状态。方程的根为

$$s_{1,2} = -\xi\omega_n \pm \omega_n \sqrt{1 - \xi^2} \tag{3-41}$$

极点分布如图 3-16a、图 3-16c 所示。系统处于负阻尼状态时,系统的极点处于 s 平面的右半平面。

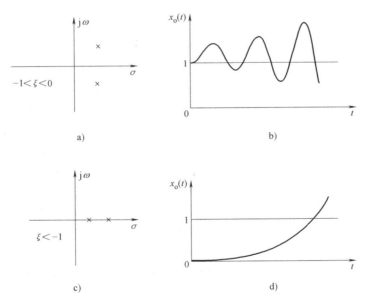

Fig. 3-16　Pole distribution and step response curve of a second-order system with $\xi<0$
($\xi<0$ 时二阶系统的极点分布及阶跃响应曲线)

二阶系统有很多实例,如前述的 RCL 网络、带有惯性载荷的助力器、质量-弹簧-阻尼机械系统等。

3.3.2　二阶系统的时间响应

1. 二阶系统的单位阶跃响应

下面讨论二阶系统不同阻尼比情况下的单位阶跃响应。单位阶跃输入信号的拉氏变换为

is, the larger the damping ratio is; the farther the root is from the negative real axis, the smaller the damping ratio is. According to equation (3-44), the first term is the steady-state term, and the second term is the transient term that is a sinusoidal oscillation function which decays with time t. The oscillation frequency is ω_d, and the amplitude decay rate depends on the time decay constant $1/(\xi\omega_n)$. From the pole distribution, the farther the pole is from the real axis, the higher the oscillation frequency is; the farther the pole is from the imaginary axis, the faster the attenuation is.

3) When $\xi = 1$, we have

$$X_o(s) = \frac{\omega_n^2}{s(s+\omega_n)^2} = \frac{1}{s} - \frac{1}{s+\omega_n} - \frac{\omega_n}{(s+\omega_n)^2}$$

Taking the inverse Laplace transform of the system, we have

$$x_o(t) = 1 - (1+\omega_n t)e^{-\omega_n t} \quad (t \geq 0) \tag{3-45}$$

The response curve is shown in Fig. 3-14b. It can be seen that the response curve of the critical damping system is a monotonously rising curve.

4) When $\xi > 1$, we have

$$X_o(s) = \frac{\omega_n^2}{s(s-s_1)(s-s_2)} = \frac{1}{s} + \frac{\omega_n^2}{s_1-s_2}\left[\frac{1}{s_1(s-s_1)} - \frac{1}{s_2(s-s_2)}\right]$$

Taking the inverse Laplace transform of the system, we have

$$x_o(t) = 1 + \frac{\omega_n}{2\sqrt{\xi^2-1}}\left(\frac{e^{s_1 t}}{s_1} - \frac{e^{s_2 t}}{s_2}\right) \quad (t \geq 0) \tag{3-46}$$

where $s_{1,2} = -\xi\omega_n \pm \omega_n\sqrt{\xi^2-1}$. The response curve is shown in Fig. 3-14b. It can be seen that the response curve of the critical damping system is also a monotonously rising curve.

5) When $\xi < 0$, the response curve is shown in Fig. 3-16b and Fig. 3-16d. The exponential term of its response expression becomes a positive exponent. So when $t \to \infty$, its output $x_o(t) \to \infty$. In other words, the unit step response of the negative damping system is a divergent or oscillatory divergence curve over time, and the system is unstable.

It can be seen from the above analysis that when the damping ratio of the second-order system $\xi > 0$, the poles of the system lie in the left half of the complex plane, and the output response of the system converges to the steady-state value, namely, it is in a stable state; when the damping ratio of the second-order system $\xi < 0$, the poles of the system lie in the right half of the complex plane, and the output response of the system doesn't converge to a steady-state value, namely, it is in an unstable state; when the damping ratio of the second-order system $\xi = 0$, the poles of the system lie on the imaginary axis of the complex plane, and the output response of the system oscillates with a constant amplitude, namely, it is in a critical steady state. Due to the fact that there always is a damping ratio and $\xi > 0$ in an actual second-order system, the second-order system is definitely a stable system.

2. The unit pulse response of a second-order system

When the input signal is a unit pulse signal, the unit pulse response of a second-order system can be obtained by differentiating the unit step response under different damping ratios.

$X_i(s) = \dfrac{1}{s}$，二阶系统单位阶跃响应的拉氏变换为

$$X_o(s) = G(s)X_i(s) = \frac{1}{s}\frac{\omega_n^2}{s^2+2\xi\omega_n s+\omega_n^2} \tag{3-42}$$

1) 当 $\xi=0$ 时，

$$X_o(s) = \frac{1}{s}\frac{\omega_n^2}{s^2+\omega_n^2} = \frac{1}{s} - \frac{s}{s^2+\omega_n^2}$$

取拉氏反变换，得

$$x_o(t) = 1-\cos\omega_n t \quad (t\geq 0) \tag{3-43}$$

响应曲线如图 3-12b，可见零阻尼系统的响应曲线为等幅振荡曲线，其振荡频率为 ω_n。

2) 当 $0<\xi<1$ 时，

$$X_o(s) = \frac{\omega_n^2}{s(s^2+2\xi\omega_n s+\omega_n^2)} = \frac{1}{s} - \frac{s+\xi\omega_n}{(s+\xi\omega_n)^2+\omega_d^2} - \frac{\omega_d}{(s+\xi\omega_n)^2+\omega_d^2}\frac{\xi\omega_n}{\omega_d}$$

式中，$\omega_d = \omega_n\sqrt{1-\xi^2}$ 称为阻尼固有频率。取拉氏反变换，得

$$x_o(t) = 1 - \frac{e^{-\xi\omega_n t}}{\sqrt{1-\xi^2}}\sin(\omega_d t+\beta) \quad (t\geq 0) \tag{3-44}$$

响应曲线如图 3-13b，可见欠阻尼系统的响应曲线为衰减振荡曲线。式（3-44）中 $\beta = \arctan(\sqrt{1-\xi^2}/\xi)$ 或者 $\beta = \arccos\xi$，如图 3-13a 所示。可以看出，根离负实轴越近，阻尼比 ξ 越大；根离负实轴越远，阻尼比 ξ 越小。由式（3-44）可知第一项是稳态项，第二项是瞬态项，是随时间 t 衰减的正弦振荡函数，振荡的频率为 ω_d，振幅衰减速度取决于时间衰减常数 $1/(\xi\omega_n)$。从极点分布图来看，极点与实轴的距离越远，则振荡频率越大；极点与虚轴的距离越远，则衰减越快。

3) 当 $\xi=1$ 时，

$$X_o(s) = \frac{\omega_n^2}{s(s+\omega_n)^2} = \frac{1}{s} - \frac{1}{s+\omega_n} - \frac{\omega_n}{(s+\omega_n)^2}$$

取拉氏反变换，得

$$x_o(t) = 1-(1+\omega_n t)e^{-\omega_n t} \quad (t\geq 0) \tag{3-45}$$

响应曲线如图 3-14b 所示，可见临界阻尼系统的响应曲线为单调上升曲线。

4) 当 $\xi>1$ 时，

$$X_o(s) = \frac{\omega_n^2}{s(s-s_1)(s-s_2)} = \frac{1}{s} + \frac{\omega_n^2}{s_1-s_2}\left[\frac{1}{s_1(s-s_1)} - \frac{1}{s_2(s-s_2)}\right]$$

取拉氏反变换，得

$$x_o(t) = 1 + \frac{\omega_n}{2\sqrt{\xi^2-1}}\left(\frac{e^{s_1 t}}{s_1} - \frac{e^{s_2 t}}{s_2}\right) \quad (t\geq 0) \tag{3-46}$$

式中，$s_{1,2} = -\xi\omega_n \pm \omega_n\sqrt{\xi^2-1}$。响应曲线如图 3-15b 所示，可见过阻尼系统响应曲线也为单调上升曲线。

1) When $\xi = 0$, we have
$$x_o(t) = \omega_n \sin\omega_n t \quad (t \geq 0) \tag{3-47}$$

2) When $0 < \xi < 1$, we have
$$x_o(t) = \frac{\omega_n}{\sqrt{1-\xi^2}} e^{-\xi\omega_n t} \sin\omega_d t \quad (t \geq 0) \tag{3-48}$$

3) When $\xi = 1$, we have
$$x_o(t) = \omega_n^2 t e^{-\omega_n t} \quad (t \geq 0) \tag{3-49}$$

4) When $\xi > 1$, we have
$$x_o(t) = \frac{\omega_n}{2\sqrt{\xi^2-1}} [e^{-(\xi-\sqrt{\xi^2-1})\omega_n t} - e^{-(\xi+\sqrt{\xi^2-1})\omega_n t}] \quad (t \geq 0) \tag{3-50}$$

The unit impulse response curve of the second-order system is shown in Fig. 3-17. As the damping ratio ξ decreases, the oscillation amplitude of the system increases. When $\xi > 0$, the response of the system approaches to zero as $t \to \infty$. It indicates that its steady-state error is zero when the system tracks the unit pulse signal.

3. The unit ramp response of a second-order system

When the input signal is a unit ramp signal, the first-order derivative of the unit ramp signal with respect to time is the unit step signal. Therefore, according to the important characteristics of a linear-invariant system, the unit ramp response of the second-order system can be obtained by integrating the unit step response of the second-order system under zero initial conditions.

1) When $\xi = 0$, we have
$$x_o(t) = t - \frac{1}{\omega_n} \sin\omega_n t \quad (t \geq 0) \tag{3-51}$$

2) When $0 < \xi < 1$, we have
$$x_o(t) = t - \frac{2\xi}{\omega_n} + \frac{e^{-\xi\omega_n t}}{\omega_d} \sin(\omega_d t + \beta) \quad (t \geq 0) \tag{3-52}$$

When $t \to \infty$, the steady-state error of the system is calculated as
$$e(\infty) = \lim_{t \to \infty} [x_i(t) - x_o(t)] = \frac{2\xi}{\omega_n}$$

The response curve is shown in Fig. 3-18, and its oscillation amplitude increases as ξ decreases.

3) When $\xi = 1$, we have
$$x_o(t) = t - \frac{2}{\omega_n} + \frac{2}{\omega_n}\left(1 + \frac{\omega_n t}{2}\right) e^{-\omega_n t} \quad (t \geq 0) \tag{3-53}$$

When $t \to \infty$, the steady-state error of the system is calculated as
$$e(\infty) = \lim_{t \to \infty} [x_i(t) - x_o(t)] = \frac{2}{\omega_n}$$

Its response curve is shown in Fig. 3-19.

4) When $\xi > 0$, we have

5) 当 $\xi<0$ 时，响应曲线如图 3-16b、3-16d 所示。其响应表达式的指数项变为正指数，故随着时间 $t\to\infty$，其输出 $x_o(t)\to\infty$，即负阻尼系统的单位阶跃响应是发散或振荡发散曲线，系统不稳定。

由以上分析可知，当二阶系统的阻尼比 $\xi>0$ 时，系统的极点位于复平面的左半平面，系统的输出响应收敛于稳态值，即系统处于稳定状态；当二阶系统的阻尼比 $\xi<0$ 时，系统的极点位于复平面的右半平面，系统的输出响应不收敛于稳态值，即系统处于不稳定状态；当二阶系统的阻尼比 $\xi=0$ 时，系统的极点位于复平面的虚轴上，系统的输出响应等幅振荡，即系统处于临界稳定状态。但实际的二阶系统总是存在阻尼比且 $\xi>0$，故二阶系统肯定是稳定的系统。

2. 二阶系统的单位脉冲响应

当输入信号为单位脉冲信号时，分别对不同阻尼比情况下的单位阶跃响应进行微分，就可以得到二阶系统的单位脉冲响应。

1) 当 $\xi=0$ 时，

$$x_o(t)=\omega_n\sin\omega_n t \quad (t\geqslant 0) \tag{3-47}$$

2) 当 $0<\xi<1$ 时，

$$x_o(t)=\frac{\omega_n}{\sqrt{1-\xi^2}}e^{-\xi\omega_n t}\sin\omega_d t \quad (t\geqslant 0) \tag{3-48}$$

3) 当 $\xi=1$ 时，

$$x_o(t)=\omega_n^2 t e^{-\omega_n t} \quad (t\geqslant 0) \tag{3-49}$$

4) 当 $\xi>1$ 时，

$$x_o(t)=\frac{\omega_n}{2\sqrt{\xi^2-1}}\left[e^{-\left(\xi-\sqrt{\xi^2-1}\right)\omega_n t}-e^{-\left(\xi+\sqrt{\xi^2-1}\right)\omega_n t}\right] \quad (t\geqslant 0) \tag{3-50}$$

二阶系统的单位脉冲响应曲线如图 3-17 所示，随着阻尼比 ξ 的减小，系统的振荡幅度加大；当 $\xi>0$ 时，随着 $t\to\infty$，系统的响应趋近于零，说明系统跟随单位脉冲信号时，其稳态误差为零。

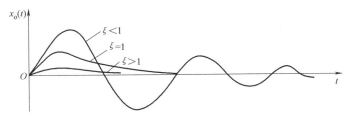

Fig. 3-17 Unit pulse response of a second-order system
（二阶系统的单位脉冲响应）

3. 二阶系统的单位斜坡响应

当输入信号是单位斜坡信号时，由于单位斜坡信号对时间的一阶微分就是单位阶跃信号，因此根据线性定常系统的重要特性，二阶系统的单位斜坡响应可以由二阶系统的单位阶跃响应在零初始条件下对其进行积分求得。

1) 当 $\xi=0$ 时，

$$x_o(t) = t - \frac{2\xi}{\omega_n} + \frac{2\xi^2 - 1 + 2\xi\sqrt{\xi^2 - 1}}{2\omega_n\sqrt{\xi^2 - 1}} e^{-(\xi - \sqrt{\xi^2 - 1})\omega_n t} - \frac{2\xi^2 - 1 - 2\xi\sqrt{\xi^2 - 1}}{2\omega_n\sqrt{\xi^2 - 1}} e^{-(\xi + \sqrt{\xi^2 - 1})\omega_n t} \quad (t \geq 0)$$

(3-54)

When $t \to \infty$, the steady-state error of the system is calculated as

$$e(\infty) = \lim_{t \to \infty} [x_i(t) - x_o(t)] = \frac{2\xi}{\omega_n}$$

Its response curve is shown in Fig. 3-20.

The above analysis shows that the second-order system has a steady-state error $2\xi/\omega_n$ when tracking the unit ramp signal, and its steady-state error becomes smaller as ξ decreases.

3.3.3 Time-domain performance indicators of second-order systems

In real-world engineering, performance indicators of a second-order system is given according to the response of the second-order element to the unit step input in the underdamping state. It takes too long time for the transition process of a completely non-oscillating second-order system. In order to obtain the shorter transition process, the system is generally allowed to be moderately damped, except for those systems that do not allow oscillation. It is the reason why the performance indicators of the second-order system is selected in the underdamping state. The unit step signal is used as the input signal based on two following aspects: one is that it is easier to generate a unit step signal, and it is also easier to obtain the response to any input based on the response of the unit step input of the system; the other is that many inputs are similar to the step input in engineering, and the step input is often the worst case in practice.

1. Rise time t_r

According to the definition of rise time, when $t = t_r$, we have $x_o(t_r) = 1$. According to equation (3-44), we have

$$1 = 1 - \frac{e^{-\xi\omega_n t_r}}{\sqrt{1-\xi^2}} \sin(\omega_d t_r + \beta)$$

Before the steady state is reached, we have $e^{-\xi\omega_n t_r} > 0$. To make the above equation true, it is required for $\sin(\omega_d t + \beta) = 0$. Considering that the rise time is the first time to reach the steady state value, we have $\omega_d t + \beta = \pi$, i.e.

$$t_r = \frac{\pi - \beta}{\omega_d} = \frac{\pi - \beta}{\omega_n\sqrt{1-\xi^2}} \quad (3-55)$$

It can be seen that when ξ is constant, the rise time is shortened as ω_n increases; and when ω_n is constant, the rise time becomes longer as ξ increases.

2. Peak time t_p

When the first peak of the response curve occurs, the change rate of the unit step response with time is zero. Take the derivative of the equation (3-44), and let $\dot{x}_o(t_p) = 0$, then we have

$$\frac{\sin(\omega_d t_p + \beta)}{\cos(\omega_d t_p + \beta)} = \frac{\sqrt{1-\xi^2}}{\xi}$$

$$x_o(t) = t - \frac{1}{\omega_n}\sin\omega_n t \quad (t \geqslant 0) \tag{3-51}$$

2) 当 $0<\xi<1$ 时，

$$x_o(t) = t - \frac{2\xi}{\omega_n} + \frac{e^{-\xi\omega_n t}}{\omega_d}\sin(\omega_d t + \beta) \quad (t \geqslant 0) \tag{3-52}$$

当 $t\to\infty$ 时，其稳态误差为

$$e(\infty) = \lim_{t\to\infty}[x_i(t) - x_o(t)] = \frac{2\xi}{\omega_n}$$

其响应曲线如图 3-18 所示，随着 ξ 的减小，其振荡幅度增大。

3) 当 $\xi=1$ 时，

$$x_o(t) = t - \frac{2}{\omega_n} + \frac{2}{\omega_n}\left(1 + \frac{\omega_n t}{2}\right)e^{-\omega_n t} \quad (t \geqslant 0) \tag{3-53}$$

当 $t\to\infty$ 时，其稳态误差为

$$e(\infty) = \lim_{t\to\infty}[x_i(t) - x_o(t)] = \frac{2}{\omega_n}$$

其响应曲线如图 3-19 所示。

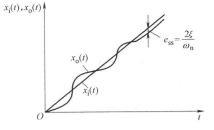

Fig. 3-18 Unit ramp response of an underdamping second-order system

（欠阻尼二阶系统单位斜坡响应曲线）

4) 当 $\xi>1$ 时，

$$x_o(t) = t - \frac{2\xi}{\omega_n} + \frac{2\xi^2 - 1 + 2\xi\sqrt{\xi^2-1}}{2\omega_n\sqrt{\xi^2-1}}e^{-(\xi-\sqrt{\xi^2-1})\omega_n t} - \frac{2\xi^2 - 1 - 2\xi\sqrt{\xi^2-1}}{2\omega_n\sqrt{\xi^2-1}}e^{-(\xi+\sqrt{\xi^2-1})\omega_n t} \quad (t \geqslant 0) \tag{3-54}$$

当 $t\to\infty$ 时，其稳态误差为

$$e(\infty) = \lim_{t\to\infty}[x_i(t) - x_o(t)] = \frac{2\xi}{\omega_n}$$

其响应曲线如图 3-20 所示。

以上分析说明，二阶系统在跟踪单位斜坡信号时存在稳态误差 $2\xi/\omega_n$，而且随着 ξ 减小，其稳态误差变小。

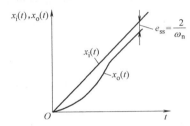

Fig. 3-19 Unit ramp response of a critical damping second-order system

（临界阻尼二阶系统单位斜坡响应曲线）

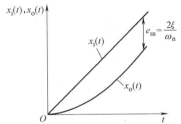

Fig. 3-20 Unit ramp response of overdamping second-order system

（过阻尼二阶系统单位斜坡响应曲线）

The equation can be rewritten as

$$\tan(\omega_d t_p + \beta) = \tan\beta$$

Then we have $\omega_d t_p = n\pi$. According to the definition of peak time, when $n = 1$, the corresponding peak time is

$$t_p = \frac{\pi}{\omega_d} = \frac{\pi}{\omega_n \sqrt{1-\xi^2}} \tag{3-56}$$

It can also be seen directly from the response curve that it takes exactly half a period to reach the peak. Since the period of the damped oscillation is $\frac{2\pi}{\omega_d}$, we have $t_p = \frac{\pi}{\omega_d}$. When ξ is constant, the peak time is shortened as ω_n increases; when ω_n is constant, the peak time becomes longer as ξ increases.

3. Maximum overshoot M_p

The maximum overshoot M_p occurs at the peak time $t = t_p$, and the value t_p is substituted into equation (3-44) to obtain $x_o(t_p)$. Let $x_o(\infty) = 1$, M_p is obtained by equation (3-22) as

$$M_p = -\frac{e^{-\xi\pi\omega_n/\omega_d}}{\sqrt{1-\xi^2}} \sin(\pi+\beta)$$

Since $\sin(\pi+\beta) = -\sin\beta = -\sqrt{1-\xi^2}$, we have

$$M_p = e^{-\xi\pi/\sqrt{1-\xi^2}} \times 100\% \tag{3-57}$$

It can be seen from the analysis, the characteristics of the second-order system performance indicators are as follows.

1) M_p only depends on the value ξ, and is independent of the natural frequency of the undamped oscillation.

2) As ξ increases, M_p gradually decreases. When $\xi \geq 1$, the step response curve rises monotonically, so $M_p = 0$.

3) As ξ increases, t_r and t_p also increase, and it will slow down the experience, According to the experience, damping ratio is better to be between 0.4 and 0.8, so that both the rapidity and stationarity of the unit step response are considered adequately. Thus, the damping ratio of the second-order system in practical applications is within this range. When $\xi = 0.707$, the maximum overshoot of the system is small ($M_p < 5\%$), and the rise time is also small. Thus, $\xi = 0.707$ is called the optimal damping ratio.

4. Settling time t_s

According to equation (3-44) and equation (3-23), considering $x_o(\infty) = 1$, we have

$$\left| \frac{e^{-\xi\omega_n t}}{\sqrt{1-\xi^2}} \sin(\omega_d t + \beta) \right| \leq \Delta$$

where Δ is the allowed error band, generally taking $\Delta = \pm 2\%$ or $\pm 5\%$. For the sake of simplification, the effect of the sine function in the above equation can be ignored. We have

3.3.3 二阶系统的时域性能指标

在实际工程中,二阶系统的性能指标是根据欠阻尼状态下的二阶环节对单位阶跃输入的响应给出的。因为完全无振荡的二阶系统过渡过程时间太长,所以除了那些不允许产生振荡的系统外,通常都允许系统有适度的阻尼,其目的是为获得较短的过渡过程。这就是选择在欠阻尼状态下研究二阶系统性能指标的原因。以单位阶跃信号作为输入信号是基于以下两个方面的考虑:一是产生单位阶跃信号比较容易,而且在获得系统单位阶跃输入的响应后,也较容易求得系统对任何其他输入的响应;二是在工程中,许多输入与阶跃输入相似,而且阶跃输入又往往是实际中最糟糕的情况。

1. 上升时间 t_r

根据上升时间定义,当 $t=t_r$ 时,$x_o(t_r)=1$,由式(3-44)得

$$1 = 1 - \frac{e^{-\xi\omega_n t_r}}{\sqrt{1-\xi^2}}\sin(\omega_d t_r + \beta)$$

且在达到稳态之前有 $e^{-\xi\omega_n t_r} > 0$。要使上式成立,即要求 $\sin(\omega_d t + \beta) = 0$。考虑到上升时间是第一次达到稳态值的时间,故取 $\omega_d t + \beta = \pi$,即

$$t_r = \frac{\pi - \beta}{\omega_d} = \frac{\pi - \beta}{\omega_n\sqrt{1-\xi^2}} \tag{3-55}$$

可知,当 ξ 一定时,ω_n 增大,上升时间缩短;而当 ω_n 一定时,ξ 越大,上升时间就越长。

2. 峰值时间 t_p

响应曲线出现第一个峰值时,单位阶跃响应随时间的变化率为零。对式(3-44)求导,并令 $\dot{x}_o(t_p) = 0$,可解得

$$\frac{\sin(\omega_d t_p + \beta)}{\cos(\omega_d t_p + \beta)} = \frac{\sqrt{1-\xi^2}}{\xi}$$

整理得

$$\tan(\omega_d t_p + \beta) = \tan\beta$$

所以有 $\omega_d t_p = n\pi$,根据峰值时间定义,$n=1$ 对应的峰值时间为

$$t_p = \frac{\pi}{\omega_d} = \frac{\pi}{\omega_n\sqrt{1-\xi^2}} \tag{3-56}$$

从响应曲线也可直接看出到达峰值的时间刚好是半个周期的长度,而有阻尼振荡的周期为 $\frac{2\pi}{\omega_d}$,所以 $t_p = \frac{\pi}{\omega_d}$。当 ξ 一定时,ω_n 增大,峰值时间就缩短;而当 ω_n 一定时,ξ 越大,峰值时间就越长。

3. 最大超调量 M_p

最大超调量 M_p 出现在峰值时间 $t=t_p$ 处,将 t_p 值代入式(3-44)求得 $x_o(t_p)$,令 $x_o(\infty)=1$,由式(3-22)可得 M_p 为

$$\frac{\mathrm{e}^{-\xi\omega_n t}}{\sqrt{1-\xi^2}} \leq \Delta \quad (t \geq t_s)$$

Arranging the above inequation, we have

$$t_s \geq \frac{1}{\xi\omega_n}\ln\frac{1}{\Delta\sqrt{1-\xi^2}}$$

Taking the approximate value of the above inequation, we have

$$t_s = \frac{4}{\xi\omega_n} \quad (\Delta = \pm 2\%) \tag{3-58}$$

$$t_s = \frac{3}{\xi\omega_n} \quad (\Delta = \pm 5\%) \tag{3-59}$$

Note that the settling time obtained here must be the settling time of the oscillation process (i.e. as for $0<\xi<1$). When $\xi \geq 1$, the step curve monotonically rises, and the settling time cannot be obtained according to the above equations.

5. The number of oscillation N

The period of damped oscillation is $T_d = \frac{2\pi}{\omega_d}$, and according to the definition of the number of oscillation, we have

$$N = \frac{t_s}{2\pi/\omega_d} = \frac{t_s\omega_d}{2\pi}$$

Since $t_s = \frac{4}{\xi\omega_n}$ or $t_s = \frac{3}{\xi\omega_n}$, and $\omega_d = \omega_n\sqrt{1-\xi^2}$, we have

$$N = \begin{cases} \dfrac{2}{\pi}\dfrac{\sqrt{1-\xi^2}}{\xi} & (\Delta = \pm 2\%) \\ \dfrac{1.5}{\pi}\dfrac{\sqrt{1-\xi^2}}{\xi} & (\Delta = \pm 5\%) \end{cases} \tag{3-60}$$

It can be seen that N only depends on the ξ, and it decreases as ξ increases. Hence, it is another indicator that reflects the damping characteristics of the system.

From the above discussion, the following conclusions can be drawn.

1) In order to obtain satisfactory dynamic performance indicators for the second-order system, the appropriate damping ratio ξ and undamped natural frequency ω_n should be selected. Increasing ω_n can improve the response speed of the second-order system, and reduce the rise time, peak time and settling time; increasing ξ can weaken the oscillation performance of the system, but increase the rise time and peak time. Since the maximum overshoot M_p only depends on the ξ, the damping ratio should be selected in terms of the allowable maximum overshoot M_p firstly when designing a system.

2) There is often a conflict between the response speed of the system and the oscillation performance. For example, for the mass-damper-spring system, according to $\omega_n = \sqrt{k/m}$, the improvement of ω_n is generally achieved by increasing the value of k; on the other hand, according to

$$M_{\mathrm{p}} = -\frac{\mathrm{e}^{-\xi\pi\omega_{\mathrm{n}}/\omega_{\mathrm{d}}}}{\sqrt{1-\xi^2}}\sin(\pi+\beta)$$

又 $\sin(\pi+\beta) = -\sin\beta = -\sqrt{1-\xi^2}$，得

$$M_{\mathrm{p}} = \mathrm{e}^{-\xi\pi/\sqrt{1-\xi^2}} \times 100\% \tag{3-57}$$

由分析可知，二阶系统性能指标具有如下特点。

1）可见 M_{p} 唯一取决于 ξ 值，而与无阻尼振动固有频率无关。

2）随着 ξ 增大，M_{p} 逐渐减小。当 $\xi \geqslant 1$ 时，阶跃响应曲线单调上升，所以 $M_{\mathrm{p}} = 0$。

3）随着 ξ 增大，t_{r} 和 t_{p} 也随之增大，这将使响应速度变慢。根据经验，阻尼比以在 0.4~0.8 之间为佳，此时单位阶跃响应的快速性和平稳性得到兼顾，故实际应用中二阶系统的阻尼比在此范围内取值。当 $\xi = 0.707$ 时，系统的最大超调量很小（$M_{\mathrm{p}} < 5\%$），上升时间 t_{r} 也很小，故 $\xi = 0.707$ 称为最佳阻尼比。

4. 调节时间 t_{s}

由式（3-44）和式（3-23），并考虑 $x_{\mathrm{o}}(\infty) = 1$，得

$$\left|\frac{\mathrm{e}^{-\xi\omega_{\mathrm{n}}t}}{\sqrt{1-\xi^2}}\sin(\omega_{\mathrm{d}}t+\beta)\right| \leqslant \Delta$$

式中，Δ 是允许的误差带，一般取 $\Delta = \pm 2\%$ 或 $\pm 5\%$。为简单起见，可忽略上式中正弦函数的影响。则有

$$\frac{\mathrm{e}^{-\xi\omega_{\mathrm{n}}t}}{\sqrt{1-\xi^2}} \leqslant \Delta \quad (t \geqslant t_{\mathrm{s}})$$

整理得

$$t_{\mathrm{s}} \geqslant \frac{1}{\xi\omega_{\mathrm{n}}}\ln\frac{1}{\Delta\sqrt{1-\xi^2}}$$

对上式近似取值得

$$t_{\mathrm{s}} = \frac{4}{\xi\omega_{\mathrm{n}}} \quad (\Delta = \pm 2\%) \tag{3-58}$$

$$t_{\mathrm{s}} = \frac{3}{\xi\omega_{\mathrm{n}}} \quad (\Delta = \pm 5\%) \tag{3-59}$$

注意，这里求得的调整时间一定是振荡过程的调节时间（即针对的是 $0 < \xi < 1$ 的情况），当 $\xi \geqslant 1$ 时，阶跃曲线单调上升，此时调节时间不能按上式求。

5. 振荡次数 N

有阻尼振荡的周期为 $T_{\mathrm{d}} = \dfrac{2\pi}{\omega_{\mathrm{d}}}$，由振荡次数的定义可求得

$$N = \frac{t_{\mathrm{s}}}{2\pi/\omega_{\mathrm{d}}} = \frac{t_{\mathrm{s}}\omega_{\mathrm{d}}}{2\pi}$$

$\xi = \dfrac{c}{2\sqrt{mk}}$, in order to increase the damping ξ, it is desirable to decrease the value of k. Therefore, both the values of ω_n and ξ should be selected appropriately to weaken the oscillation performance and to have a certain response speed of the system.

Example 3-4 The block diagram of a system is shown in Fig. 3-21. If the unit step response indicator is required to be $M_p = 20\%$, $t_p = 1\text{s}$, try to determine the values of K and K_t of the system.

Solution: The closed-loop transfer function of the system is

$$G(s) = \dfrac{K}{s^2 + (1+KK_t)s + K}$$

Comparing it with the standard form of the second-order system transfer function, we have

$$\begin{cases} \omega_n = \sqrt{K} \quad \text{rad/s} \\ 2\xi\omega_n = 1 + KK_t \end{cases} \Rightarrow \xi = \dfrac{1+KK_t}{2\sqrt{K}}$$

According to $M_p = 20\%$, we have

$$20\% = e^{-\dfrac{\xi\pi}{\sqrt{1-\xi^2}}} \times 100\%$$

$$\xi = 0.456$$

According to $t_p = 1\text{s}$, we have

$$1 = \dfrac{\pi}{\omega_n \sqrt{1-\xi^2}}$$

So, we have

$$\omega_n = 3.53 \text{rad/s}$$

$$K = \omega_n^2 = 12.5$$

$$K_t = \dfrac{2\xi\omega_n - 1}{K} = 0.178$$

Example 3-5 The unit step response of a second-order control system is shown in Fig. 3-22. Figure out the damping ratio ξ and the undamped natural frequency ω_n of the system.

Solution: It can be seen from Fig. 3-22 that the steady-state value of the unit step response is 3, so the open-loop gain of the system is 3, and then the transfer function is

$$G(s) = \dfrac{3\omega_n^2}{s^2 + 2\xi\omega_n s + \omega_n^2}$$

According to Fig. 3-22, we have $t_p = 0.1\text{s}$. According to the equation of the damping ratio ξ, we have

$$M_p = e^{-\xi\pi/\sqrt{1-\xi^2}} = \dfrac{x_o(t_p) - x_o(\infty)}{x_o(\infty)} = \dfrac{4-3}{3} = 33.3\%$$

So we have $\xi = 0.33$.

According to the equation of the undamped natural frequency, we have

$$t_p = \dfrac{\pi}{\omega_n\sqrt{1-\xi^2}} \Rightarrow \omega_n = \dfrac{\pi}{t_p\sqrt{1-\xi^2}} = 33.2\text{rad/s}$$

因为 $t_s = \dfrac{4}{\xi\omega_n}$ 或 $t_s = \dfrac{3}{\xi\omega_n}$，$\omega_d = \omega_n\sqrt{1-\xi^2}$，所以

$$N = \begin{cases} \dfrac{2}{\pi}\dfrac{\sqrt{1-\xi^2}}{\xi} & (\Delta = \pm 2\%) \\ \dfrac{1.5}{\pi}\dfrac{\sqrt{1-\xi^2}}{\xi} & (\Delta = \pm 5\%) \end{cases} \qquad (3\text{-}60)$$

由此可见，N 只取决于 ξ，且随着 ξ 的增大而减小，所以 N 是反映系统阻尼特性的另一个指标。

由以上讨论，可得出如下结论。

1) 为使二阶系统具有满意的动态性能指标，必须选择合适的阻尼比 ξ 和无阻尼固有频率 ω_n。提高 ω_n 可以提高二阶系统的响应速度，减少上升时间、峰值时间和调整时间；增大 ξ 可以减弱系统的振荡性能，但增大上升时间和峰值时间。由于最大超调量 M_p 由 ξ 唯一决定，因此在设计系统时应首先根据允许的超调量 M_p 来选择阻尼比 ξ。

2) 系统的响应速度与振荡性能之间往往存在矛盾。譬如，对于质量-阻尼-弹簧系统，根据 $\omega_n = \sqrt{k/m}$，ω_n 的提高一般通过提高 k 值来实现；另一方面，根据 $\xi = \dfrac{c}{2\sqrt{mk}}$，为增大阻尼 ξ，希望减小 k 值。因此，既要减弱系统的振荡性能，又要系统具有一定的响应速度，就应同时选取合适的 ξ 和 ω_n 值。

例 3-4 已知系统的框图如图 3-21 所示，若要求单位阶跃响应指标满足 $M_p = 20\%$，$t_p = 1\text{s}$，试确定系统的 K 和 K_t 之值。

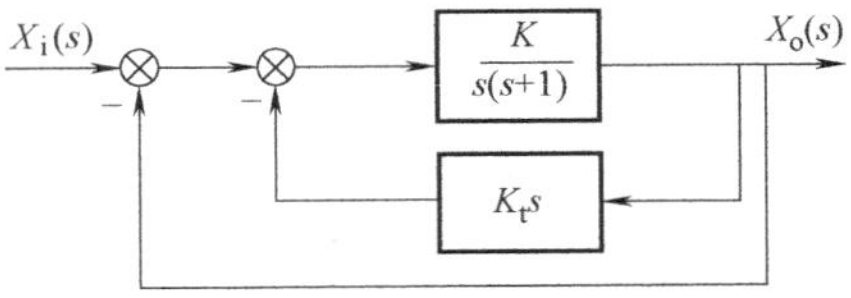

解：系统的闭环传递函数为

$$G(s) = \dfrac{K}{s^2 + (1+KK_t)s + K}$$

Fig. 3-21 Block diagram of example 3-4
（例 3-4 的框图）

与二阶系统传递函数的标准形式比较，得

$$\begin{cases} \omega_n = \sqrt{K}\,\text{rad/s} \\ 2\xi\omega_n = 1 + KK_t \end{cases} \Rightarrow \xi = \dfrac{1+KK_t}{2\sqrt{K}}$$

因为要求 $M_p = 20\%$，故

$$20\% = e^{-\dfrac{\xi\pi}{\sqrt{1-\xi^2}}} \times 100\%$$

$$\xi = 0.456$$

由峰值时间 $t_p = 1\text{s}$ 得

$$1 = \dfrac{\pi}{\omega_n\sqrt{1-\xi^2}}$$

解得

$$\omega_n = 3.53\,\text{rad/s}$$

Example 3-6 As is shown in Fig. 3-23a, after applying the step force $x_i(t) = 3N$ on the mass m, the time response of the system is shown in Fig. 3-23b. Try to figure out the spring stiffness k, the mass m, and the damping coefficient c.

Solution: According to Newton's law, establishing the dynamic differential equation of the mechanical system, we have the transfer function of the systems

$$G(s) = \frac{X(s)}{F(s)} = \frac{1}{ms^2 + cs + k} = \frac{\frac{1}{k} \cdot \frac{k}{m}}{s^2 + \frac{c}{m}s + \frac{k}{m}}$$

Comparing it with the standard form of the second-order system transfer function, we have

$$\omega_n = \sqrt{\frac{k}{m}}, \quad \xi = \frac{c}{2\sqrt{mk}}$$

(1) Calculating k according to the steady-state value (1cm) of the response curve. Since the step force is $F(t) = 3N$, its Laplace transform is $F(s) = 3/s$. Then we have

$$X(s) = \frac{1}{ms^2 + cs + k} F(s) = \frac{3}{(ms^2 + cs + k)s}$$

The steady-state value of $x(t)$ can be obtained in terms of the final value theorem of the Laplace transform as

$$x(t)\big|_{t \to \infty} = \lim_{s \to 0} sX(s) = \frac{3}{k} = 1 \text{ (cm)}$$

So, $k = 3\text{N/cm} = 300\text{N/m}$.

(2) Calculating ξ and ω_n while $M_p = 0.095$, $t_p = 2s$ according to the response curve. According to $M_p = e^{-\xi\pi/\sqrt{1-\xi^2}} \times 100\% = 0.095$, we have

$$\xi = 0.6, \quad t_p = \frac{\pi}{\omega_n \sqrt{1-\xi^2}} = 2s$$

So, we have $\omega_n = 1.96 \text{rad/s}$.

(3) Calculating m and c from ω_n and ξ. Substituting $\omega_n = 1.96 \text{rad/s}$ and $\xi = 0.6$ into $\omega_n = \sqrt{\frac{k}{m}}$ and $\xi = \frac{c}{2\sqrt{mk}}$, we have

$$m = 78.09 \text{kg}, \quad c = 180 \text{N} \cdot \text{s/m}$$

According to $\omega_n = \sqrt{\frac{k}{m}}$ and $\xi = \frac{c}{2\sqrt{mk}}$, in order to make the system response smoothly, the value of ξ should be increased. So, the damping coefficient c should be increased and the mass m should be reduced. In order to make the system respond quickly, the value of ω_n should be increased. So, the mass m should be reduced. The stiffness k of the spring is generally determined by the steady-state value. In order to make the system have good transient response performance, the mass m should be reduced, and the damping coefficient c should be increased. In engineering, lightweight materials or hollow structures are often used to reduce the mass.

$$K = \omega_n^2 = 12.5$$

$$K_t = \frac{2\xi\omega_n - 1}{K} = 0.178$$

例 3-5 某二阶控制系统的单位阶跃响应如图 3-22 所示，求系统的阻尼比 ξ 和无阻尼固有频率 ω_n。

解：由图 3-22 可知，系统的单位阶跃响应稳态值是 3，故其开环增益为 3，所以其传递函数为

$$G(s) = \frac{3\omega_n^2}{s^2 + 2\xi\omega_n s + \omega_n^2}$$

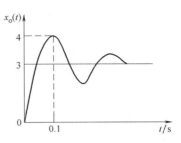

Fig. 3-22 Unit step response curve of example 3-5
（例 3-5 的单位阶跃响应曲线）

又由图 3-22 可知 $t_p = 0.1\text{s}$，根据阻尼比的公式，有

$$M_p = e^{-\xi\pi/\sqrt{1-\xi^2}} = \frac{x_o(t_p) - x_o(\infty)}{x_o(\infty)} = \frac{4-3}{3} = 33.3\%$$

可解得 $\xi = 0.33$。

无阻尼固有频率为

$$t_p = \frac{\pi}{\omega_n\sqrt{1-\xi^2}} \Rightarrow \omega_n = \frac{\pi}{t_p\sqrt{1-\xi^2}} = 33.2\text{rad/s}$$

例 3-6 如图 3-23a 所示的机械系统，在质量块 m 上施加 $F = 3N$ 的阶跃力后，系统的时间响应如图 3-23b 所示。试求弹簧刚度 k、质量 m 和阻尼系数 c。

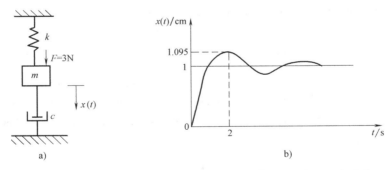

Fig. 3-23 System structure and time response diagram of example 3-6
（例 3-6 的系统结构及其时间响应图）

解：根据牛顿定律，建立机械系统的动力学微分方程，得系统传递函数为

$$G(s) = \frac{X(s)}{F(s)} = \frac{1}{ms^2 + cs + k} = \frac{\frac{1}{k} \cdot \frac{k}{m}}{s^2 + \frac{c}{m}s + \frac{k}{m}}$$

与二阶系统传递函数的标准形式比较可知

$$\omega_n = \sqrt{\frac{k}{m}}, \quad \xi = \frac{c}{2\sqrt{mk}}$$

3.4 Time Response and Performance Analysis of Higher-Order Systems

Third-order and above systems are called higher-order systems. In fact, a large number of systems, especially engineering systems, generally are higher-order systems. A higher-order system can always be decomposed into a combination of several zero-order, first-order and second-order elements, and it can also include delay elements. The time response of the higher-order system is composed of the superposition of the response functions of these elements.

3.4.1 Time response of higher-order systems

In Chapter 2, the differential equation of a linear time-invariant system with n-order is defined as

$$a_n x_o^{(n)}(t) + a_{n-1} x_o^{(n-1)}(t) + \cdots + a_1 \dot{x}_o(t) + a_0 x_o(t)$$
$$= b_m x_i^{(m)}(t) + b_{m-1} x_i^{(m-1)}(t) + \cdots + b_1 \dot{x}_i(t) + b_0 x_i(t) \quad (n \geq m) \tag{3-61}$$

The transfer function of the system is

$$G(s) = \frac{b_m s^m + b_{m-1} s^{m-1} + \cdots + b_1 s + b_0}{a_n s^n + a_{n-1} s^{n-1} + \cdots + a_1 s + a_0} \quad (n \geq m) \tag{3-62}$$

The characteristic equation of the system is

$$a_n s^n + a_{n-1} s^{n-1} + \cdots + a_1 s + a_0 = 0 \tag{3-63}$$

There are n characteristic roots in the characteristic equation, where the number of real roots is n_1, the number of conjugate complex roots is n_2, and $n = n_1 + 2n_2$. Thus, the characteristic equation can be decomposed into the product of n_1 one-time factors $s + p_j$ ($j = 1, 2, \cdots, n_1$) and n_2 quadratic factors $s^2 + 2\xi_k \omega_{nk} s + \omega_{nk}^2$ ($0 < \xi_k < 1$; $k = 1, 2, \cdots, n_2$). In other words, there are n_1 real poles $-p_j$ and n_2 conjugate complex poles $-\xi_k \omega_{nk} \pm j\omega_{nk}\sqrt{1-\xi_k^2}$ for the transfer function of the system, where $-\xi_k \omega_{nk}$ is the real part of the conjugate complex pole. Assuming that there are m zeros $-z_i$ ($i = 1, 2, \cdots, m$) in the transfer function of the system, the transfer function of the system can be written as

$$G(s) = \frac{K_g \prod_{i=1}^{m}(s + z_i)}{\prod_{j=1}^{n_1}(s + p_j) \prod_{k=1}^{n_2}(s^2 + 2\xi_k \omega_{nk} s + \omega_{nk}^2)} \tag{3-64}$$

where K_g is the transfer coefficient or root-locus gain of the system.

The output of the system under the unit step input $X_i(s) = \dfrac{1}{s}$ is

$$X_o(s) = G(s) \frac{1}{s} = \frac{K_g \prod_{i=1}^{m}(s + z_i)}{s \prod_{j=1}^{n_1}(s + p_j) \prod_{k=1}^{n_2}(s^2 + 2\xi_k \omega_{nk} s + \omega_{nk}^2)}$$

(1) 由响应曲线的稳态值（1cm）求 k　由于阶跃力 $F = 3$N，其拉氏变换为 $F(s) = 3/s$，故

$$X(s) = \frac{1}{ms^2+cs+k}F(s) = \frac{3}{(ms^2+cs+k)s}$$

由拉氏变换的终值定理，可求得 $x(t)$ 的稳态值为

$$x(t)\big|_{t\to\infty} = \lim_{s\to 0}sX(s) = \frac{3}{k} = 1\,(\text{cm})$$

因此，$k = 3\text{N/cm} = 300\text{N/m}$。

(2) 由响应曲线可知 $M_p = 0.095$，$t_p = 2$s，求 ξ 和 ω_n　由 $M_p = \mathrm{e}^{-\xi\pi/\sqrt{1-\xi^2}} \times 100\% = 0.095$，解得

$$\xi = 0.6，\quad t_p = \frac{\pi}{\omega_n\sqrt{1-\xi^2}} = 2\,(\text{s})$$

因此，$\omega_n = 1.96\text{rad/s}$。

(3) 由 ω_n 和 ξ 求 m 和 c　将 $\omega_n = 1.96\text{rad/s}$ 和 $\xi = 0.6$ 代入 $\omega_n = \sqrt{\dfrac{k}{m}}$ 和 $\xi = \dfrac{c}{2\sqrt{mk}}$ 求得

$$m = 78.09\text{kg}，\quad c = 180\text{N}\cdot\text{s/m}$$

根据 $\omega_n = \sqrt{\dfrac{k}{m}}$ 和 $\xi = \dfrac{c}{2\sqrt{mk}}$ 可知，要使系统响应平稳，应增大 ξ 值，故应增大阻尼系数 c，减小质量 m；要使系统响应快速，应增大 ω_n 值，故应减小质量 m。弹簧的刚度 k 一般由稳态值决定。为使系统具有良好的暂态响应性能，应该减小质量块 m 的质量，增大阻尼系数 c 值，在工程中经常采用轻型材料或空心结构减小质量。

3.4 高阶系统的时间响应与性能分析

三阶及三阶以上的系统称为高阶系统。实际上大量的系统特别是工程系统，一般都属于高阶系统。高阶系统可以分解成若干零阶、一阶和二阶环节等的组合，而且也可以包含延时环节。高阶系统的时间响应即由这些环节的响应函数叠加组成。

3.4.1 高阶系统的时间响应

在第2章中定义的线性定常 n 阶系统的微分方程式为

$$\begin{aligned}&a_n x_o^{(n)}(t) + a_{n-1} x_o^{(n-1)}(t) + \cdots + a_1 \dot{x}_o(t) + a_0 x_o(t) \\ &= b_m x_i^{(m)}(t) + b_{m-1} x_i^{(m-1)}(t) + \cdots + b_1 \dot{x}_i(t) + b_0 x_i(t) \quad (n \geq m)\end{aligned} \quad (3\text{-}61)$$

系统的传递函数为

$$G(s) = \frac{b_m s^m + b_{m-1} s^{m-1} + \cdots + b_1 s + b_0}{a_n s^n + a_{n-1} s^{n-1} + \cdots + a_1 s + a_0} \quad (n \geq m) \quad (3\text{-}62)$$

The above equation can be expanded by the partial fractions as

$$X_o(s) = \frac{A_o}{s} + \sum_{j=1}^{n_1} \frac{A_j}{s+p_j} + \sum_{k=1}^{n_2} \frac{B_k s + C_k}{s^2 + 2\xi_k \omega_{nk} s + \omega_{nk}^2} \tag{3-65}$$

where A_0, A_j ($j=1, 2, \cdots, n_1$), B_k, C_k ($k=1, 2, \cdots, n_2$) are coefficients which are obtained from the partial fraction expansion.

By taking the Laplace inverse transformation on equation (3-65), the unit step response of the higher-order system can be obtained as

$$x_o(t) = A_0 + \sum_{j=1}^{n_1} A_j e^{-p_j t} + \sum_{k=1}^{n_2} D_k e^{-\xi_k \omega_{nk} t} \sin(\omega_{dk} t + \beta_k) \tag{3-66}$$

where

$$\omega_{dk} = \omega_{nk}\sqrt{1-\xi_k^2} \quad (k=1,2,\cdots,n_2)$$

$$\beta_k = \arctan \frac{B_k \omega_{dk}}{C_k - \xi_k \omega_{nk} B_k} \quad (k=1,2,\cdots,n_2)$$

$$D_k = \sqrt{B_k^2 + \left(\frac{C_k - \xi_k \omega_{nk} B_k}{\omega_{dk}}\right)^2} \quad (k=1,2,\cdots,n_2)$$

In equation (3-66), the first term is the steady-state component; the second term is the sum of the transient components of the first-order elements, which depends on n_1 real poles $-p_j$ ($j=1, 2, \cdots, n_1$) of the system; The third term is the sum of the transient components of the second-order elements, which depends on the real parts $-\xi_k \omega_{nk}$ ($k=1, 2, \cdots, n_2$) of n_2 conjugate complex poles. The poles of the system are directly related to the inherent characteristics of the system, thus the sum of the second and third terms is the free motion response of the system (also called the free motion modal).

Similarly, when the input signal is a unit pulse signal ($X_i(s)=1$), the unit impulse response of the higher-order system can be obtained as

$$x_o(t) = \sum_{j=1}^{n_1} A_j e^{-p_j t} + \sum_{k=1}^{n_2} D_k e^{-\xi_k \omega_{nk} t} \sin(\omega_{dk} t + \beta_k) \tag{3-67}$$

The above equation shows that the steady-state component of the time response of the system is zero with the unit pulse signal input, and there is only a free motion response synthesized by the first-order and second-order elements.

To sum up, the free motion response of a higher-order linear time-invariant system can be regarded as the superposition of time responses with respect to several first-order and second-order elements. The time responses of the first-order and second-order elements depend on the variables $-p_j$, ξ_k, ω_{nk} and the coefficients A_j, D_k, which are related to the distribution of zeros and poles of the system. Therefore, understanding the distribution of zeros and poles enables us to do qualitative analysis of the performance of the system.

系统的特征方程为

$$a_n s^n + a_{n-1} s^{n-1} + \cdots + a_1 s + a_0 = 0 \tag{3-63}$$

特征方程有 n 个特征根，设其中有实数根 n_1 个，共轭复数根 n_2 对，且 $n = n_1 + 2n_2$。由此，特征方程可以分解为 n_1 个一次因式 $s + p_j$（$j = 1, 2, \cdots, n_1$）及 n_2 个二次因式 $s^2 + 2\xi_k \omega_{nk} s + \omega_{nk}^2$（$0 < \xi_k < 1$；$k = 1, 2, \cdots, n_2$）的乘积。换言之，系统的传递函数有 n_1 个实数极点 $-p_j$ 和 n_2 对共轭复数极点 $-\xi_k \omega_{nk} \pm j\omega_{nk}\sqrt{1-\xi_k^2}$，$-\xi_k \omega_{nk}$ 为极点的实部。设系统传递函数的 m 个零点为 $-z_i$（$i = 1, 2, \cdots, m$），则系统的传递函数可写为

$$G(s) = \frac{K_g \prod_{i=1}^{m}(s + z_i)}{\prod_{j=1}^{n_1}(s + p_j) \prod_{k=1}^{n_2}(s^2 + 2\xi_k \omega_{nk} s + \omega_{nk}^2)} \tag{3-64}$$

式中，K_g 为控制系统的传递系数或根轨迹增益。

系统在单位阶跃输入 $X_i(s) = \dfrac{1}{s}$ 作用下，输出为

$$X_o(s) = G(s) \frac{1}{s} = \frac{K_g \prod_{i=1}^{m}(s + z_i)}{s \prod_{j=1}^{n_1}(s + p_j) \prod_{k=1}^{n_2}(s^2 + 2\xi_k \omega_{nk} s + \omega_{nk}^2)}$$

对上式按部分分式展开得

$$X_o(s) = \frac{A_o}{s} + \sum_{j=1}^{n_1} \frac{A_j}{s + p_j} + \sum_{k=1}^{n_2} \frac{B_k s + C_k}{s^2 + 2\xi_k \omega_{nk} s + \omega_{nk}^2} \tag{3-65}$$

式中，A_0，A_j（$j = 1, 2, \cdots, n_1$），B_k，C_k（$k = 1, 2, \cdots, n_2$）是由部分分式展开式获得的系数。

对式（3-65）进行拉氏反变换，可得到高阶系统的单位阶跃响应为

$$x_o(t) = A_0 + \sum_{j=1}^{n_1} A_j e^{-p_j t} + \sum_{k=1}^{n_2} D_k e^{-\xi_k \omega_{nk} t} \sin(\omega_{dk} t + \beta_k) \tag{3-66}$$

式中，

$$\omega_{dk} = \omega_{nk} \sqrt{1-\xi_k^2} \quad (k = 1, 2, \cdots, n_2)$$

$$\beta_k = \arctan \frac{B_k \omega_{dk}}{C_k - \xi_k \omega_{nk} B_k} \quad (k = 1, 2, \cdots, n_2)$$

$$D_k = \sqrt{B_k^2 + \left(\frac{C_k - \xi_k \omega_{nk} B_k}{\omega_{dk}}\right)^2} \quad (k = 1, 2, \cdots, n_2)$$

式（3-66）中第一项为稳态分量；第二项为一阶环节的瞬态分量和，取决于系统的 n_1

3.4.2 Performance analysis of higher-order systems

1. The free motion modal composition of a higher-order system

Whether it is an external input signal (including the disturbance signal) or the initial state of the system, it can stimulate the free motion determined by system poles. Therefore, the free motion of the system is the inherent motion property of the system, which is independent of the external input signal. In general, the form of the system poles determine the specific form of the system's free motion modal.

1) When the poles are unequal real roots, such as $-p_j$ ($j=1, 2, \cdots, n_1$), the free motion modal form of the system is

$$e^{-p_j t} \quad (j=1, 2, \cdots, n)$$

which is similar to the second term of equation (3-66) or the first term of equation (3-67).

2) When the poles are several conjugate complex roots, such as $\sigma_i \pm j\omega_i$ ($i=1, 2, \cdots, n$), the free motion modal form of the system is

$$e^{\sigma_i t}\cos\omega_i t \quad \text{or} \quad e^{\sigma_i t}\sin\omega_i t \quad (i=1, 2, \cdots, n)$$

which is similar to the third term of equation (3-66) or the second term of equation (3-67).

3) When the poles are repeated real roots, such as m repeated $-p_j$ ($j=1, 2, \cdots, n$), the free motion modal form of the system is

$$e^{-p_j t}, \ t e^{-p_j t}, \ \cdots, \ t^{m-1} e^{-p_j t} \quad (j=1, 2, \cdots, n)$$

4) When the poles are repeated conjugate complex roots, such as m repeated $\sigma_i \pm j\omega_i$ ($i=1, 2, \cdots, n$), the free motion modal form of the system is

$$e^{\sigma_i t}\cos\omega_i t, \ t e^{\sigma_i t}\cos\omega_i t, \ \cdots, \ t^{m-1} e^{\sigma_i t}\cos\omega_i t \quad (i=1, 2, \cdots, n)$$

or

$$e^{\sigma_i t}\sin\omega_i t, \ t e^{\sigma_i t}\sin\omega_i t, \ \cdots, \ t^{m-1} e^{\sigma_i t}\sin\omega_i t \quad (i=1, 2, \cdots, n)$$

5) When the poles are not only distinct real roots, conjugate complex roots, but also repeated real roots, repeated conjugate complex roots, the free motion modal form of the system is a linear combination of the above forms.

2. The effect of system poles on the dynamic performance

(1) The necessary and sufficient condition for the stability of a higher-order system   It can be seen from the above analysis that the free motion modal composition of the system includes various exponential functions $e^{-p_j t}$ or $e^{\sigma_i t}$ related to the real pole $-p_j$ of the system or the real part of the complex pole $\sigma_i \pm j\omega_i$. If all the real poles or the real part of complex poles of the system are less than zero, the modal amplitude of the system will converge and approach to zero as $t \to \infty$. In other words, when the closed-loop poles of the system are all on the left half of the s-plane, the characteristic root (pole) has a negative real part, there is $-p_j < 0$ ($j=1, 2, \cdots, n_1$) and $-\xi_k \omega_{nk} < 0$ ($k=1, 2, \cdots, n_2$). When $t \to \infty$, it can be known from equations (3-66) and equation (3-67) that $\sum_{j=1}^{n_1} A_j e^{-p_j t}$ and $\sum_{k=1}^{n_2} D_k e^{-\xi_k \omega_{nk} t} \sin(\omega_{dk} t + \beta_k)$ in the free motion response of the system will decay

个实数极点 $-p_j$ ($j=1, 2, \cdots, n_1$)；第三项为二阶环节瞬态分量和，取决于 n_2 对共轭复数极点的实部 $-\xi_k\omega_{nk}$ ($k=1, 2, \cdots, n_2$)。系统的极点与系统固有特性直接相关，故第二项与第三项之和为系统的自由运动响应（也称自由运动模态）。

同理，当输入信号为单位脉冲信号 ($X_i(s)=1$) 时，可得高阶系统的单位脉冲响应为

$$x_o(t) = \sum_{j=1}^{n_1} A_j e^{-p_j t} + \sum_{k=1}^{n_2} D_k e^{-\xi_k \omega_{nk} t} \sin(\omega_{dk} t + \beta_k) \tag{3-67}$$

此式说明，系统在单位脉冲信号作用下，其时间响应的稳态分量为零，只存在由一阶环节和二阶环节合成的自由运动响应。

综上可知，一个高阶线性定常系统的自由运动响应可以看成是若干个一阶环节和二阶环节时间响应的叠加。一阶环节及二阶环节的时间响应取决于变量 $-p_j$、ξ_k、ω_{nk} 及系数 A_j、D_k，即与系统的零、极点分布有关。因此了解零、极点的分布情况就可以对系统的性能进行定性分析。

3.4.2 高阶系统的性能分析

1. 高阶系统的自由运动模态组成形式

无论是外部输入信号（包括扰动信号），还是系统初始状态，都可激发出由系统极点决定的自由运动，因此系统的自由运动是系统的固有运动属性，而与外部输入信号无关。一般而言，系统极点的形式决定了系统自由运动模态的具体形式。

1) 当极点为互不相等的实数根时，如 $-p_j$ ($j=1, 2, \cdots, n$)，系统的自由运动模态形式为

$$e^{-p_j t} \quad (j=1,2,\cdots,n)$$

这类似于式（3-66）的第二项或式（3-67）的第一项。

2) 当极点为若干共轭复数根，如 $\sigma_i \pm j\omega_i$ ($i=1, 2, \cdots, n$) 时，系统的自由运动模态形式为

$$e^{\sigma_i t}\cos\omega_i t \quad \text{或} \quad e^{\sigma_i t}\sin\omega_i t \quad (i=1,2,\cdots,n)$$

这类似于式（3-66）的第三项或式（3-67）中的第二项。

3) 当极点为若干实数重根，如 m 重实数根 $-p_j$ ($j=1, 2, \cdots, n$) 时，系统的自由运动模态形式将出现

$$e^{-p_j t}, \; te^{-p_j t}, \cdots, t^{m-1}e^{-p_j t} \quad (j=1,2,\cdots,n)$$

4) 当极点为若干复数重根，如 m 重共轭复数根 $\sigma_i \pm j\omega_i$ ($i=1, 2, \cdots, n$) 时，系统的自由运动模态形式将出现

$$e^{\sigma_i t}\cos\omega_i t, \; te^{\sigma_i t}\cos\omega_i t, \cdots, t^{m-1}e^{\sigma_i t}\cos\omega_i t \quad (i=1,2,\cdots,n)$$

或

$$e^{\sigma_i t}\sin\omega_i t, \; te^{\sigma_i t}\sin\omega_i t, \cdots, t^{m-1}e^{\sigma_i t}\sin\omega_i t \quad (i=1,2,\cdots,n)$$

5) 当极点既有互异的实数根、共轭复数根，又有重实数根、重共轭复数根时，系统的自由运动模态形式将是上述几种形式的线性组合。

as time t increases. They will eventually approach to zero and disappear, and the complete response of the system output will converge to the steady state component. Namely, the unit step steady-state response of the system is

$$\lim_{t \to \infty} x_o(t) = x_o(\infty) = A_0 \qquad (3\text{-}68)$$

The unit pulse steady-state response of the system is

$$\lim_{t \to \infty} x_o(t) = x_o(\infty) = 0 \qquad (3\text{-}69)$$

Obviously, as long as one or several closed-loop poles of the system are in the right half of the s-plane, the corresponding characteristic roots (poles) have positive real parts, and the corresponding modal will tend to infinity as time t increases. In other words, the output response $x_o(t)$ of the system will diverge when $t \to \infty$, which makes the system in an unstable state. When the system has a closed-loop pole on the imaginary axis of the s-plane, the corresponding modal displays persistent oscillation as time t increases. The output response $x_o(t)$ of the system will continue to oscillate with constant amplitude when $t \to \infty$, which makes the system in a critical stable state.

According to the above analysis, the necessary and sufficient condition for the stability of a linear control system is that all the roots of the system characteristic equation have negative real parts, or the characteristic roots are all in the left half of the s-plane. If the system has a root in the right half of the s-plane (i.e. the real part is positive), the system is in an unstable state. If there is a pure virtual root in the characteristic roots and the other roots are in the left half of the s-plane, the system is in a critical stable state.

(2) Effects of system poles on the dynamic performance of the system The dynamic response performance of a stable control system is related to the degree of amplitude attenuation and oscillation frequency of each component of the output response, and mainly depends on the distance between the poles and the imaginary axis and the real axis. The greater the absolute values of $-p_j$ and $-\xi_k \omega_{nk}$ are, the farther the poles are from the imaginary axis, and the faster the amplitude attenuation of the output response component is; the larger the value of $\omega_{dk} = \omega_{nk}\sqrt{1-\xi_k^2}$ is, the farther the poles are from the real axis, and the higher the oscillation frequency of the output response component is. The farther the position of the system pole is from the origin, the smaller the amplitude of the corresponding item is and the less the influence on the system transition process is. The output response curves at different pole positions on the s-plane is shown in Fig. 3-24.

If the real part of the pole that is the closest one to the imaginary axis in the higher-order system is actually less than 1/5 of the real part of the other poles, and there is no zero nearby, it can be considered that the dynamic response of the system is mainly determined by the pole, which is called the dominant pole. By using the concept of dominant poles, a higher-order system whose dominant poles are conjugate complex poles can be treated approximately as a second-order system.

Fig. 3-25 shows the relation between the pole positions and the corresponding unit step response curves. A system is provided, and the poles' distribution of its transfer function on the s-plane is shown in Fig. 3-25a. The distance between the pole s_3 and the imaginary axis is no less than 5 times of the distance between the conjugate complex poles s_1, s_2 and the imaginary axis, i.e. $|\text{Re} s_3| \geqslant 5$

2. 极点对系统动态性能的影响

（1）高阶系统稳定的充分必要条件　由上述分析可知，系统的自由运动模态组成包含与系统实数极点 $-p_j$ 或复数极点 $\sigma_i \pm j\omega_i$ 的实部有关的指数函数 $e^{-p_j t}$ 或 $e^{\sigma_i t}$。若系统的所有实数极点或复数极点的实部均小于零，则随着 $t \to \infty$，系统的模态幅值将收敛并趋近于零。换言之，当系统的闭环极点全部在复平面的左半平面时，其特征根（极点）具有负实部，有 $-p_j < 0$（$j = 1, 2, \cdots, n_1$）和 $-\xi_k \omega_{nk} < 0$（$k = 1, 2, \cdots, n_2$）。在 $t \to \infty$ 时，由式（3-66）和式（3-67）可知，系统的自由运动响应中，$\sum_{j=1}^{n_1} A_j e^{-p_j t}$ 及 $\sum_{k=1}^{n_2} D_k e^{-\xi_k \omega_{nk} t} \sin(\omega_{dk} t + \beta_k)$ 将随着时间 t 的增长而衰减，最终趋于零而消失，系统输出全响应将收敛于稳态分量。即，系统的单位阶跃稳态响应为

$$\lim_{t \to \infty} x_o(t) = x_o(\infty) = A_0 \tag{3-68}$$

系统的单位脉冲稳态响应为

$$\lim_{t \to \infty} x_o(t) = x_o(\infty) = 0 \tag{3-69}$$

很显然系统的所有闭环极点中只要有一个或几个在复平面的右半平面，那么相应的特征根（极点）就具有正实部，对应的模态随着时间 t 的增加将趋于无穷大，即系统的输出响应 $x_o(t)$ 随着 $t \to \infty$，将发散或振荡发散，使系统处于不稳定状态。而当系统有闭环极点在复平面的虚轴上时，对应的模态随着时间 t 的增加而呈现等幅振荡，系统输出响应 $x_o(t)$ 随着时间 $t \to \infty$，也将持续等幅振荡，使系统处于临界稳定状态。

由以上分析可知，线性控制系统稳定的充分必要条件是系统特征方程的所有根具有负的实部，或特征根全部在复平面的左半平面。如果系统有一个根在复平面的右半平面（即实部为正），则系统处于不稳定状态。若特征根中有纯虚根，其余根均在复平面左半平面，则系统处于临界稳定状态。

（2）系统极点对系统动态性能的影响　一个稳定的控制系统，其动态响应性能与输出响应各分量的幅值衰减快慢程度及振荡频率等因素有关，主要取决于极点到虚轴和实轴的距离。$-p_j$ 和 $-\xi_k \omega_{nk}$ 的绝对值越大，极点与虚轴的距离越远，输出响应分量的幅值衰减越快；$\omega_{dk} = \omega_{nk} \sqrt{1 - \xi_k^2}$ 越大，极点与实轴的距离越远，输出响应分量的振荡频率越高。系统极点位置距原点越远，则对应项的幅值就越小，对系统过渡过程的影响就越小。复平面上不同位置极点的输出响应曲线如图3-24所示。

如果高阶系统中离虚轴最近的极点，其实部小于其他极点实部的1/5，并且附近不存在零点，则可以认为系统的动态响应主要由这一极点决定，其称为主导极点。利用主导极点的概念，可将主导极点为共轭复数极点的高阶系统降阶而近似为二阶系统来处理。

图3-25所示为极点位置与单位阶跃响应曲线间的对应关系。设有一系统，其传递函数极点在复平面上的分布如图3-25a所示。极点 s_3 到虚轴距离不小于共轭复数极点 s_1、s_2 到虚轴距离的5倍，即 $|\mathrm{Re} s_3| \geq 5|\mathrm{Re} s_1| = 5\xi \omega_n$（此处 ξ，ω_n 对应于极点 s_1、s_2）；同时，极点 s_1、s_2 的附近不存在系统的零点。由以上条件可算出与极点 s_3 所对应的过渡过程分量的调整时间为

$|\text{Re}s_1|=5\xi\omega_n$ (here ξ and ω_n correspond to the poles s_1 and s_2); at the same time, there is no system zero near the poles s_1 and s_2. According to the above conditions, the settling time of the transition process component corresponding to the pole s_3 can be calculated as

$$t_{s3} \leqslant \frac{1}{5} \times \frac{4}{\xi\omega_n} = \frac{1}{5} t_{s1}$$

where t_{s1} is the settling time of the transition process corresponding to the poles s_1 and s_2.

Fig. 3-25b shows the curves of the components of the unit impulse response function determined by the poles shown in Fig. 3-25a. It can be seen from Fig. 3-25b that the component determined by the conjugate complex poles s_1 and s_2 decays the slowest, and they play a dominant role in the unit step response function of the system, i.e. the dominant pole. While the unit step responses corresponding to the poles s_3, s_4, and s_5 who are far away from the imaginary axis decay faster, and they only have a certain influence in a very short time. Therefore, the influence of these components on the transition process of the higher-order system can be ignored in the approximate analysis of the transition process.

3. The effect of zeros and gains on the dynamic performance of the system

The system modal is also related to the amplitude coefficient A_j of each component as well as the coefficients B_k, C_k in D_k, the magnitude of which depends on the position distribution of system zeros and poles. Therefore, the system zeros will have an impact on the transition process. When the poles and the zeros are very close, the amplitude of the corresponding term is also small. In other words, poles and zeros have little influence on the transition process of the system. In addition, those components with large coefficients will play a dominant role in the dynamic process.

Some proof is as follows that the magnitude of the corresponding term is small when the poles and the zeros are close to each other.

Assuming that the poles of the system are real roots, then we have

$$G(s) = \frac{K_g \prod_{i=1}^{m}(s+z_i)}{\prod_{j=1}^{n}(s+p_j)}$$

$$A_j = \lim_{s \to -p_j} \frac{1}{s} K_g \frac{\prod_{i=1}^{m}(s+z_i)}{\prod_{j=1}^{n}(s+p_j)}(s+p_j) = \frac{K_g}{-p_j} \frac{\prod_{i=1}^{m}(-p_j+z_i)}{\prod_{\substack{i=1 \\ i \neq j}}^{n}(-p_j+p_i)} \tag{3-70}$$

According to equation (3-70), $-p_j+z_i$ is the distance between the pole-p_j and the zero $-z_i$. If the distance is small or even zero, the value A_j is small or even zero. Even if the component corresponding to the pole-p_j decays slowly, it has little effect on the dynamic performance of the system response. It can be seen that the zero of the system determines the "proportion" of each modal in the response, and thus affects the shape of the system response curve, and also affects the rapidity of the system response. Generally speaking, when the zero is far from the pole, the proportion corre-

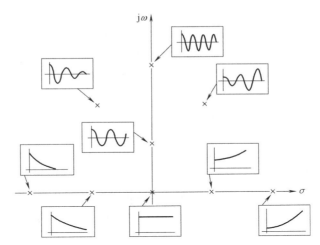

Fig. 3-24　The output response curves at different pole positions on the s-plane
（复平面上不同位置极点的输出响应曲线）

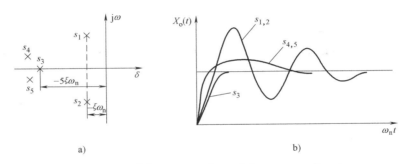

Fig. 3-25　The positions of the system poles and the corresponding step responses
（系统极点的位置与阶跃响应的关系）

$$t_{s3} \leq \frac{1}{5} \times \frac{4}{\xi\omega_n} = \frac{1}{5}t_{s1}$$

式中，t_{s1} 是极点 s_1、s_2 所对应过渡过程的调节时间。

图 3-25b 以曲线的形式表示了图 3-25a 所示极点对应的单位脉冲响应函数的分量。由图 3-25b 可知，由共轭复数极点 s_1、s_2 确定的分量衰减得最慢，它们在该系统的单位阶跃响应函数中起主导作用，即为主导极点。其他远离虚轴的极点，即 s_3、s_4、s_5 所对应的单位阶跃响应衰减较快，它们仅在极短时间内产生一定的影响。因此，对高阶系统过渡过程进行近似分析时，可以忽略这些分量对系统过渡过程的影响。

3. 零点及增益对系统动态性能的影响

系统模态还与各分量幅值系数 A_j 以及 D_k 中的系数 B_k、C_k 有关，这些系数的大小取决于系统零、极点的位置分布。因此，系统零点将对过渡过程产生影响，当极点和零点靠得很近时，对应项的幅值也很小，换言之，零、极点对系统过渡过程影响将很小。此外，系数大的分量将在动态过程起主导作用。

下面简单证明当极点和零点靠得很近时，对应项的幅值很小。

假设系统极点都是实根，则

sponding to the pole modal is large; when the zero is close to the pole, the proportion corresponding to the pole modal is small; when the zero coincides with the pole, the pole-zero cancellation phenomenon occurs, and the modal corresponding to the pole disappears (actually, the proportion of the modal is zero). Therefore, the zero has the effect of blocking the generation of the pole modal.

Besides the influence of zero and pole position distribution on the transition process of the system response, according to equation (3-70), the transfer coefficient K_g of the system is also a parameter that directly determines the magnitude of the modal amplitude. The larger the transfer coefficient K_g is, the larger the modal amplitude A_j is, and the greater the impact on the dynamic performance of the system response is, and vice versa. In addition, the steady-state gain K of the system is related to the transfer coefficient K_g, zero-z_i and pole-p_j of the system, i.e.

$$K = K_g \frac{\prod_{i=1}^{m} z_i}{\prod_{j=1}^{n} p_j} \quad (n \geq m) \tag{3-71}$$

It can be seen that the steady-state transfer performance (control accuracy) of the system is also related to the transfer coefficients, zeros and poles of the system. The larger the value K is, the higher the steady-state accuracy of the system is.

$$G(s) = \frac{K_g \prod\limits_{i=1}^{m}(s+z_i)}{\prod\limits_{j=1}^{n}(s+p_j)}$$

$$A_j = \lim_{s \to -p_j} \frac{1}{s} K_g \frac{\prod\limits_{i=1}^{m}(s+z_i)}{\prod\limits_{j=1}^{n}(s+p_j)}(s+p_j) = \frac{K_g}{-p_j} \frac{\prod\limits_{i=1}^{m}(-p_j+z_i)}{\prod\limits_{\substack{i=1\\i \neq j}}^{n}(-p_j+p_i)} \tag{3-70}$$

由式（3-70）可见，$-p_j+z_i$ 是极点 $-p_j$ 与零点 $-z_i$ 之间的距离。如果这个距离很小甚至为零，则 A_j 数值就很小甚至为零。与极点 $-p_j$ 对应的分量即使衰减很慢，对系统的动态响应性能影响也很小。由此可知，系统的零点决定了各模态在响应中所占的"比重"，进而影响系统响应的曲线形状，因此会影响系统响应的快速性。一般来讲，零点距离极点较远时，该极点模态所占的比重较大，距离极点较近时，该极点模态所占的比重较小。当零点与极点重合时，出现零、极点对消现象，此时该极点的模态消失（实际是该模态比重为零）。因此零点有阻断极点模态产生的作用。

除零、极点位置分布对系统响应的过渡过程产生影响外，由式（3-70）还可看出，系统的传递系数 K_g 也直接决定模态幅值大小，传递系数 K_g 越大，模态幅值 A_j 就越大，对系统响应的动态性能影响越大，反之亦然。此外，系统的稳态增益 K 与传递系数 K_g、系统的零点 $-z_i$、极点 $-p_j$ 之间的关系为

$$K = K_g \frac{\prod\limits_{i=1}^{m} z_i}{\prod\limits_{j=1}^{n} p_j} \quad (n \geq m) \tag{3-71}$$

由此可见，系统的稳态传递性能（控制精度）也与系统的传递系数、零点和极点相关，K 越大，系统的稳态精度越高。

Chapter 4　Frequency Characteristic Analysis of Control Systems

第4章　控制系统的频率特性分析

The time-domain analysis introduced in the previous chapter is a direct method to analyze a control system, which is relatively intuitive and accurate. But it is usually difficult to study higher-order systems. Because the solving process of differential equations becomes more complex as their orders increase. In this chapter, a different method for analyzing control systems is described, which is the frequency-domain analysis method. The frequency characteristics of a control system can be taken as mathematical model elements in the frequency domain, and it is not necessary to solve the differential equations of the system. It uses plots as analysis tools to reveal the system performance and indicate the direction to improve system performance. Therefore, it is appropriate to use the frequency-domain analysis method for the performance analysis of higher-order systems.

4.1 Basic Concepts of Frequency Characteristics

4.1.1 Frequency response

The frequency response is the steady-state response of a linear time-invariant system to a sinusoidal signal (or harmonic signal). The response of the linear time-invariant system to a sinusoidal signal is the same as that of other typical signals, including the transient response and steady-state response. The transient response is not a sinusoidal signal while the steady-state response is a sinusoidal signal with the same frequency as the input signal. The amplitude and phase angle of the steady-state response are generally different from the input signal.

Example 4-1 Fig. 4-1 shows a passive RC circuit, where $u_i(t)$ is the input voltage, $u_o(t)$ is the output voltage, $i(t)$ is the current, R is the resistance, and C is the capacitance. When the input voltage is $u_i(t) = U_i \sin\omega t$, figure out the steady-state response of the output voltage.

Solution: The transfer function of the RC circuit is

$$G(s) = \frac{U_o(s)}{U_i(s)} = \frac{1}{Ts+1}$$

where T is the time constant of the circuit and $T = RC$.

Since the input voltage is a sinusoidal signal, i.e. $u_i(t) = U_i \sin\omega t$, and its Laplace transform is

$$U_i(s) = \frac{U_i \omega}{s^2 + \omega^2}$$

Then the Laplace transform of the output $u_o(t)$ for the RC circuit is

$$U_o(s) = \frac{1}{Ts+1} \frac{U_i \omega}{s^2 + \omega^2}$$

According to the inverse Laplace transform, the full response of the output for the circuit is

$$u_o(t) = \frac{U_i T\omega}{1+T^2\omega^2} e^{-\frac{t}{T}} + \frac{U_i}{\sqrt{1+T^2\omega^2}} \sin(\omega t - \arctan\omega T)$$

where the first term on the right is the transient component of the input voltage (the free motion mo-

第4章 控制系统的频率特性分析

上一章的时域分析法是分析控制系统的直接方法，比较直观、精确，但分析高阶系统则较为困难。因为微分方程的求解过程将随着微分方程阶数的增大而变得复杂。本章将介绍另一种分析控制系统的方法，即频域分析法。在频域内，可将控制系统的频率特性作为数学模型，而无需求解系统微分方程。它以图表作为分析工具来揭示系统性能，并指明改进系统性能的方向。因此，对于高阶系统的性能分析，使用频域分析法是比较方便的。

4.1 频率特性的基本概念

4.1.1 频率响应

频率响应是指线性定常系统对正弦信号（或谐波信号）的稳态响应。线性定常系统对正弦信号的响应与对其他典型信号的响应一样，包含瞬态响应和稳态响应，其瞬态响应不是正弦信号，而稳态响应是与输入信号频率相同的正弦信号。稳态响应的幅值和相位一般与输入信号不同。

例 4-1 图 4-1 为无源 RC 电路，$u_i(t)$ 为输入电压，$u_o(t)$ 为输出电压，$i(t)$ 为电流，R 为电阻，C 为电容。当输入电压 $u_i(t) = U_i \sin\omega t$ 时，求输出电压 $u_o(t)$ 的稳态响应。

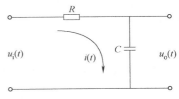

Fig. 4-1　Passive RC circuit
（无源 RC 电路）

解：该 RC 电路的传递函数为

$$G(s) = \frac{U_o(s)}{U_i(s)} = \frac{1}{Ts+1}$$

式中，T 为电路的时间常数，$T=RC$。

输入电压为正弦信号 $u_i(t) = U_i \sin\omega t$，其拉氏变换为

$$U_i(s) = \frac{U_i \omega}{s^2 + \omega^2}$$

则 RC 电路输出电压 $u_o(t)$ 的拉氏变换为

$$U_o(s) = \frac{1}{Ts+1} \frac{U_i \omega}{s^2 + \omega^2}$$

根据拉氏反变换，得电路的输出全响应为

$$u_o(t) = \frac{U_i T\omega}{1+T^2\omega^2} e^{-\frac{t}{T}} + \frac{U_i}{\sqrt{1+T^2\omega^2}} \sin(\omega t - \arctan\omega T)$$

式中，右边第一项为输入电压的瞬态分量（系统的自由运动模态），取决于初始条件和系统的极点 $-\frac{1}{T}$；第二项为稳态分量，取决于输入量的受控项。随着 $t \to \infty$，瞬态分量将衰减为零，所以系统的稳态响应为

dal of the system), which depends on the initial conditions and the pole $-1/T$ of the system; the second term is the steady-state component, and is the controlled term that depends on the input. When $t \to \infty$, the transient component will decay to zero. Therefore, the steady-state response of the system is

$$u_o(t) = \frac{U_i}{\sqrt{1+T^2\omega^2}} \sin(\omega t - \arctan \omega T) = U_i A(\omega) \sin[\omega t + \phi(\omega)]$$

where $A(\omega) = \dfrac{1}{\sqrt{1+T^2\omega^2}}$ is the amplitude ratio of the steady-state output to the input of the sinusoidal signal (or harmonic signal) for the RC circuit; $\varphi(\omega) = -\arctan \omega T$ is the phase difference between the steady-state output and the sinusoidal (or harmonic) input signal for the RC circuit. They separately reflect the amplitude and phase angle changes of the steady-state response of the RC circuit under the sinusoidal signal. Moreover, they are functions of frequency ω in the input sinusoidal signal.

According to the results of the example 4-1, the frequency response is a special case of the time response. In order to study the variation of the system with frequency, the concept of frequency characteristics is introduced.

4.1.2 Frequency characteristics and solving methods

It can be seen from the above that under the harmonic input, the amplitude ratio of the steady-state output to the input of a linear time-invariant system is a function with respect to frequency ω, which is called the amplitude frequency characteristic of the system and is denoted as $A(\omega)$. The phase difference between the steady-state output and the signal input of the system is a function with respect to frequency ω, which is called the phase frequency characteristic of the system and is denoted as $\varphi(\omega)$. When $\varphi(\omega) > 0$, the phase of the output signal is ahead of the phase of the input signal; when $\varphi(\omega) < 0$, the phase of the output signal lags behind the phase of the input signal. Taking the amplitude frequency characteristic $A(\omega)$ as the amplitude (or module) and the phase frequency characteristic as the phase angle, the frequency characteristic of the system described by the vector form is

$$G(j\omega) = A(\omega) e^{j\varphi(\omega)} \tag{4-1}$$

The frequency characteristics can generally be obtained by the following three methods.

1) Firstly, according to the differential equation or transfer function of the system, figure out the steady-state response of the system with the sinusoidal signal as the input signal; then figure out the frequency characteristic according to the definition of frequency characteristic (as is the method used in the example 4-1).

2) Figure out the frequency characteristics according to the transfer function of the system directly. When the system is at a steady state, substituting $s = j\omega$ into the system transfer function $G(s)$, the frequency characteristics of the system can be directly obtained.

The transfer function of a linear time-invariant system is

$$u_o(t) = \frac{U_i}{\sqrt{1+T^2\omega^2}}\sin(\omega t - \arctan\omega T) = U_i A(\omega)\sin[\omega t + \varphi(\omega)]$$

式中，$A(\omega) = \dfrac{1}{\sqrt{1+T^2\omega^2}}$ 为 RC 电路的稳态输出与正弦信号（或谐波信号）输入的幅值比；$\varphi(\omega) = -\arctan\omega T$ 为 RC 电路的稳态输出与正弦（或谐波）输入信号的相位差。它们分别反映 RC 电路在正弦信号作用下，稳态响应的幅值和相位的变化，并且它们都是关于输入正弦信号的频率 ω 的函数。

由例 4-1 结果可知，频率响应是时间响应的一种特例。为了研究系统随频率变化的情况，引入频率特性的概念。

4.1.2 频率特性及其求取方法

由上可知，线性定常系统在谐波输入作用下，其稳态输出与输入信号的幅值比是关于频率 ω 的函数，称为系统的幅频特性，记为 $A(\omega)$。该系统的稳态输出与输入信号的相位差也是关于频率 ω 的函数，称为系统的相频特性，记为 $\varphi(\omega)$。当 $\varphi(\omega)>0$ 时，输出信号的相位超前于输入信号的相位；当 $\varphi(\omega)<0$ 时，输出信号的相位滞后于输入信号的相位。以幅频特性 $A(\omega)$ 为幅值（或模），相频特性 $\varphi(\omega)$ 为相角，采用向量形式描述的系统的频率特性为

$$G(j\omega) = A(\omega)e^{j\varphi(\omega)} \tag{4-1}$$

频率特性一般可通过如下三种方法得到。

1) 首先根据已知系统的微分方程或传递函数，以正弦信号为输入信号，求系统的稳态响应；然后根据频率特性的定义，求得频率特性（例 4-1 中所用的方法）。

2) 直接根据系统的传递函数求得频率特性。在稳态系统时将 $s=j\omega$ 代入系统传递函数 $G(s)$ 中，就可以直接得到系统的频率特性。

线性定常系统的传递函数为

$$G(s) = \frac{L[x_o(t)]}{L[x_i(t)]} = \frac{X_o(s)}{X_i(s)} = \frac{b_m s^m + b_{m-1}s^{m-1} + \cdots + b_1 s + b_0}{a_n s^n + a_{n-1}s^{n-1} + \cdots + a_1 s + a_0} \quad (n \geq m) \tag{4-2}$$

在系统稳态时，将 $s=j\omega$ 代入式（4-2），则有

$$G(j\omega) = \frac{X_o(j\omega)}{X_i(j\omega)} = \frac{b_m(j\omega)^m + b_{m-1}(j\omega)^{m-1} + \cdots + b_1(j\omega) + b_0}{a_n(j\omega)^n + a_{n-1}(j\omega)^{n-1} + \cdots + a_1(j\omega) + a_0} \quad (n \geq m) \tag{4-3}$$

系统的频率特性 $G(j\omega)$ 是一个复变函数，可在复平面上用复数表示，如图 4-2 所示。将其分解为实部和虚部，即

$$G(j\omega) = U(\omega) + jV(\omega) \tag{4-4}$$

式中，$U(\omega)$ 是 $G(j\omega)$ 的实部，称为实频特性；$V(\omega)$ 是 $G(j\omega)$ 的虚部，称为虚频特性。$G(j\omega)$ 的模、幅频、相频、实频、虚频之间的关系为

$$\begin{cases} A(\omega) = |G(j\omega)| = \sqrt{[U(\omega)]^2 + [V(\omega)]^2} \\ \varphi(\omega) = \angle G(j\omega) = \arctan\dfrac{V(\omega)}{U(\omega)} \end{cases} \tag{4-5}$$

$$G(s) = \frac{L[x_o(t)]}{L[x_i(t)]} = \frac{X_o(s)}{X_i(s)} = \frac{b_m s^m + b_{m-1} s^{m-1} + \cdots + b_1 s + b_0}{a_n s^n + a_{n-1} s^{n-1} + \cdots + a_1 s + a_0} \quad (n \geq m) \tag{4-2}$$

When the system is at a steady state, substituting $s = j\omega$ into the above equation, we have

$$G(j\omega) = \frac{X_o(j\omega)}{X_i(j\omega)} = \frac{b_m(j\omega)^m + b_{m-1}(j\omega)^{m-1} + \cdots + b_1(j\omega) + b_0}{a_n(j\omega)^n + a_{n-1}(j\omega)^{n-1} + \cdots + a_1(j\omega) + a_0} \quad (n \geq m) \tag{4-3}$$

The frequency characteristic $G(j\omega)$ of the system is a complex function, which can be represented by a complex number on the complex plane, as is shown in Fig. 4-2. It can be decomposed into the real and imaginary parts, i. e.

$$G(j\omega) = U(\omega) + jV(\omega) \tag{4-4}$$

where $U(\omega)$ is the real part of $G(j\omega)$, which is called the real frequency characteristic; $V(\omega)$ is the imaginary part of $G(j\omega)$, which is called the virtual frequency characteristic.

The relation among the module, amplitude frequency, phase frequency, real frequency and virtual frequency with regard to $G(j\omega)$ can be described as

$$\begin{cases} A(\omega) = |G(j\omega)| = \sqrt{[U(\omega)]^2 + [V(\omega)]^2} \\ \varphi(\omega) = \angle G(j\omega) = \arctan \dfrac{V(\omega)}{U(\omega)} \end{cases} \tag{4-5}$$

$$\begin{cases} U(\omega) = \mathrm{Re}\, G(j\omega) = A(\omega) \cos\varphi(\omega) \\ V(\omega) = \mathrm{Im}\, G(j\omega) = A(\omega) \sin\varphi(\omega) \end{cases} \tag{4-6}$$

$$G(j\omega) = A(\omega) e^{j\varphi(\omega)} = A(\omega)[\cos\varphi(\omega) + j\sin\varphi(\omega)] \tag{4-7}$$

According to equation (4-7), we have

$$e^{j\varphi(\omega)} = \cos\varphi(\omega) + j\sin\varphi(\omega) \tag{4-8}$$

3) Figure out the frequency characteristics based on the experimental method. The experimental method is common and useful to obtain the frequency characteristics of an actual system. Because the above two methods cannot be used to calculate the frequency characteristics without knowing the mathematical models of a system like differential equations or transfer functions, in this case, the frequency characteristics can be only obtained by the experimental method, and then the transfer function can also be obtained.

According to the definition of the frequency characteristics, firstly we can change frequency ω of the input harmonic signal, and measure the amplitude and phase angle of the corresponding steady-state response. Then, a function curve of the amplitude ratio of output to input with frequency ω can be drawn, which is the amplitude frequency characteristic curve; a function curve of the phase ratio of output to input with frequency ω can be drawn, which is the phase frequency characteristic curve.

Like the transfer function and differential equation, the frequency characteristic characterizes the motion law of the system. Thus, the frequency characteristic is also a mathematical model. In fact, it is the mathematical model of the system in the frequency domain.

$$\begin{cases} U(\omega) = \mathrm{Re}\,G(\mathrm{j}\omega) = A(\omega)\cos\varphi(\omega) \\ V(\omega) = \mathrm{Im}\,G(\mathrm{j}\omega) = A(\omega)\sin\varphi(\omega) \end{cases} \quad (4\text{-}6)$$

$$G(\mathrm{j}\omega) = A(\omega)\mathrm{e}^{\mathrm{j}\varphi(\omega)} = A(\omega)[\cos\varphi(\omega) + \mathrm{j}\sin\varphi(\omega)] \quad (4\text{-}7)$$

根据式（4-7），可得

$$\mathrm{e}^{\mathrm{j}\phi(\omega)} = \cos\varphi(\omega) + \mathrm{j}\sin\varphi(\omega) \quad (4\text{-}8)$$

3) 基于试验方法求得频率特性。利用试验获得实际系统频率特性是一种常用而又重要的方法。因为不知道系统的微分方程或传递函数等数学模型就无法用上面两种方法求取频率特性，所以在

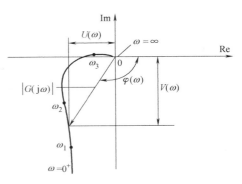

Fig. 4-2　Polar plot of frequency characteristics
（频率特性的极坐标图）

这样的情况下，只有通过试验求取频率特性，然后也能求出传递函数。

根据频率特性的定义，首先通过改变输入谐波信号的频率 ω，测出稳态响应的幅值和相位。然后，作出输出与输入幅值比关于频率 ω 的函数曲线，即为幅频特性曲线；作出输出与输入相位比关于频率 ω 的函数曲线，即为相频特性曲线。

频率特性与传递函数、微分方程一样，表征系统的运动规律。因此频率特性也是一种数学模型，事实上，它是系统在频域内的数学模型。

4.1.3　频率特性的图形表示法

1. 极坐标图（奈奎斯特图）

前面已说明系统的频率特性 $G(\mathrm{j}\omega)$ 是一个向量，其端点的轨迹即为频率特性的极坐标图，如图 4-2 所示。频率特性的极坐标图也称为奈奎斯特图或幅相频特性图，当给定频率 ω 时，可以算出相应幅频特性和相频特性的值，进而在极坐标复平面上画出 $\omega = -\infty \to +\infty$ 的 $G(\mathrm{j}\omega)$ 向量曲线图。完整的幅相频特性图是以实轴（横坐标轴）为对称轴的封闭曲线，即 $G(\mathrm{j}\omega)$ 向量曲线图在 $\omega = 0^- \to -\infty$ 时的曲线与 $G(\mathrm{j}\omega)$ 在 $\omega = 0^+ \to +\infty$ 时的曲线完全对称于实轴，因此只需要画出 $\omega = 0 \to \infty$ 时的极坐标曲线，就可以根据对称性获得完整的极坐标曲线。

2. 对数频率特性图（伯德图）

对数频率特性图又称为伯德图，伯德图是由对数幅频特性图和对数相频特性图组成。对数幅频特性和对数相频特性定义为

$$\begin{cases} L(\omega) = 20\lg A(\omega) \\ \varphi(\omega) = \arctan\dfrac{V(\omega)}{U(\omega)} \end{cases} \quad (4\text{-}9)$$

对数幅频特性图的横坐标是频率 ω，采用对数分度。频率变化 10 倍称为一个十倍频程，对应横坐标的间隔距离为一个单位。频率 ω 的单位为弧度/秒（rad/s），也可表示为 s^{-1}。对数幅频特性图的纵坐标以对数幅值 $L(\omega) = 20\lg A(\omega)$ 表示，单位是分贝（dB）；对数相频特性图的横坐标与对数幅频特性图的横坐标相同，对数相频特性图的纵坐标采用线性分度，用 $\varphi(\omega)$ 表示，单位是度（°），如图 4-3 所示。

4.1.3 Graphical representation of frequency characteristics

1. Polar plot (Nyquist plot)

It has been explained above that the frequency characteristic $G(j\omega)$ of the system is a vector, and the trajectory of the endpoint is the polar plot of the frequency characteristic, as is shown in Fig. 4-2. The polar plot of the frequency characteristic is also called Nyquist plot or the amplitude-phase frequency characteristic plot. When the frequency ω is given, the corresponding amplitude frequency characteristic and phase frequency characteristic value can be calculated. In this way, the vector curve $G(j\omega)$ can be drawn on the complex plane of the polar coordinate when $\omega = -\infty \rightarrow +\infty$. A complete amplitude frequency characteristic plot is a closed curve with the real axis (the horizontal axis) as the symmetry axis. That is, the curve of the vector graph $G(j\omega)$ at $\omega = 0^- \rightarrow -\infty$ and the curve of the vector graph $G(j\omega)$ at $\omega = 0^+ \rightarrow +\infty$ are completely symmetrical to the real axis. So it is only necessary to draw the polar curve of $\omega = 0 \rightarrow \infty$, and then to obtain the complete polar curve according to the symmetry.

2. Logarithmic frequency characteristics plot (Bode plot)

The logarithmic frequency characteristics plot is also called the Bode plot.

The Bode plot consists of a logarithmic amplitude frequency characteristic plot and a logarithmic phase frequency characteristic plot. The logarithmic amplitude frequency characteristic and the logarithmic phase frequency characteristic are defined as

$$\begin{cases} L(\omega) = 20\lg A(\omega) \\ \varphi(\omega) = \arctan \dfrac{V(\omega)}{U(\omega)} \end{cases} \qquad (4\text{-}9)$$

The horizontal axis of the logarithmic amplitude frequency characteristic plot is the frequency ω, where the logarithmic scale is used. When the frequency changes ten times, it is called a decade, and the distance corresponding to the horizontal axis is one unit. The unit of frequency ω is radians/second (rad/s). The vertical axis of the logarithmic amplitude frequency characteristic plot is expressed with the logarithmic amplitude $L(\omega) = 20\lg A(\omega)$, and the unit is decibel (dB); the horizontal axis of the logarithmic phase frequency characteristic plot is the same as the horizontal axis of the logarithmic amplitude frequency characteristic plot; the vertical axis of the logarithmic phase frequency characteristic plot is linearly divided, and it is represented by $\varphi(\omega)$, the unit is degree (°), as is shown in Fig. 4-3.

3. The correspondence between polar plot and Bode plot

As is shown in Fig. 4-4, the correspondence between polar plot and Bode plot can be described as follows.

1) The unit circle $[A(\omega) = 1]$ on the polar plot corresponds to the 0dB line of the Bode plot, which is the horizontal axis of the logarithmic amplitude frequency characteristic plot as $L(\omega) = 20\lg A(\omega) = 0$; the inside of the unit circle $[A(\omega) < 1]$ corresponds to the part of the Bode plot where $L(\omega) = 20\lg A(\omega) < 0$; the outside of the unit circle $[A(\omega) > 1]$ corresponds to the part of the Bode plot where $L(\omega) = 20\lg A(\omega) > 0$. The frequency ω_c at which the polar coordinate curve G

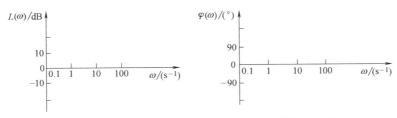

Fig. 4-3　Coordinate system of Bode plot（伯德图坐标系）

3. 极坐标图与伯德图之间的对应关系

如图 4-4 所示，不难看出极坐标图与伯德图之间有如下关系。

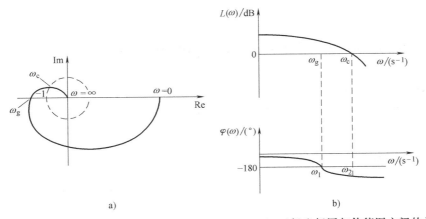

Fig. 4-4　The relationship between Nyquist plot and Bode plot（极坐标图与伯德图之间的关系）

1）极坐标图上的单位圆 $[A(\omega)=1]$ 对应于伯德图的 0dB 线，即对数幅频特性图的横轴 $[L(\omega)=20\lg A(\omega)=0]$；单位圆内 $[A(\omega)<1]$ 对应于伯德图 $L(\omega)=20\lg A(\omega)<0$ 的部分；单位圆外 $[A(\omega)>1]$ 对应于伯德图 $L(\omega)=20\lg A(\omega)>0$ 的部分。定义极坐标曲线 $G(j\omega)$ 与其单位圆交点处的频率 ω_c 为幅值穿越频率或截止频率，此处，$L(\omega_c)=0$ 或 $A(\omega_c)=1$。

2）极坐标图上的负实轴对应于伯德图上的 $-180°$ 相位线。定义极坐标曲线 $G(j\omega)$ 与负实轴相交处的频率 ω_g，或伯德图中相频特性曲线与 $-180°$ 水平线交点处的频率为相频穿越频率，即 $\varphi(\omega_g)=-180°$。

3）伯德图只对应于 $\omega=0^+\to+\infty$ 的极坐标图，而极坐标图可以绘出 $\omega=-\infty\to+\infty$ 的封闭完整的极坐标曲线，$\omega=0\to+\infty$ 的极坐标曲线与 $\omega=0\to-\infty$ 的极坐标曲线完全对称于横坐标轴。

4.2　典型环节的频率特性

控制系统通常由若干典型环节组成，故系统频率特性也是由典型环节的频率特性组成的。下面就介绍一些典型环节的频率特性。

4.2.1　比例环节

比例环节的频率特性与传递函数一样，为

($j\omega$) intersects its unit circle is defined as the amplitude crossing frequency or cutoff frequency, where $L(\omega_c) = 0$ or $A(\omega_c) = 1$.

2) The negative real axis on the polar plot corresponds to the $-180°$ phase line on the Bode plot. The frequency ω_g at which the polar coordinate curve $G(j\omega)$ intersects the negative real axis, or the frequency at which the phase frequency characteristic curve intersects the $-180°$ horizontal line in the Bode plot, is defined as the phase crossing frequency, i.e. $\varphi(\omega_g) = -180°$.

3) The Bode plot only corresponds to the polar plot of $\omega = 0^+ \to +\infty$ while the polar plot can draw the closed complete polar coordinate curve of $\omega = -\infty \to +\infty$. The polar coordinate curve of $\omega = 0 \to \infty$ and the polar coordinate curve of $\omega = 0 \to -\infty$ are completely symmetrical to the horizontal axis.

4.2 Frequency Characteristics of Typical Elements

A control system usually consists of several typical elements, so the system frequency characteristic is also composed of typical element frequency characteristics. The frequency characteristics of some typical elements are described as below.

4.2.1 Proportion element

The frequency characteristic of the proportion element is the same as the transfer function, i.e.

$$G(j\omega) = K \tag{4-10}$$

where K is a proportional coefficient. According to equation (4-10), the amplitude frequency characteristics, logarithmic amplitude frequency characteristics and phase frequency characteristics of the proportion element are respectively

$$\begin{cases} A(\omega) = |G(j\omega)| = K \\ L(\omega) = 20\lg A(\omega) = 20\lg K \\ \varphi(\omega) = \angle G(j\omega) = 0° \end{cases} \tag{4-11}$$

The polar plot of the proportion element is a fixed point on the real axis, as is shown in Fig. 4-5.

The logarithmic amplitude frequency characteristic of the proportion element is a horizontal straight line and its amplitude is equal to $20\lg K$ (dB); the logarithmic phase frequency characteristic plot is a straight line that coincides with the 0dB line, as is shown in Fig. 4-6.

4.2.2 Integral element and differential element

1. Integral element

The frequency characteristic of the integral element is

$$G(j\omega) = \frac{1}{j\omega} = -j\frac{1}{\omega} \tag{4-12}$$

Accordingly, the amplitude frequency characteristic, logarithmic amplitude frequency characteristic and phase frequency characteristic of the integral element are respectively

$$G(j\omega) = K \tag{4-10}$$

式中，K 为比例系数。依据式（4-10），比例环节的幅频特性、对数幅频特性和相频特性分别为

$$\begin{cases} A(\omega) = |G(j\omega)| = K \\ L(\omega) = 20\lg A(\omega) = 20\lg K \\ \varphi(\omega) = \angle G(j\omega) = 0° \end{cases} \tag{4-11}$$

比例环节的极坐标图为实轴上的一个定点，如图 4-5 所示。

比例环节的对数幅频特性图为一条水平直线，其幅值等于 $20\lg K$（dB）；对数相频特性图是与 0dB 线重合的一条直线，如图 4-6 所示。

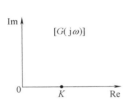

Fig. 4-5 Polar plot of proportion element
（比例环节的极坐标图）

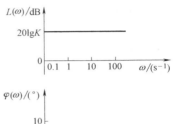

Fig. 4-6 Bode plot of proportion element
（比例环节的伯德图）

4.2.2 积分环节和微分环节

1. 积分环节

积分环节的频率特性为

$$G(j\omega) = \frac{1}{j\omega} = -j\frac{1}{\omega} \tag{4-12}$$

相应地，积分环节的幅频特性、对数幅频特性和相频特性分别为

$$\begin{cases} A(\omega) = |G(j\omega)| = \frac{1}{\omega} \\ L(\omega) = 20\lg A(\omega) = -20\lg\omega \\ \varphi(\omega) = \angle G(j\omega) = -90° \end{cases} \tag{4-13}$$

积分环节的极坐标图为负虚轴，且由负无穷远处指向原点，如图 4-7 所示。

当 $\omega = 1$ 时，$L(\omega) = -20\lg\omega = -20\lg 1 = 0$。由此可见，积分环节的对数幅频特性图是一条斜率为 -20dB/dec 的直线，并与 0dB 线相交于 $\omega = 1$；对数相频特性图为 -90° 水平直线，如图 4-8 所示。

2. 微分环节

微分环节的频率特性为

$$G(j\omega) = j\omega \tag{4-14}$$

则对应的幅频特性、对数幅频特性和相频特性分别为

$$\begin{cases} A(\omega) = |G(j\omega)| = \omega \\ L(\omega) = 20\lg A(\omega) = 20\lg\omega \\ \varphi(\omega) = \angle G(j\omega) = 90° \end{cases} \tag{4-15}$$

$$\begin{cases} A(\omega) = |G(j\omega)| = \dfrac{1}{\omega} \\ L(\omega) = 20\lg A(\omega) = -20\lg\omega \\ \varphi(\omega) = \angle G(j\omega) = -90° \end{cases} \quad (4\text{-}13)$$

The polar plot of the integral element is the negative imaginary axis and points from the negative infinity to the origin, as is shown in Fig. 4-7.

When $\omega = 1$, we have $L(\omega) = -20\lg\omega = -20\lg 1 = 0$. It can be seen that the logarithmic amplitude frequency characteristic of the integral element is a straight line with a slope of -20 dB/dec and intersects the 0dB-line at $\omega = 1$; the logarithmic phase frequency characteristic is a horizontal line with $-90°$, as is shown in Fig. 4-8.

2. Differential element

The frequency characteristic of the differential element is

$$G(j\omega) = j\omega \quad (4\text{-}14)$$

Accordingly, the amplitude frequency characteristic, logarithmic amplitude frequency characteristic and phase frequency characteristic of the integral element are respectively

$$\begin{cases} A(\omega) = |G(j\omega)| = \omega \\ L(\omega) = 20\lg A(\omega) = 20\lg\omega \\ \varphi(\omega) = \angle G(j\omega) = 90° \end{cases} \quad (4\text{-}15)$$

The polar plot of the integral element is the positive imaginary axis and points from the origin to the positive infinity, as is shown in Fig. 4-9.

When $\omega = 1$, we have $L(\omega) = 20\lg\omega = 20\lg 1 = 0$. It can be seen that the logarithmic amplitude frequency characteristic of the differential element is a straight line with a slope of 20dB/dec and intersects the 0dB-line at $\omega = 1$; the logarithmic phase frequency characteristic is a horizontal line with $90°$, as is shown in Fig. 4-10.

4.2.3 Inertial element and first-order differential element

1. Inertial element

The transfer function of the inertia element and the corresponding frequency characteristic are

$$\begin{cases} G(s) = \dfrac{1}{Ts+1} \\ G(j\omega) = \dfrac{1}{j\omega T+1} = \dfrac{1}{1+\omega^2 T^2} - j\dfrac{\omega T}{1+\omega^2 T^2} \end{cases} \quad (4\text{-}16)$$

The corresponding real frequency characteristic and virtual frequency characteristic are respectively

$$\begin{cases} U(\omega) = \dfrac{1}{1+\omega^2 T^2} \\ V(\omega) = -\dfrac{\omega T}{1+\omega^2 T^2} \end{cases} \quad (4\text{-}17)$$

The amplitude frequency characteristic, logarithmic amplitude frequency characteristic and phase frequency characteristic are respectively

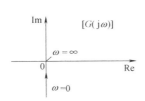

Fig. 4-7　Polar plot of integral element
（积分环节的极坐标图）

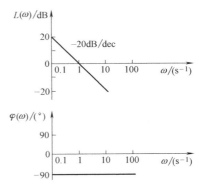

Fig. 4-8　Bode plot of integral element
（积分环节的伯德图）

微分环节的极坐标图为正虚轴，且由原点指向正无穷远处，如图 4-9 所示。

当 $\omega=1$ 时，$L(\omega)=20\lg\omega=20\lg1=0$。由此可见，微分环节的对数幅频特性图是一条斜率为 20dB/dec 的直线，并与 0dB 线相交于 $\omega=1$；对数相频特性图为 90°水平直线，如图 4-10 所示。

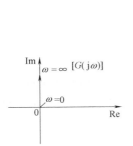

Fig. 4-9　Polar plot of differential element
（微分环节的极坐标图）

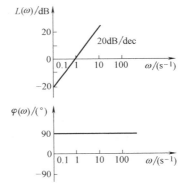

Fig. 4-10　Bode plot of differential element
（微分环节的伯德图）

4.2.3　惯性环节和一阶微分环节

1. 惯性环节

惯性环节的传递函数和相应的频率特性为

$$\begin{cases} G(s)=\dfrac{1}{Ts+1} \\ G(\mathrm{j}\omega)=\dfrac{1}{\mathrm{j}\omega T+1}=\dfrac{1}{1+\omega^2 T^2}-\mathrm{j}\dfrac{\omega T}{1+\omega^2 T^2} \end{cases} \qquad (4\text{-}16)$$

则对应的实频特性和虚频特性分别为

$$\begin{cases} U(\omega)=\dfrac{1}{1+\omega^2 T^2} \\ V(\omega)=-\dfrac{\omega T}{1+\omega^2 T^2} \end{cases} \qquad (4\text{-}17)$$

$$\begin{cases} A(\omega) = |G(j\omega)| = \dfrac{1}{\sqrt{1+\omega^2 T^2}} \\ L(\omega) = 20\lg A(\omega) = 20\lg \dfrac{1}{\sqrt{T^2\omega^2+1}} \\ \varphi(\omega) = \angle G(j\omega) = -\arctan\omega T \end{cases} \quad (4\text{-}18)$$

When $\omega = 0$, $A(\omega) = 1$, the phase angle is $\varphi(\omega) = 0°$; when $\omega = 1/T$, $A(\omega) = 1/\sqrt{2}$, the phase angle is $\varphi(\omega) = -45°$; when $\omega = \infty$, $A(\omega) = 0$, the phase angle is $\varphi(\omega) = -90°$.

In addition, the real frequency characteristic and the virtual frequency characteristic satisfy a relation as

$$\left[U(\omega) - \dfrac{1}{2}\right]^2 + V^2(\omega) = \left(\dfrac{1}{2}\right)^2$$

The above equation shows that when $\omega = 0 \to \infty$, the polar plot of the inertia element is a semicircle whose circle center is at $(1/2, j0)$ and radius is $1/2$, as is shown in Fig. 4-11.

When $\omega = 0 \to \infty$, the corresponding $L(\omega)$ and $\varphi(\omega)$ can be calculated, and then the logarithmic amplitude frequency characteristic plot and the logarithmic phase frequency characteristic plot can be drawn. In engineering, the approximate drawing method is often used. That is, the logarithmic amplitude frequency characteristic plot is represented by asymptotic curves. Usually, the frequency $\omega_T = 1/T$ at which the two adjacent asymptotic curves intersect is called the corner frequency.

When $\omega \ll \omega_T$, $L(\omega) = 20\lg A(\omega) = -20\lg\sqrt{T^2\omega^2+1} \approx -20\lg 1 = 0\text{dB}$. It means that the logarithmic amplitude frequency characteristic is approximated to 0dB line in the low frequency band, which is called the low frequency asymptotic curve.

When $\omega \gg \omega_T$, $L(\omega) = 20\lg A(\omega) = -20\lg\sqrt{T^2\omega^2+1} \approx -20\lg\omega T$. It means that the logarithmic amplitude frequency characteristic is approximate to a straight line with a slope of -20dB/dec passing through the point of $(\omega_T, 0)$ in the high frequency band, which is called the high frequency asymptotic curve.

Therefore, the piecewise function of the asymptotic lines regarding the inertia element can be expressed as

$$L(\omega) = -20\lg\sqrt{T^2\omega^2+1} \approx \begin{cases} 0 & \left(0 < \omega < \dfrac{1}{T}\right) \\ 20\lg\dfrac{1}{\omega T} & \left(\omega > \dfrac{1}{T}\right) \end{cases} \quad (4\text{-}19)$$

Since the phase angle is represented by the inverse tangent function in the logarithmic phase frequency characteristic plot of the inertia element, the logarithmic phase frequency characteristic plot is an arctangent curve passing through point of $(1/T, -45°)$, as is shown in Fig. 4-12.

Drawing with asymptotic curves is simple and convenient, and it is close to its exact curve, which is often used in the preliminary design stage of a system. If an precise logarithmic amplitude frequency characteristic curve is required, the asymptotic curve can be corrected by referring to the

幅频特性、对数幅频特性、相频特性分别为

$$\begin{cases} A(\omega) = |G(j\omega)| = \dfrac{1}{\sqrt{1+\omega^2 T^2}} \\ L(\omega) = 20\lg A(\omega) = 20\lg \dfrac{1}{\sqrt{T^2\omega^2+1}} \\ \varphi(\omega) = \angle G(j\omega) = -\arctan\omega T \end{cases} \quad (4\text{-}18)$$

当 $\omega=0$ 时，$A(\omega)=1$，相位角 $\varphi(\omega)=0°$；当 $\omega=\dfrac{1}{T}$ 时，$A(\omega)=\dfrac{1}{\sqrt{2}}$，相位角 $\varphi(\omega)=-45°$；当 $\omega=\infty$ 时，$A(\omega)=0$，相位角 $\varphi(\omega)=-90°$。

此外，实频特性和虚频特性满足

$$\left[U(\omega)-\dfrac{1}{2}\right]^2 + V^2(\omega) = \left(\dfrac{1}{2}\right)^2$$

上式表明，当 $\omega=0\rightarrow\infty$ 时，惯性环节的极坐标图为一个圆心在 (1/2, j0)、半径为 1/2 的半圆，如图 4-11 所示。

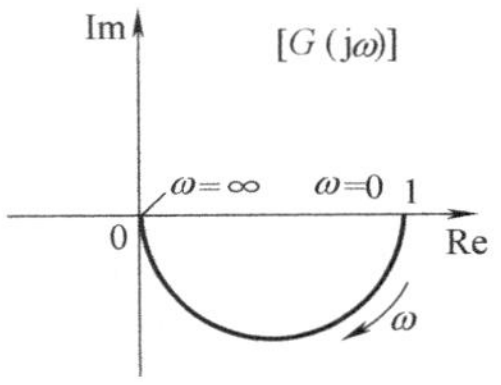

当 $\omega=0\rightarrow\infty$ 时，可以计算相应的 $L(\omega)$ 和 $\varphi(\omega)$，画出对数幅频特性图和对数相频特性图。在工程上常采用近似作图法，即用渐近线表示对数幅频特性曲线。通常将相邻两条渐近线相交处的频率 $\omega_T=1/T$ 称为转折频率。

当 $\omega\ll\omega_T$ 时，$L(\omega)=20\lg A(\omega)=-20\lg\sqrt{T^2\omega^2+1}\approx-20\lg 1=0\text{dB}$，即对数幅频特性曲线在低频段近似为 0dB 线，其称为低频渐近线；当 $\omega\gg\omega_T$ 时，$L(\omega)=20\lg A(\omega)=-20\lg\sqrt{T^2\omega^2+1}\approx-20\lg\omega T$，即对数幅频特性曲线在高频段近似为一条过点 $(\omega_T, 0)$、斜率为 -20dB/dec 的直线，其称为高频渐近线。因此，惯性环节的渐近线分段函数可表示为

Fig. 4-11 Polar plot of inertial element
（惯性环节的极坐标图）

$$L(\omega) = -20\lg\sqrt{T^2\omega^2+1} \approx \begin{cases} 0 & \left(0<\omega\leqslant\dfrac{1}{T}\right) \\ 20\lg\dfrac{1}{\omega T} & \left(\omega>\dfrac{1}{T}\right) \end{cases} \quad (4\text{-}19)$$

在惯性环节的对数相频特性图中，相角是用反正切函数来表示的，所以对数相频特性图是一条过点 $(1/T, -45°)$ 的反正切曲线，如图 4-12 所示。

用渐近线作图简单方便，且接近其精确曲线，在系统初步设计阶段经常采用。若需要精确的对数幅频特性曲线，可以参照图 4-13 的误差曲线对渐近线进行修正。最大误差在转折频率 $\omega=\omega_T$ 处，其误差值约为 3dB。

由对数幅频特性图可看出：惯性环节具有低通滤波特性，对于高频信号，其对数幅值迅速衰减。

2. 一阶微分环节

一阶微分环节的传递函数和相应的频率特性为

$$\begin{cases} G(s) = 1+Ts \\ G(j\omega) = 1+j\omega T \end{cases} \quad (4\text{-}20)$$

error curve in Fig. 4-13. The maximum error is at the corner frequency $\omega = \omega_T$, and the error value is approximate to 3dB.

According to the logarithmic amplitude frequency characteristic plot, it can be seen that the inertia element has low-pass filtering characteristics, and that the logarithmic amplitude of the inertia element is rapidly attenuated for high-frequency signals.

2. First-order differential element

The transfer function of the first-order differential element and the corresponding frequency characteristic are

$$\begin{cases} G(s) = 1+Ts \\ G(j\omega) = 1+j\omega T \end{cases} \quad (4-20)$$

Accordingly, the amplitude frequency characteristic, logarithmic amplitude frequency characteristic and phase frequency characteristic are respectively

$$\begin{cases} A(\omega) = |G(j\omega)| = \sqrt{1+\omega^2 T^2} \\ L(\omega) = 20\lg A(\omega) = 20\lg\sqrt{T^2\omega^2+1} \\ \varphi(\omega) = \angle G(j\omega) = \arctan\omega T \end{cases} \quad (4-21)$$

According to the equation of the frequency characteristic, when $\omega = 0$, the amplitude is $A(\omega) = 1$ and the phase angle is $\varphi(\omega) = 0°$; when $\omega = \infty$, the amplitude is $A(\omega) = \infty$ and the phase angle is $\varphi(\omega) = 90°$.

When $\omega = 0 \to \infty$, the polar plot of the first-order differential element starts at the point of (1, j0), and is a vertical line in the first quadrant, which is parallel to the imaginary axis, as is shown in Fig. 4-14.

The asymptotic piecewise function of the logarithmic amplitude frequency characteristic of the first-order differential element is

$$L(\omega) = 20\lg\sqrt{T^2\omega^2+1} \approx \begin{cases} 0 & \left(0<\omega \leq \dfrac{1}{T}\right) \\ 20\lg\omega T & \left(\omega > \dfrac{1}{T}\right) \end{cases} \quad (4-22)$$

Obviously, there is only one sign difference between the logarithmic frequency characteristics of the first-order differential element and the inertial element. Thus, their Bode plots are symmetric about ω axis, as is shown in Fig. 4-15.

4.2.4 Oscillation element and second-order differential element

1. Oscillation element

The transfer function of the second-order oscillation element and the corresponding frequency characteristic are

$$\begin{cases} G(s) = \dfrac{\omega_n^2}{s^2+2\xi\omega_n s+\omega_n^2} \quad (0<\xi<1) \\ G(j\omega) = \dfrac{1}{(j\omega/\omega_n)^2+2\xi(j\omega/\omega_n)+1} = \dfrac{1}{1-(\omega/\omega_n)^2+j2\xi\omega/\omega_n} \end{cases} \quad (4-23)$$

对应的幅频特性、对数幅频特性及相频特性为

$$\begin{cases} A(\omega) = |G(j\omega)| = \sqrt{1+\omega^2 T^2} \\ L(\omega) = 20\lg A(\omega) = 20\lg\sqrt{T^2\omega^2+1} \\ \varphi(\omega) = \angle G(j\omega) = \arctan\omega T \end{cases} \quad (4-21)$$

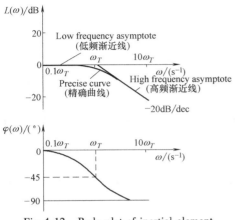

Fig. 4-12 Bode plot of inertial element
（惯性环节的伯德图）

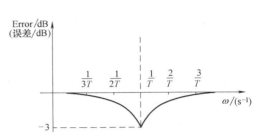

Fig. 4-13 Error curve of inertial element
（惯性环节的误差曲线）

根据频率特性公式，当 $\omega = 0$ 时，幅值为 $A(\omega) = 1$，相位角 $\varphi(\omega) = 0°$；当 $\omega = \infty$ 时，幅值 $A(\omega) = \infty$，相位角 $\varphi(\omega) = 90°$。

当 $\omega = 0 \to \infty$ 时，一阶微分环节的极坐标图为始于点 (1, j0)，在第一象限平行于虚轴的一条竖线，如图 4-14 所示。

一阶微分环节的对数幅频特性渐近线分段函数为

$$L(\omega) = 20\lg\sqrt{T^2\omega^2+1} \approx \begin{cases} 0 & \left(0<\omega\leqslant\dfrac{1}{T}\right) \\ 20\lg\omega T & \left(\omega>\dfrac{1}{T}\right) \end{cases} \quad (4-22)$$

显然，一阶微分环节与惯性环节的对数频率特性相比，仅相差一个符号。所以它们的伯德图关于 ω 轴成对称关系，如图 4-15 所示。

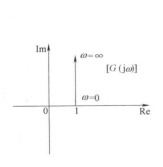

Fig. 4-14 Polar plot of first-order differential element
（一阶微分环节的极坐标图）

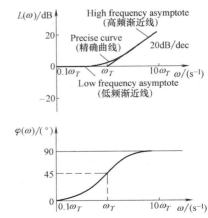

Fig. 4-15 Bode plot of first-order differential element
（一阶微分环节的伯德图）

Accordingly, the amplitude frequency characteristic, logarithmic amplitude frequency characteristic and phase frequency characteristic are respectively

$$\begin{cases} A(\omega) = |G(j\omega)| = \dfrac{1}{\sqrt{[1-(\omega/\omega_n)^2]^2 + (2\xi\omega/\omega_n)^2}} \\ L(\omega) = -20\lg\sqrt{[1-(\omega/\omega_n)^2]^2 + (2\xi\omega/\omega_n)^2} \\ \varphi(\omega) = \angle G(j\omega) = -\arctan\dfrac{2\xi\omega/\omega_n}{1-(\omega/\omega_n)^2} \end{cases} \qquad (4\text{-}24)$$

According to the equations of the frequency characteristic, when $\omega = 0$, the amplitude is $A(\omega) = 1$ and the phase angle is $\varphi(\omega) = 0°$; when $\omega = \omega_n$, the amplitude is $A(\omega) = 1/(2\xi)$ and the phase angle is $\varphi(\omega) = -90°$; when $\omega = \infty$, the amplitude is $A(\omega) = 0$ and the phase angle is $\varphi(\omega) = -180°$.

The polar plot of the second-order oscillation is related to the damping ratio ξ. Considering different values of the damping ratio, a cluster of polar plots can be formed, as is shown in Fig. 4-16. It can be seen from the figure, when $\omega = 0 \to \infty$, regardless of the value of ξ, the polar plot starts from the point of (0, j0) and ends at the point of (0, j0), and the phase angle changes from $0°$ to $-180°$; when $\omega = \omega_n$, all the polar plots intersect the negative imaginary axis. The phase angle is $-90°$ and the amplitude is $1/(2\xi)$. For an underdamped system ($\xi < 1$), the system will have a resonant peak, and it is denoted as M_r. The corresponding frequency at which the resonant peak appears is called the resonant frequency ω_r.

The asymptotic piecewise function of the logarithmic amplitude frequency characteristic of the second-order oscillation element is

$$L(\omega) = 20\lg\dfrac{1}{\sqrt{[1-(\omega/\omega_n)^2]^2 + (2\xi\omega/\omega_n)^2}} \approx \begin{cases} 0 & \left(0 < \omega < \dfrac{1}{T}\right) \\ 20\lg\dfrac{1}{\omega^2 T^2} & \left(\omega > \dfrac{1}{T}\right) \end{cases} \qquad (4\text{-}25)$$

where $\omega_n = 1/T = \omega_T$, i.e. the corner frequency according to which the asymptotic curve of the logarithmic amplitude frequency characteristic plot can be calculated.

When $\omega \ll \omega_n$, $L(\omega) = 20\lg A(\omega) \approx -20\lg 1 = 0\text{dB}$. It means that the logarithmic amplitude frequency characteristic plot is approximate to 0dB line in the low frequency band; when $\omega \gg \omega_n$, $L(\omega) = 20\lg A(\omega) \approx -20\lg\dfrac{\omega^2}{\omega_n^2} = 40\lg\omega_n - 40\lg\omega$. It means that the logarithmic amplitude frequency characteristic plot is approximate to an oblique line with a slope of -40dB/dec passing through the point of (ω_n, 0) in the high frequency band. Since the phase angle is represented by the inverse tangent function, the logarithmic phase frequency characteristic plot of the second-order oscillation element is an arctangent curve passing through the point of (ω_n, $-90°$), as is shown in Fig. 4-17.

When the logarithmic amplitude frequency characteristic plot of the second-order oscillation element is approximated by the asymptotic curves, an error is inevitable. When $\omega = \omega_n$, the error value is $-20\lg 2\xi$. According to different values of the damping ratio ξ, the maximum errors are listed in

4.2.4 振荡环节和二阶微分环节

1. 振荡环节

二阶振荡环节的传递函数和相应的频率特性为

$$\begin{cases} G(s) = \dfrac{\omega_n^2}{s^2 + 2\xi\omega_n s + \omega_n^2} & (0<\xi<1) \\ G(j\omega) = \dfrac{1}{(j\omega/\omega_n)^2 + 2\xi(j\omega/\omega_n) + 1} = \dfrac{1}{1-(\omega/\omega_n)^2 + j2\xi\omega/\omega_n} \end{cases} \quad (4\text{-}23)$$

对应的幅频特性、对数幅频特性及相频特性为

$$\begin{cases} A(\omega) = |G(j\omega)| = \dfrac{1}{\sqrt{[1-(\omega/\omega_n)^2]^2 + (2\xi\omega/\omega_n)^2}} \\ L(\omega) = -20\lg\sqrt{[1-(\omega/\omega_n)^2]^2 + (2\xi\omega/\omega_n)^2} \\ \varphi(\omega) = \angle G(j\omega) = -\arctan\dfrac{2\xi\omega/\omega_n}{1-(\omega/\omega_n)^2} \end{cases} \quad (4\text{-}24)$$

根据频率特性公式,当 $\omega=0$ 时,幅值 $A(\omega)=1$,相位角 $\varphi(\omega)=0°$;当 $\omega=\omega_n$ 时,幅值 $A(\omega)=1/(2\xi)$,相位角 $\varphi(\omega)=-90°$;当 $\omega=\infty$ 时,幅值 $A(\omega)=0$,相位角 $\varphi(\omega)=-180°$。

二阶振荡环节的极坐标图与阻尼比 ξ 有关。考虑不同的阻尼比,可形成一簇极坐标曲线,如图 4-16 所示。由图可知,当 $\omega=0\to\infty$ 时,不论 ξ 值如何,极坐标曲线均从点 (1,j0) 开始,到 (0,j0) 结束,相位角相应从 $0°$ 变到 $-180°$。当 $\omega=\omega_n$ 时,极坐标曲线均交于负虚轴,其相位角为 $-90°$,幅值为 $1/(2\xi)$。对于欠阻尼系统($\xi<1$),系统会出现谐振峰值,将其记作 M_r,出现该谐振峰值的对应频率称为谐振频率 ω_r。

二阶振荡环节的对数幅频特性渐近线分段函数为

$$L(\omega) = 20\lg\dfrac{1}{\sqrt{[1-(\omega/\omega_n)^2]^2 + (2\xi\omega/\omega_n)^2}} \approx \begin{cases} 0 & \left(0<\omega\leq\dfrac{1}{T}\right) \\ 20\lg\dfrac{1}{\omega^2 T^2} & \left(\omega>\dfrac{1}{T}\right) \end{cases} \quad (4\text{-}25)$$

式中,$\omega_n = 1/T = \omega_T$,即为转折频率,可由此频率算出其对数幅频特性图的渐近线。

当 $\omega\ll\omega_n$ 时,$L(\omega) = 20\lg A(\omega) \approx -20\lg 1 = 0\text{dB}$,这表示对数幅频特性曲线在低频段近似为 0dB 线;当 $\omega\gg\omega_n$ 时,$L(\omega) = 20\lg A(\omega) \approx -20\lg\dfrac{\omega^2}{\omega_n^2} = 40\lg\omega_n - 40\lg\omega$,即对数幅频特性曲线在高频段近似是一条过点 $(\omega_n, 0)$,斜率为 -40dB/dec 的斜直线。因为相角是用反正切函数来表示,二阶振荡环节的对数相频特性图是一条过点 $(\omega_n, -90°)$ 的反正切曲线,如图 4-17 所示。

用渐近线近似表示二阶振荡环节的对数幅频特性图时,必然会产生误差。在频率 $\omega=\omega_n$ 处,其误差值为 $-20\lg 2\xi$。根据阻尼比 ξ 的不同取值,最大误差见表 4-1。由此可见,当 $0.4\leq\xi\leq0.7$ 时,最大误差小于 3dB,可以不对渐近线进行修正;当 $\xi<0.4$ 或 $\xi>0.7$ 时,误差较大,必须对渐近线进行修正。

Table 4-1. It can be seen that the maximum error is less than 3dB when $0.4 \leqslant \omega \leqslant 0.7$, so it is not necessary to do correction for asymptotic curves; when $\xi < 0.4$ or $\xi > 0.7$, the error is increased, and correction for asymptotic curves is a must.

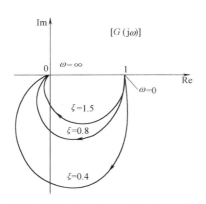

Fig. 4-16 Polar plot of second-order oscillation element （二阶振荡环节的极坐标图）

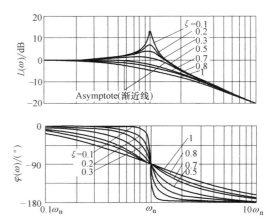

Fig. 4-17 Bode plot of second-order oscillation element （二阶振荡环节的伯德图）

Table 4-1 Error correction table for $\omega = \omega_n$ （$\omega = \omega_n$ 时的误差修正表）

The damping ratio ξ（阻尼比 ξ）	0.1	0.15	0.2	0.25	0.3	0.4	0.5	0.6	0.7	0.8	1.0
The maximum error/dB（最大误差/dB）	14.0	10.4	8.0	6.0	4.4	2.0	0	-1.6	-3.0	-4.0	-6.0

2. Second-order differential element

The transfer function of the second-order differential element and the corresponding frequency characteristics are

$$\begin{cases} G(s) = \dfrac{s^2}{\omega_n^2} + 2\xi \dfrac{s}{\omega_n} + 1 \\ G(j\omega) = \left(1 - \dfrac{\omega^2}{\omega_n^2}\right) + j2\xi \dfrac{\omega}{\omega_n} \end{cases} \quad (4\text{-}26)$$

Accordingly, the amplitude frequency characteristic, logarithmic amplitude frequency characteristic and phase frequency characteristic are respectively

$$\begin{cases} A(\omega) = \sqrt{[1-(\omega/\omega_n)^2]^2 + [2\xi\omega/\omega_n]^2} \\ L(\omega) = 20\lg\sqrt{[1-(\omega/\omega_n)^2]^2 + (2\xi\omega/\omega_n)^2} \\ \varphi(\omega) = \arctan\dfrac{2\xi(\omega/\omega_n)}{1-(\omega/\omega_n)^2} \end{cases} \quad (4\text{-}27)$$

According to the equation of the frequency characteristic, when $\omega = 0$, the amplitude is $A(\omega) = 1$ and the phase angle is $\varphi(\omega) = 0°$; when $\omega = \omega_n$, the amplitude is $A(\omega) = 2\xi$ and the phase angle is $\varphi(\omega) = 90°$; when $\omega = \infty$, the amplitude is $A(\omega) = \infty$ and the phase angle is $\varphi(\omega) = 180°$.

2. 二阶微分环节

二阶微分环节的传递函数和相应的频率特性为

$$\begin{cases} G(s) = \dfrac{s^2}{\omega_n^2} + 2\xi\dfrac{s}{\omega_n} + 1 \\ G(j\omega) = \left(1 - \dfrac{\omega^2}{\omega_n^2}\right) + j2\xi\dfrac{\omega}{\omega_n} \end{cases} \tag{4-26}$$

对应的幅频特性、对数幅频特性及相频特性为

$$\begin{cases} A(\omega) = \sqrt{[1-(\omega/\omega_n)^2]^2 + (2\xi\omega/\omega_n)^2} \\ L(\omega) = 20\lg\sqrt{[1-(\omega/\omega_n)^2]^2 + (2\xi\omega/\omega_n)^2} \\ \varphi(\omega) = \arctan\dfrac{2\xi\omega/\omega_n}{1-(\omega/\omega_n)^2} \end{cases} \tag{4-27}$$

根据频率特性公式，当 $\omega=0$ 时，幅值 $A(\omega)=1$，相位角 $\varphi(\omega)=0°$；当 $\omega=\omega_n$ 时，幅值 $A(\omega)=2\xi$，相位角 $\varphi(\omega)=90°$；当 $\omega=\infty$ 时，幅值 $A(\omega)=\infty$，相位角 $\varphi(\omega)=180°$。

二阶微分环节的极坐标图也与阻尼比 ξ 有关。考虑不同的阻尼比，可形成一簇极坐标曲线，如图 4-18 所示。由图可知，当 ω 从 $0\to\infty$ 时，不论 ξ 值如何，极坐标曲线均从点 $(1,j0)$ 开始，指向无穷远处，相位角相应从 $0°$ 变到 $180°$。当 $\omega=\omega_n$ 时，极坐标曲线均交于正虚轴，其相位角为 $90°$，幅值为 2ξ。

二阶微分环节的对数幅频特性为

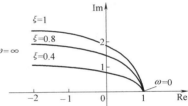

Fig. 4-18 Polar plot of second-order differential element
（二阶微分环节的极坐标图）

$$L(\omega) = 20\lg\sqrt{\left(1-\dfrac{\omega^2}{\omega_n^2}\right)^2 + \left(2\xi\dfrac{\omega}{\omega_n}\right)^2} \approx \begin{cases} 0 & \left(0<\omega<\dfrac{1}{T}=\omega_n\right) \\ 20\lg\dfrac{\omega^2}{\omega_n^2} & \left(\omega>\dfrac{1}{T}\right) \end{cases} \tag{4-28}$$

显然，二阶微分环节与振荡环节的对数频率特性相比，仅相差一个符号，所以它们的伯德图关于 ω 轴成对称关系，如图 4-19 所示。

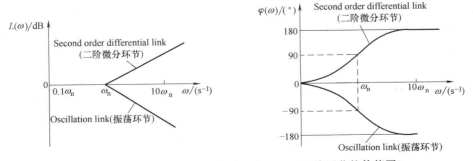

Fig. 4-19 Bode plot of second-order element（二阶环节的伯德图）

The polar plot of the second-order differential element is also related to the damping ratio ξ. Considering different values of the damping ratio, a cluster of polar plots can be formed, as is shown in Fig. 4-18. It can be seen from the figure, when $\omega=0\to\infty$, regardless of the value of ξ, the polar plot starts from the point of (1, j0) to infinite, and the phase angle changes from 0° to 180°; when $\omega=\omega_n$, all the polar plots intersect the positive imaginary axis, and the phase angle is 90° and the amplitude is 2ξ.

The logarithmic amplitude frequency characteristic of the second-order differential element is

$$L(\omega)=20\lg\sqrt{\left(1-\frac{\omega^2}{\omega_n^2}\right)^2+\left(2\xi\frac{\omega}{\omega_n}\right)^2}\approx\begin{cases}0 & \left(0<\omega<\frac{1}{T}=\omega_n\right)\\ 20\lg\frac{\omega^2}{\omega_n^2} & \left(\omega>\frac{1}{T}\right)\end{cases} \quad (4\text{-}28)$$

Obviously, there is only one sign difference between the logarithmic frequency characteristics of the second differential element and the oscillation element. Thus, their Bode plots are symmetric about ω axis, as is shown in Fig. 4-19.

4.2.5 Time delay element

The transfer function and frequency characteristic of the time delay element are respectively

$$\begin{cases}G(s)=e^{-\tau s}\\ G(j\omega)=e^{-j\omega\tau}=\cos\omega\tau+j(-\sin\omega\tau)\end{cases} \quad (4\text{-}29)$$

The corresponding real frequency characteristic and virtual frequency characteristic are

$$\begin{cases}U(\omega)=\cos\omega\tau\\ V(\omega)=-\sin\omega\tau\end{cases} \quad (4\text{-}30)$$

Accordingly, the amplitude frequency characteristic, logarithmic amplitude frequency characteristic and phase frequency characteristic are

$$\begin{cases}A(\omega)=|G(j\omega)|=1\\ L(\omega)=20\lg A(\omega)=0\\ \varphi(\omega)=\angle G(j\omega)=-\tau\omega\end{cases} \quad (4\text{-}31)$$

The polar plot of the time delay element is the unit circle. The amplitude is always equal to 1 and the phase changes proportionally with the frequency ω in the clockwise direction. In other words, the endpoint of the polar plot goes around the unit circle indefinitely, as is shown in Fig. 4-20. The logarithmic amplitude frequency characteristic plot is a 0dB line, and the logarithmic phase frequency characteristic plot increases exponentially as the frequency ω increases, as is shown in Fig. 4-21.

4.3 Open-loop Frequency Characteristics of Control Systems

4.3.1 The open-loop polar plot of control systems

Any system is composed of typical elements. Next, detailed analysis of the frequency character-

4.2.5 延时环节

延时环节的传递函数和频率特性分别为

$$\begin{cases} G(s) = e^{-\tau s} \\ G(j\omega) = e^{-j\omega\tau} = \cos\omega\tau + j(-\sin\omega\tau) \end{cases} \quad (4\text{-}29)$$

其相应的实频特性和虚频特性为

$$\begin{cases} U(\omega) = \cos\omega\tau \\ V(\omega) = -\sin\omega\tau \end{cases} \quad (4\text{-}30)$$

相应地，其幅频特性、对数幅频特性和相频特性为

$$\begin{cases} A(\omega) = |G(j\omega)| = 1 \\ L(\omega) = 20\lg A(\omega) = 0 \\ \varphi(\omega) = \angle G(j\omega) = -\tau\omega \end{cases} \quad (4\text{-}31)$$

延时环节的极坐标图为单位圆，其幅值恒等于1，而相位 $\varphi(\omega)$ 则随频率 ω 的增加而按顺时针方向成正比变化，换言之，极坐标图端点在单位圆上无限循环，如图4-20所示。其对数幅频特性图为一条0dB线，对数相频特性图为 $\varphi(\omega)$ 随着频率 ω 增加而指数增加，如图4-21所示。

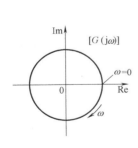

Fig. 4-20　Polar plot of time delay element
（延时环节的极坐标图）

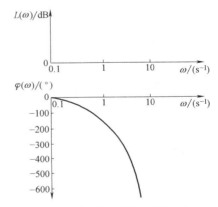

Fig. 4-21　Bode plot of time delay element
（延时环节的伯德图）

4.3　控制系统的开环频率特性

4.3.1　控制系统的开环极坐标图

任何一个系统都是由典型环节组成的，下面就典型环节组成的各种较为复杂的系统进行频率特性极坐标图分析。绘制系统极坐标图的一般步骤如下。

1）根据系统的传递函数写出系统的频率特性。
2）由系统的频率特性求出其实频特性、虚频特性、幅频特性、相频特性的表达式。
3）分别求出若干个特征点，如起点（$\omega=0$）、终点（$\omega=\infty$）、与实轴的交点、与虚轴

istics for various complex systems composed of typical elements will be introduced. The general steps for plotting a system polar plot are described as follows.

1) Write the frequency characteristics of the system according to the transfer function of the system.

2) Figure out the expressions of the actual frequency characteristic, the virtual frequency characteristic, the amplitude frequency characteristic, and the phase frequency characteristic according to the frequency characteristics of the system.

3) Find several characteristic points, such as the starting point ($\omega = 0$), the ending point ($\omega = \infty$), the point of intersection with the real axis, and the point of intersection with the imaginary axis, etc., and mark them on the polar plot.

4) Add the necessary characteristic points, and draw the approximate shape of the polar plot according to the change laws of the known points, the real frequency characteristic, the virtual frequency characteristic, the amplitude frequency characteristic and the phase frequency characteristic.

5) Connect discontinuous points and draw a complete polar plot.

1. Frequency characteristics of different types of systems

If the open-loop transfer function of a control system is

$$G_K(s) = K \frac{\prod_{k=1}^{p}(T_k s + 1) \prod_{l=1}^{q}(T_l^2 s^2 + 2\xi_l T_l s + 1)}{s^v \prod_{i=1}^{g}(T_i s + 1) \prod_{j=1}^{h}(T_j^2 s^2 + 2\xi_j T_j s + 1)} \quad (p + 2q = m, v + g + 2h = n, n \geq m)$$

(4-32)

where K is an open-loop gain coefficient; v is the number of serial integral elements.

Then substituting $s = j\omega$ into equation (4-32), the open-loop frequency characteristic of the system is

$$G_K(j\omega) = \frac{K \prod_{k=1}^{p}(1 + j\omega T_k) \prod_{l=1}^{q}(1 - T_l^2 \omega^2 + j2\xi_l \omega T_l)}{(j\omega)^v \prod_{i=1}^{g}(1 + j\omega T_i) \prod_{j=1}^{h}(1 - T_j^2 \omega^2 + j2\xi_j \omega T_j)} \quad (p + 2q = m, v + g + 2h = n, n \geq m)$$

(4-33)

The type of a control system is defined by the above equation according to v, and the general shape of the open-loop polar plot for different systems is introduced as follows.

(1) Type of systems

1) When $v = 0$, the system is called type-0 system.

When $\omega = 0$, we have $\angle G_K(j\omega) = 0°$, $|G_K(j\omega)| = K$.

When $\omega = \infty$, we have $\angle G_K(j\omega) = (m-n) \times 90°$, $|G_K(j\omega)| = \begin{cases} 0 & (n > m) \\ \text{const} & (n = m) \end{cases}$.

的交点等,并标注在极坐标图上。

4)补充必要的特征点,根据已知点和实频特性、虚频特性、幅频特性、相频特性的变化规律,绘制极坐标图的大致形状。

5)补充不连续点的极坐标曲线,绘出完整的极坐标曲线图。

1. 不同类型系统的频率特性

若控制系统的开环传递函数为

$$G_K(s) = K \frac{\prod_{k=1}^{p}(T_k s + 1)\prod_{l=1}^{q}(T_l^2 s^2 + 2\xi_l T_l s + 1)}{s^v \prod_{i=1}^{g}(T_i s + 1)\prod_{j=1}^{h}(T_j^2 s^2 + 2\xi_j T_j s + 1)} \quad (p+2q=m, v+g+2h=n, n \geq m)$$

(4-32)

式中,K 为开环增益;v 为串联积分环节的个数。

将 $s=\mathrm{j}\omega$ 代入式(4-32),可得系统的开环频率特性为

$$G_k(\mathrm{j}\omega) = \frac{K\prod_{k=1}^{p}(1+\mathrm{j}\omega T_k)\prod_{l=1}^{q}(1-T_l^2\omega^2+\mathrm{j}2\xi_l\omega T_l)}{(\mathrm{j}\omega)^v \prod_{i=1}^{g}(1+\mathrm{j}\omega T_i)\prod_{j=1}^{h}(1-T_j^2\omega^2+\mathrm{j}2\xi_j\omega T_j)}$$

$$(p+2q=m, v+g+2h=n, n \geq m)$$

(4-33)

由上式对控制系统的类型加以定义,它们的开环极坐标图的一般形状介绍如下。

(1)系统类型

1)$v=0$ 时,系统称为 0 型系统。

当 $\omega=0$ 时,$\angle G_K(\mathrm{j}\omega)=0°$,$|G_K(\mathrm{j}\omega)|=K$。

当 $\omega=\infty$ 时,$\angle G_K(\mathrm{j}\omega)=(m-n)\times 90°$,$|G_K(\mathrm{j}\omega)|=\begin{cases}0 & (n>m)\\ \mathrm{const} & (n=m)\end{cases}$。

由此可见,0 型系统极坐标图始于正实轴上坐标为 K 的点,在 $n>m$ 的高频段趋于原点,在哪个象限趋于原点取决于 $(m-n)\times 90°$;在 $n=m$ 的高频段趋于实轴上某点。

2)$v>0$ 时,系统称为 I 型系统($v=1$)、II 型系统($v=2$)及以上的系统(v 型系统,$v>2$)。

当 $\omega=0$ 时,$\angle G_K(\mathrm{j}\omega)=-v\times 90°$,$|G_K(\mathrm{j}\omega)|=\infty$。

当 $\omega=\infty$ 时,$\angle G_K(\mathrm{j}\omega)=(m-n)\times 90°$,$|G_K(\mathrm{j}\omega)|=\begin{cases}0 & (n>m)\\ \mathrm{const} & (n=m)\end{cases}$。

由此可见,I 型及以上的系统极坐标图在低频段的渐近线与角度为 $\angle G_K(\mathrm{j}\omega)=-v\times 90°$ 的虚轴或实轴平行,在高频段趋于原点($n>m$)或实轴上某点($n=m$),在哪个象限趋于原点取决于 $(m-n)\times 90°$。

综上所述,开环系统极坐标图的低频部分是由因式 $K/(\mathrm{j}\omega)^v$ 确定的。对于 0 型系统,$G_K(\mathrm{j}0)=K\angle 0°$,而对于 I 型、II 型及以上的系统,$G_K(\mathrm{j}0)=\infty\angle -90°v$,如图 4-22a 所示;对于开环系统的高频部分,在 $n>m$ 时,当 $\omega\to\infty$,$G_K(\mathrm{j}\infty)=0\angle -90°(n-m)$,$G_K(\mathrm{j}\omega)$ 曲线以顺时针方向按 $-90°(n-m)$ 的角度趋向于坐标原点。如果 $n-m$ 是偶数,则曲线与横轴相切;反之,若 $n-m$ 是奇数,则曲线与虚轴相切,如图 4-22b 所示。

It can be seen that the polar plot of the type-0 system starts from the point of value K on the positive real axis. When $n > m$, the high-frequency band approaches to the origin. Which quadrant the plot approaches to the origin is dependent of $(m-n) \times 90°$; When $n = m$, the high-frequency band tends to a certain point on the real axis.

2) When $v > 0$, the system is called type-I system ($v = 1$), type-II system ($v = 2$) and the above system (type-v system, $v = 2$).

When $\omega = 0$, we have $\angle G_K(j\omega) = -v \times 90°$, $|G_K(j\omega)| = \infty$.

When $\omega = \infty$, we have $\angle G_K(j\omega) = (m-n) \times 90°$, $|G_K(j\omega)| = \begin{cases} 0 & (n > m) \\ \text{const} & (n = m) \end{cases}$.

It can be seen that asymptotic lines of the polar plot of the type-I and above system are parallel to the imaginary axis or the real axis of the angle $\angle G_K(j\omega) = -v \times 90°$ in the low frequency band, and approaches to the origin ($n > m$) or a point on the real axis ($n = m$) in the high frequency band, and which quadrant the plot approaches to the origin is dependent of $(m-n) \times 90°$.

In summary, the low frequency part of the polar plot of the open-loop system is determined by the factor $K/(j\omega)^v$. We have $G_K(j0) = K \angle 0°$ for type-0 systems while we have $G_K(j0) = \infty \angle -90°$ v for type-I, type-II and above systems, as is shown in Fig. 4-22a; for the high frequency part of the open-loop system, when $n > m$, we have $G_K(j\infty) = 0 \angle -90°(n-m)$ at $\omega \to \infty$, the curve $G_K(j\omega)$ approaches to the origin of the coordinates in terms of the angle $-90°(n-m)$ in the clockwise direction. If $n-m$ is even, the curve is tangent to the horizontal axis; conversely, if $n-m$ is odd, the curve is tangent to the imaginary axis, as is shown in Fig. 4-22b.

(2) The polar plot of the system at $\omega = 0$ when $v > 0$ The open-loop polar plot $G_K(j\omega)$ corresponding to $\omega = -\infty \to +\infty$ must be a complete closed curve. However, when there is an integral element in the open-loop transfer function, i.e. $v \neq 0$, its amplitude frequency characteristic is $|G_K(j0)| \to \infty$ at $\omega = 0$. And then the trajectory of the polar plot $G_K(j\omega)$ will be discontinuous. Therefore, it is necessary to add the discontinuous part of the polar trajectory to form the complete closed curve.

The open-loop frequency characteristics of the system is expressed as

$$G_K(j\omega) = \frac{K}{(j\omega)^v} G_0(j\omega) \tag{4-34}$$

where

$$G_0(j\omega) = \frac{\prod_{k=1}^{p}(1 + j\omega T_k) \prod_{l=1}^{q}(1 - T_l^2 \omega^2 + j2\xi_l \omega T_l)}{\prod_{i=1}^{g}(1 + j\omega T_i) \prod_{j=1}^{h}(1 - T_j^2 \omega^2 + j2\xi_j \omega T_j)}$$

The frequency characteristic of the system at $\omega = 0$ is

$$G_K(j0) = \lim_{\omega \to 0} \frac{K}{(j\omega)^v} G_0(j\omega) = \lim_{\omega \to 0} \frac{K}{(j\omega)^v} \tag{4-35}$$

Thus, the open-loop amplitude frequency characteristic and the phase frequency characteristic

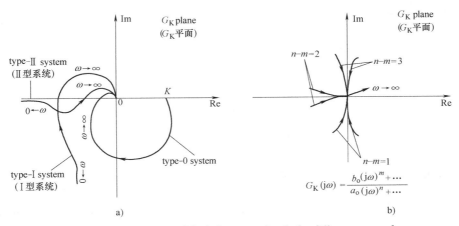

Fig. 4-22 Polar plots of low and high frequency bands for different types of systems
（不同类型系统低频段和高频段的极坐标图）

（2） $v>0$ 时系统在 $\omega=0$ 处的极坐标图　对应于 $\omega=-\infty \to +\infty$ 的开环极坐标曲线 $G_K(j\omega)$ 必须为全封闭曲线。但当开环传递函数中存在积分环节时，即 $v\neq 0$ 时，在 $\omega=0$ 处，其幅频特性 $|G_K(j0)|\to\infty$，于是极坐标曲线 $G_K(j\omega)$ 轨迹将不连续。因此，必须补充不连续部分的极坐标轨迹线以构成完整的封闭曲线。

将系统的开环频率特性表示为

$$G_K(j\omega)=\frac{K}{(j\omega)^v}G_0(j\omega) \tag{4-34}$$

式中

$$G_0(j\omega)=\frac{\prod_{k=1}^{p}(1+j\omega T_k)\prod_{l=1}^{q}(1-T_l^2\omega^2+j2\xi_l\omega T_l)}{\prod_{i=1}^{g}(1+j\omega T_i)\prod_{j=1}^{h}(1-T_j^2\omega^2+j2\xi_j\omega T_j)}$$

系统在 $\omega=0$ 处的频率特性为

$$G_K(j0)=\lim_{\omega\to 0}\frac{K}{(j\omega)^v}G_0(j\omega)=\lim_{\omega\to 0}\frac{K}{(j\omega)^v} \tag{4-35}$$

由此可得 $\omega=0^-\to 0^+$ 处的开环幅频特性和相频特性分别为

$$A(0)=|G_K(j0)|=\left|\frac{K}{(j\omega)^v}\right|_{\omega=0^-\to 0^+}=\infty$$

$$\varphi(0)=\angle G_K(j0)=\begin{cases} v\times 90° & (\omega\to 0^-) \\ -v\times 90° & (\omega\to 0^+) \end{cases} \tag{4-36}$$

相应地，当 $\omega=0^-\to 0^+$ 时，$G_K(j\omega)$ 曲线的相角变化量为

$$\Delta\arg G_K(j0)=\angle G_K(j0^+)-\angle G_K(j0^-)=\varphi(0^+)-\varphi(0^-)=-v\frac{\pi}{2}-v\frac{\pi}{2}=-v\pi \tag{4-37}$$

由以上分析可知，当 $v\neq 0$ 时，系统在 $\omega=0$（即 $\omega=0^-\to 0^+$）处的开环极坐标曲线 $G_K(j\omega)$ 是半径（幅值）为 ∞、相角为 $-v\pi$ 的圆弧轨迹线。对于 I 型系统和 II 型系统，开环

at $\omega = 0^- \to 0^+$ are respectively

$$A(0) = |G_K(j0)| = \left|\frac{K}{(j\omega)^v}\right|_{\omega=0^-\to 0^+} = \infty$$

$$\varphi(0) = \angle G_K(j0) = \begin{cases} v \times 90° & (\omega \to 0^-) \\ -v \times 90° & (\omega \to 0^+) \end{cases} \tag{4-36}$$

Accordingly, the phase angle change of the curve $G_K(j\omega)$ at $\omega = 0^- \to 0^+$ is

$$\Delta\arg G_K(j0) = \angle G_K(j0^+) - \angle G_K(j0^-) = \varphi(0^+) - \varphi(0^-) = -v\frac{\pi}{2} - v\frac{\pi}{2} = -v\pi \tag{4-37}$$

It can be seen from the above analysis that the open-loop polar plot $G_K(j\omega)$ of the system at $\omega = 0$ (i.e. $\omega = 0^- \to 0^+$) is an arc trajectory with a radius (amplitude) of infinity and a phase angle of $-v\pi$ when $v \neq 0$. For the open-loop amplitude frequency characteristics of type-I system and type-II system, the changes of their phase angles at $\omega = 0$ are shown in Fig. 4-23.

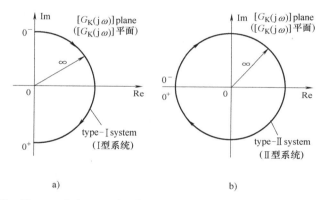

Fig. 4-23 Change of phase angle of open-loop amplitude-frequency characteristics
（开环幅频特性的相角变化情况）

Example 4-2 The open-loop transfer function of a type-0 system is $G_K(s) = \dfrac{10}{(s+1)(0.1s+1)}$, try to draw the polar plot.

Solution: According to the transfer function, the frequency characteristics of the system are

$$\begin{cases} A(\omega) = |G_K(j\omega)| = \dfrac{10}{\sqrt{1+\omega^2}\sqrt{1+(0.1\omega)^2}} \\ \varphi(\omega) = \angle G_K(j\omega) = -\arctan\omega - \arctan 0.1\omega \\ U(\omega) = \dfrac{10(1-0.1\omega^2)}{(1+\omega^2)(1+0.01\omega^2)} \\ V(\omega) = -\dfrac{11\omega}{(1+\omega^2)(1+0.01\omega^2)} \end{cases}$$

1) When $\omega = 0$, we have $A(\omega) = 10$, $\varphi(\omega) = 0°$, $U(\omega) = 10$, $V(\omega) = 0$.

幅频特性在 $\omega=0$ 时的相角变化情况如图 4-23 所示。

例 4-2 某 0 型系统开环传递函数为 $G_K(s)=\dfrac{10}{(s+1)(0.1s+1)}$，试绘制其极坐标图。

解：由传递函数求得系统频率特性为

$$\begin{cases} A(\omega)=|G_K(j\omega)|=\dfrac{10}{\sqrt{1+\omega^2}\sqrt{1+(0.1\omega)^2}} \\ \varphi(\omega)=\angle G_K(j\omega)=-\arctan\omega-\arctan 0.1\omega \\ U(\omega)=\dfrac{10(1-0.1\omega^2)}{(1+\omega^2)(1+0.01\omega^2)} \\ V(\omega)=-\dfrac{11\omega}{(1+\omega^2)(1+0.01\omega^2)} \end{cases}$$

1) 当 $\omega=0$ 时，$A(\omega)=10$，$\varphi(\omega)=0°$，$U(\omega)=10$，$V(\omega)=0$。
2) 当 $\omega=\infty$ 时，$A(\omega)=0$，$\varphi(\omega)=-180°$，$U(\omega)=0$，$V(\omega)=0$。
3) 当 $U(\omega)=0$ 时，$1-0.1\omega^2=0$，解得 $\omega=\pm\sqrt{10}$，$V(\pm\sqrt{10})=\pm 2.9$。

绘出极坐标图，如图 4-24 所示。

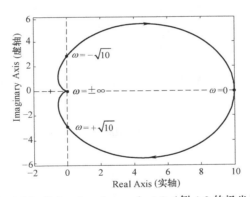

Fig. 4-24 Polar plot of example 4-2（例 4-2 的极坐标图）

例 4-3 某 I 型系统开环传递函数为 $G_K(s)=\dfrac{10}{s(s+1)}$，试绘制其极坐标图。

解：由传递函数求得系统频率特性为

$$\begin{cases} A(\omega)=|G_K(j\omega)|=\dfrac{10}{\omega\sqrt{1+\omega^2}} \\ \varphi(\omega)=\angle G_K(j\omega)=-90°-\arctan\omega \\ U(\omega)=-\dfrac{10}{1+\omega^2} \\ V(\omega)=-\dfrac{10}{\omega(1+\omega^2)} \end{cases}$$

1) 当 $\omega=0$ 时，$A(\omega)=\infty$，$\varphi(\omega)=-90°$，$U(\omega)=-10$，$V(\omega)=-\infty$。
2) 当 $\omega=\infty$ 时，$A(\omega)=0$，$\varphi(\omega)=-180°$，$U(\omega)=0$，$V(\omega)=0$。

2) When $\omega = \infty$, we have $A(\omega) = 0$, $\varphi(\omega) = -180°$, $U(\omega) = 0$, $V(\omega) = 0$.

3) When $U(\omega) = 0$, we have $1 - 0.1\omega^2 = 0$. Then $\omega = \pm\sqrt{10}$, $V(\pm\sqrt{10}) = \pm 2.9$.

The polar plot is drawn as Fig. 4-24.

Example 4-3 The open-loop transfer function of a type-I system is $G_K(s) = \dfrac{10}{s(s+1)}$, try to draw the polar plot.

Solution: According to the transfer function, the frequency characteristics of the system are

$$\begin{cases} A(\omega) = |G_K(j\omega)| = \dfrac{10}{\omega\sqrt{1+\omega^2}} \\ \varphi(\omega) = \angle G_K(j\omega) = -90° - \arctan\omega \\ U(\omega) = -\dfrac{10}{1+\omega^2} \\ V(\omega) = -\dfrac{10}{\omega(1+\omega^2)} \end{cases}$$

1) When $\omega = 0$, we have $A(\omega) = \infty$, $\varphi(\omega) = -90°$, $U(\omega) = -10$, $V(\omega) = -\infty$.
2) When $\omega = \infty$, we have $A(\omega) = 0$, $\varphi(\omega) = -180°$, $U(\omega) = 0$, $V(\omega) = 0$.
3) When $\omega = 0^- \to 0^+$, we have $\Delta \arg G_K(j0) = \varphi(0^+) - \varphi(0^-) = -180°$.

The polar plot is drawn as Fig. 4-25.

Example 4-4 The open-loop transfer function of a type-II system is $G_K(s) = \dfrac{10}{s^2(s+1)}$, try to plot the polar plot.

Solution: According to the transfer function, the frequency characteristics of the system are

$$\begin{cases} A(\omega) = |G_K(j\omega)| = \dfrac{10}{\omega^2\sqrt{1+\omega^2}} \\ \varphi(\omega) = \angle G_K(j\omega) = -180° - \arctan\omega \\ U(\omega) = -\dfrac{10}{\omega^2(1+\omega^2)} \\ V(\omega) = \dfrac{10}{\omega(1+\omega^2)} \end{cases}$$

1) When $\omega = 0$, we have $A(\omega) = \infty$, $\varphi(\omega) = -180°$, $U(\omega) = -\infty$, $V(\omega) = \infty$.
2) When $\omega = \infty$, we have $A(\omega) = 0$, $\varphi(\omega) = -270°$, $U(\omega) = 0$, $V(\omega) = 0$.
3) When $\omega = 0^- \to 0^+$, we have $\Delta \arg G_K(j0) = \varphi(0^+) - \varphi(0^-) = -360°$.

The polar plot is drawn as Fig. 4-26.

2. Frequency characteristics of non-minimum phase system

(1) Minimum phase system and non-minimum phase system If an element of a system has zeros located in the right half of the s-plane, the element is called a non-minimum phase element. If the transfer function does not contain non-minimum phase elements, in other words, all the zeros and poles of the transfer function are all located in the left half of the s-plane, it is called the mini-

3) 当 $\omega = 0^- \to 0^+$ 时，$\Delta \arg G_K(j0) = \varphi(0^+) - \varphi(0^-) = -180°$。

绘出极坐标图，如图 4-25 所示。

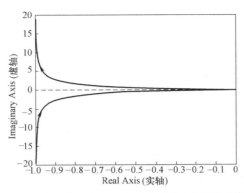

Fig. 4-25 Polar plot of example 4-3（例 4-3 的极坐标图）

例 4-4 某 II 型系统开环传递函数为 $G_K(s) = \dfrac{10}{s^2(s+1)}$，试绘制其极坐标图。

解： 由传递函数求得系统频率特性为

$$\begin{cases} A(\omega) = |G_K(j\omega)| = \dfrac{10}{\omega^2\sqrt{1+\omega^2}} \\ \varphi(\omega) = \angle G_K(j\omega) = -180° - \arctan\omega \\ U(\omega) = -\dfrac{10}{\omega^2(1+\omega^2)} \\ V(\omega) = \dfrac{10}{\omega(1+\omega^2)} \end{cases}$$

1) 当 $\omega = 0$ 时，$A(\omega) = \infty$，$\varphi(\omega) = -180°$，$U(\omega) = -\infty$，$V(\omega) = \infty$。

2) 当 $\omega = \infty$ 时，$A(\omega) = 0$，$\varphi(\omega) = -270°$，$U(\omega) = 0$，$V(\omega) = 0$。

3) 当 $\omega = 0^- \to 0^+$ 时，$\Delta \arg G_K(j0) = \varphi(0^+) - \varphi(0^-) = -360°$。

绘出极坐标图，如图 4-26 所示。

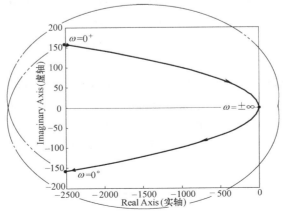

Fig. 4-26 Polar plot of example 4-4（例 4-4 的极坐标图）

mum phase transfer function. A system with a minimum phase open-loop transfer function is called a minimum phase system. If the transfer function contains a non-minimum phase element, in other words, there is one or more poles or zeros in the right half of the s-plane, it is called a non-minimum phase transfer function. A system with a non-minimum phase open-loop transfer function is called a non-minimum phase system.

Sometimes the amplitude frequency characteristics of two systems are identical, but their phase frequency characteristics are different. Among systems with the identical amplitude frequency characteristics, when $w = 0 \to \infty$, the phase angle range of the minimum phase system is the smallest, and the phase angle range of any non-minimum phase system is greater than that of the minimum phase transfer function. Moreover, there is a certain single-valued correspondence between the logarithmic amplitude frequency characteristic and the logarithmic phase frequency characteristic for the minimum phase system, but there is no correspondence for the non-minimum phase system. Therefore, according to the asymptotic curves of the logarithmic amplitude frequency characteristic of the minimum phase system, the phase frequency characteristic and the transfer function can be figured out, and vice versa. When analyzing and compensating a minimum phase system using Bode plot, we only need to draw a logarithmic amplitude frequency characteristic plot (or a logarithmic phase frequency characteristic plot).

For example, the transfer functions of two systems are respectively

$$G_1(s) = \frac{1+\tau s}{1+Ts}, \quad G_2(s) = \frac{1-\tau s}{1+Ts} \quad (0 < \tau < T)$$

The two systems have the same amplitude frequency characteristics but different phase frequency characteristics, as is shown in Fig. 4-27. Obviously, according to the definition of the minimum phase system, the system with $G_1(s)$ is a minimum phase system, and its phase angle range is minimum. The system with $G_2(s)$ is a non-minimum phase system.

In order to determine whether a system is a minimum phase system, it is necessary to check the slope of the asymptotic curve in high frequency band of the logarithmic amplitude frequency characteristic plot, and also check the phase angle at $\omega = \infty$. When $\omega = \infty$, if the slope of the logarithmic amplitude frequency characteristic plot is $20 \times (m-n)$ dB/dec (where n and m are the orders of the denominator and numerator polynomial of the transfer function, respectively), and the phase angle is equal to $90° \times (m-n)$, then the system is surely a minimum phase system, otherwise it is a non-minimum phase system.

(2) Common non-minimum phase elements The non-minimum phase system is generally generated by two causes. One is that the system contains non-minimum phase elements (such as time delay elements), and the other one is that the inner loop of the system is unstable. Transmitting time delay is a non-minimum phase characteristic, which may cause severe phase lag at high frequency.

The time delay element $G(s) = e^{-\tau s}$ can be expanded into power series, i.e.

$$G(s) = e^{-\tau s} = 1 - \tau s + \frac{1}{2}\tau^2 s^2 - \frac{1}{3}\tau^3 s^3 + \cdots$$

Since some of the factors in the above equation have a negative coefficient, it can be written as

2. 非最小相位系统的频率特性

（1）最小相位系统与非最小相位系统　若系统的某个环节在复平面右半面存在零点，此环节称为非最小相位环节。传递函数若不包含非最小相位环节，即所有零点和极点全部位于复平面的左半平面，则称为最小相位传递函数。具有最小相位开环传递函数的系统，称为最小相位系统。传递函数若包含非最小相位环节，即在复平面的右半平面上有一个或多个极点或零点，则称为非最小相位传递函数。具有非最小相位开环传递函数的系统，称为非最小相位系统。

有时两个系统的幅频特性完全相同，而相频特性却相异。在具有相同幅频特性的系统中，当 $\omega=0\to\infty$，最小相位系统的相角范围在所有这类系统中是最小的，且任何非最小相位系统的相角范围都大于最小相位传递函数的相角范围。此外，最小相位系统的对数幅频特性与对数相频特性之间具有确定的单值对应关系，而非最小相位系统却没有这种对应关系。因此，根据最小相位系统对数幅频特性的渐近线，就能确定其相频特性和传递函数，反之亦然。当利用伯德图对最小相位系统进行分析和校正时，只需画出对数幅频特性图（或对数相频特性图）。

例如，两个系统的传递函数分别为

$$G_1(s)=\frac{1+\tau s}{1+Ts},\quad G_2(s)=\frac{1-\tau s}{1+Ts}\quad(0<\tau<T)$$

这两个系统具有相同的幅频特性，却有不同的相频特性，如图 4-27 所示。显然，根据最小相位系统的定义，$G_1(s)$ 表达的系统为最小相位系统，它的相角变化范围最小。$G_2(s)$ 表达的系统为非最小相位系统。

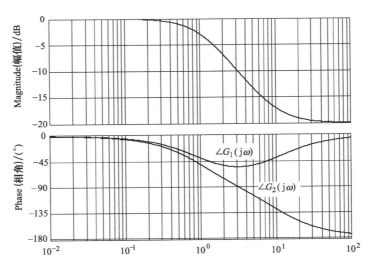

Fig. 4-27　Bode plots of minimum phase system and non-minimum phase system
（最小相位系统与非最小相位系统的伯德图）

为了确定一个系统是否为最小相位系统，需要检查对数幅频特性图高频段渐近线的斜率，也要检查在 $\omega=\infty$ 时的相角。当 $\omega=\infty$ 时，若对数幅频特性图的斜率为 $20\times(m-n)$ dB/dec（其中 n、m 分别为传递函数分母、分子多项式的阶数），其相角等于 $90°\times(m-n)$，则

$$G(s) = (s+a)(s-b)(s+c)\cdots$$

where a, b, c, $\cdots$ are positive values. If the time delay element is connected in series in the system, the numerator of the transfer function $G(s)$ has a positive root. It indicates that the time delay element causes the system to have a zero in the right half of the s-plane. So, the system is a non-minimum phase system. Other unstable first-order and second-order elements are listed in Table 4-2.

Example 4-5 The transfer functions of two systems are respectively, $G_1(s) = \dfrac{1+s}{1+2s}$ and $G_2(s) = \dfrac{1-s}{1+2s}$. Try to plot the polar plots of the two systems.

Solution: According to the definition, the system corresponding to $G_1(s)$ is a minimum phase system, and the system corresponding to $G_2(s)$ is a non-minimum phase system. The corresponding frequency characteristics are respectively

$$G_1(j\omega) = \frac{1+j\omega}{1+2j\omega}, \quad G_2(j\omega) = \frac{1-j\omega}{1+2j\omega}$$

Accordingly, the corresponding amplitude frequency characteristics and phase frequency characteristics are respectively

$$\begin{cases} A_1(\omega) = |G_1(j\omega)| = \sqrt{\dfrac{1+\omega^2}{1+4\omega^2}} \\ \varphi_1(\omega) = \angle G_1(j\omega) = \arctan\omega - \arctan 2\omega \end{cases}$$

$$\begin{cases} A_2(\omega) = |G_2(j\omega)| = \sqrt{\dfrac{1+\omega^2}{1+4\omega^2}} \\ \varphi_2(\omega) = \angle G_2(j\omega) = \arctan(-\omega) - \arctan 2\omega \end{cases}$$

When $\omega = 0$, we have $A_1(\omega) = 1$, $\varphi_1(\omega) = 0°$.

When $\omega = \infty$, we have $A_1(\omega) = 0.5$, $\varphi_1(\omega) = 0°$.

When $\omega = 0$, we have $A_2(\omega) = 1$, $\varphi_2(\omega) = 0°$.

When $\omega = \infty$, we have $A_2(\omega) = 0.5$, $\varphi_2(\omega) = -180°$.

The amplitude frequency characteristics of the two systems are identical, and the phase frequency characteristics are different. $G_1(s)$ has a smaller phase angle than $G_2(s)$. The polar plots of the two systems are shown in Fig. 4-28.

Example 4-6 The open-loop transfer function of a system is $G_K(s) = \dfrac{K(s+3)}{s(s-1)}$. Try to plot the polar plot.

Solution: According to the system open-loop transfer function, it can be observed that the system is a non-minimum phase type-I system, and its frequency characteristic is

$$G_K(j\omega) = \frac{K(j\omega+3)}{j\omega(j\omega-1)} = -\frac{4K}{1+\omega^2} + j\frac{3-\omega^2}{\omega(1+\omega^2)}$$

The real frequency characteristic and the virtual frequency characteristic are respectively

可确定其为最小相位系统，否则为非最小相位系统。

（2）常见非最小相位环节　非最小相位系统一般由两种情况产生，一是系统内包含非最小相位环节（如延迟因子），二是系统内环不稳定。传递延迟是一种非最小相位特性，高频时将造成严重的相位滞后。

可将延时环节 $G(s) = e^{-\tau s}$ 展开成幂级数，即

$$G(s) = e^{-\tau s} = 1 - \tau s + \frac{1}{2}\tau^2 s^2 - \frac{1}{3}\tau^3 s^3 + \cdots$$

因为上式中有些项的系数为负，故可写成

$$G(s) = (s+a)(s-b)(s+c)\cdots$$

式中，a，b，c，$\cdots$，均为正值。若延时环节串联在系统中，则传递函数 $G(s)$ 的分子有正根，表示延时环节使系统有零点位于复平面的右半平面，因此，该系统为非最小相位系统。其他不稳定的一阶环节和二阶环节见表 4-2。

Table 4-2　Phase angle change of unstable elements（不稳定环节的相角变化）

Transfer function(传递函数)	Phase angle $\omega = 0 \to \infty$（相角 $\omega = 0 \to \infty$）
$Ts - 1$	$-180° \to -270°$
$1 - Ts$	$0° \to -90°$
$\dfrac{1}{Ts - 1}$	$-180° \to -90°$
$\dfrac{1}{1 - Ts}$	$0° \to 90°$
$s^2/\omega_n^2 - 2\xi s/\omega_n + 1$	$0° \to -180°$
$\dfrac{1}{s^2/\omega_n^2 - 2\xi s/\omega_n + 1}$	$0° \to 180°$
$-K$	$-180°$

例 4-5　已知两个系统的传递函数分别为 $G_1(s) = \dfrac{1+s}{1+2s}$ 和 $G_2(s) = \dfrac{1-s}{1+2s}$，试绘制这两个系统的极坐标图。

解：由定义知 $G_1(s)$ 对应的系统为最小相位系统，$G_2(s)$ 对应的系统为非最小相位系统。其频率特性分别为

$$G_1(j\omega) = \dfrac{1+j\omega}{1+2j\omega}, \quad G_2(j\omega) = \dfrac{1-j\omega}{1+2j\omega}$$

则对应的幅频特性、相频特性分别为

$$\begin{cases} A_1(\omega) = |G_1(j\omega)| = \sqrt{\dfrac{1+\omega^2}{1+4\omega^2}} \\ \varphi_1(\omega) = \angle G_1(j\omega) = \arctan\omega - \arctan 2\omega \end{cases}$$

$$\begin{cases} A_2(\omega) = |G_2(j\omega)| = \sqrt{\dfrac{1+\omega^2}{1+4\omega^2}} \\ \varphi_2(\omega) = \angle G_2(j\omega) = \arctan(-\omega) - \arctan 2\omega \end{cases}$$

当 $\omega = 0$ 时，有 $A_1(\omega) = 1$，$\varphi_1(\omega) = 0°$。

$$U(\omega) = -\frac{4K}{1+\omega^2}, \quad V(\omega) = \frac{3-\omega^2}{\omega(1+\omega^2)}$$

The amplitude frequency characteristic and the phase frequency characteristic are respectively

$$|G_K(j\omega)| = \frac{\sqrt{9+\omega^2}}{\omega\sqrt{1+\omega^2}}, \quad \angle G_K(j\omega) = \arctan\frac{\omega}{3} - 90° - \arctan\frac{\omega}{-1}$$

When $\omega = 0$, we have $|G_K(j\omega)| = \infty$, $\angle G_K(j\omega) = -270°$, $U(\omega) = -4K$, $V(\omega) = +\infty$.

When $\omega = \infty$, we have $|G_K(j\omega)| = 0$, $\angle G_K(j\omega) = 270°$, $U(\omega) = 0$, $V(\omega) = 0$.

When $V(\omega) = 0$, we have $3 - 0.1\omega^2 = 0$. Then $\omega = \pm\sqrt{10}$, so $U(\pm\sqrt{3}) = \pm K$.

The polar plot is plotted as Fig. 4-29.

It can be seen that the open-loop polar plot of the system is parallel to the positive imaginary axis in the low frequency band, and approaches to the origin along the negative imaginary axis in the high frequency band.

4.3.2 The open-loop Bode plot of control systems

If the open-loop transfer function of the system is composed of several elements in series, the corresponding logarithmic amplitude frequency characteristics and phase frequency characteristics are respectively

$$\begin{cases} L(\omega) = 20\lg|G_K(j\omega)| = 20\lg|G_{K1}(j\omega)| + 20\lg|G_{K2}(j\omega)| + \cdots + 20\lg|G_{Kn}(j\omega)| \\ \quad\quad = L_1(\omega) + L_2(\omega) + \cdots + L_n(\omega) \\ \varphi(\omega) = \angle G_K(j\omega) = \angle G_{K1}(j\omega) + \angle G_{K2}(j\omega) + \cdots + \angle G_{Kn}(j\omega) \end{cases} \quad (4-38)$$

Therefore, if the logarithmic amplitude frequency and phase frequency characteristic curves of each element contained in $G_K(j\omega)$ are plotted, the Bode plot of the open-loop system can be obtained by adding the curves algebraically.

1. Basic steps for plotting the Bode plot of control systems

1) Write down frequency characteristic according to the transfer function of the control system, and convert the frequency characteristic into the form in which the frequency characteristics of several typical elements are multiplied.

2) Calculate the corner frequency of each typical element.

3) Draw asymptotic curves of the logarithmic amplitude frequency characteristic of each typical element.

4) Correct the asymptotic curves to obtain an accurate curve of the logarithmic amplitude frequency characteristic of each element according to the error correction value.

5) Superimpose the logarithmic amplitude frequency characteristic of each element to obtain the logarithmic amplitude frequency characteristic of the system.

6) Calculate the logarithmic phase frequency characteristic of each typical element, and then superimpose them to obtain the logarithmic phase frequency characteristic of the system.

The Bode plot of a linear time-invariant system has the following geometric characteristics.

1) There are certain corresponding relations between the asymptotic curves of the initial seg-

当 $\omega = \infty$ 时,有 $A_1(\omega) = 0.5$, $\varphi_1(\omega) = 0°$。
当 $\omega = 0$ 时,有 $A_2(\omega) = 1$, $\varphi_2(\omega) = 0°$。
当 $\omega = \infty$ 时,有 $A_2(\omega) = 0.5$, $\varphi_2(\omega) = -180°$。

两系统的幅频特性是相同的,相频特性是不同的,且 $G_1(s)$ 比 $G_2(s)$ 有更小的相位角。两系统的极坐标图如图 4-28。

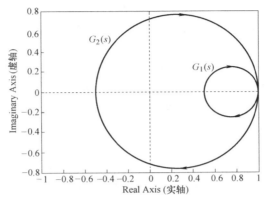

Fig. 4-28　Polar plot of example 4-5(例 4-5 的极坐标图)

例 4-6　某系统的开环传递函数为 $G_K(s) = \dfrac{K(s+3)}{s(s-1)}$,试绘制其极坐标图。

解：由系统开环传递函数可知此系统为非最小相位的 I 型系统,其频率特性为

$$G_K(j\omega) = \frac{K(j\omega+3)}{j\omega(j\omega-1)} = -\frac{4K}{1+\omega^2} + j\frac{3-\omega^2}{\omega(1+\omega^2)}$$

实频特性和虚频特性分别为

$$U(\omega) = -\frac{4K}{1+\omega^2}, \quad V(\omega) = \frac{3-\omega^2}{\omega(1+\omega^2)}$$

幅频特性和相频特性分别为

$$|G_K(j\omega)| = \frac{\sqrt{9+\omega^2}}{\omega\sqrt{1+\omega^2}}, \quad \angle G_K(j\omega) = \arctan\frac{\omega}{3} - 90° - \arctan\frac{\omega}{-1}$$

当 $\omega = 0$ 时, $|G_K(j\omega)| = \infty$, $\angle G_K(j\omega) = -270°$, $U(\omega) = -4K$, $V(\omega) = +\infty$。
当 $\omega = \infty$ 时, $|G_K(j\omega)| = 0$, $\angle G_K(j\omega) = 270°$, $U(\omega) = 0$, $V(\omega) = 0$。
当 $V(\omega) = 0$ 时,有 $3 - \omega^2 = 0$,解得 $\omega = \pm\sqrt{3}$,$U(\pm\sqrt{3}) = -K$。

绘制极坐标曲线,如图 4-29 所示。

由此可见,此系统开环极坐标图在低频段渐近线与正虚轴平行,在高频段沿负虚轴趋于原点。

4.3.2　控制系统的开环伯德图

如果系统的开环传递函数由若干环节串联而成,则对应的对数幅频特性和相频特性分别为

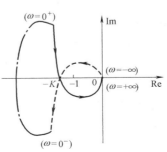

Fig. 4-29　Polar plot of example 4-6(例 4-6 的极坐标图)

ment of the logarithmic amplitude frequency characteristic plots and the system types. The system whose initial asymptotic curve is a horizontal line corresponds to type-0 system; the system whose initial segment asymptote slope is -20dB/dec corresponds to the type-I system; the system whose initial segment asymptotic slope is -40dB/dec corresponds to the type-II system.

2) The slope change of each asymptotic curve in the amplitude frequency characteristic plot depends on the typical element of the corresponding corner frequency. If it is an inertia element, the slope of the asymptotic curve drops by 20dB/dec after the corner frequency; if it is an oscillation element, the slope of the asymptotic curve drops by 40dB/dec after the corner frequency; if it is an first-order differential element, the slope of the asymptotic curve increases by 20dB/dec after the corner frequency; if it is an second-order differential element, the slope of the asymptotic curve increases by 40dB/dec after the corner frequency.

Example 4-7 The open-loop transfer function of a system is $G_K(s) = \dfrac{1000}{s(s+10)}$. Try to plot its Bode plot.

Solution: 1) The system is a type-I system, and the frequency characteristic of the system can be written as a form of product of frequency characteristics of several typical elements, i.e.

$$G_K(j\omega) = \frac{1000}{j\omega(j\omega+10)} = \frac{100}{j\omega(j0.1\omega+1)}$$

It can be seen that the open-loop transfer function of the system consists of a proportion element, an integral element and an inertia element.

2) Calculate the corner frequency of each typical element The corner frequency of the inertia element is $\omega_{T1} = 10$.

3) Draw asymptotic curves of the logarithmic amplitude frequency characteristics of each typical element.

4) Correct the asymptotic curves according to the error correction values, and draw the exact curve of the logarithmic amplitude frequency characteristic of each element (the step is omitted for this example).

5) Superimpose the logarithmic amplitude frequency characteristic of each element to obtain the logarithmic amplitude frequency characteristics of the system, as is shown in Fig. 4-30.

6) Draw the logarithmic phase frequency characteristic plot of each typical element, and then superimpose them to obtain the logarithmic phase frequency characteristic plot of the system, as is shown in Fig. 4-30.

Example 4-8 An open-loop transfer function of a feedback control system is $G_K(s) = \dfrac{10(1+0.1s)}{s(1+0.5s)}$. Try to plot the Bode plot of the open-loop system.

Solution: The open loop frequency characteristic of the system is

$$G_K(j\omega) = \frac{10\left(1+j\dfrac{\omega}{10}\right)}{j\omega\left(1+j\dfrac{\omega}{2}\right)}$$

$$\begin{cases} L(\omega) = 20\lg|G_K(j\omega)| = 20\lg|G_{K1}(j\omega)| + 20\lg|G_{K2}(j\omega)| + \cdots + 20\lg|G_{Kn}(j\omega)| \\ \qquad = L_1(\omega) + L_2(\omega) + \cdots + L_n(\omega) \\ \varphi(\omega) = \angle G_K(j\omega) = \angle G_{K1}(j\omega) + \angle G_{K2}(j\omega) + \cdots + \angle G_{Kn}(j\omega) \end{cases} \quad (4\text{-}38)$$

因此，若绘制出 $G_K(j\omega)$ 所包含各环节的对数幅频和相频特性曲线，然后对它们进行代数相加，最终就能求得开环系统的伯德图。

1. 绘制控制系统伯德图的基本步骤

1) 根据控制系统的传递函数写出其频率特性，并化为若干典型环节频率特性相乘的形式。
2) 确定各典型环节的转折频率。
3) 作出各典型环节的对数幅频特性的渐近线。
4) 根据误差修正值对渐近线进行修正，得出各环节的对数幅频特性的精确曲线。
5) 将各环节的对数幅频特性叠加，得到系统的对数幅频特性。
6) 计算各典型环节的对数相频特性，叠加得到系统的对数相频特性。

线性定常系统伯德图有如下几何特点。

1) 对数幅频特性图的起始段渐近线与系统类型之间存在一定的对应关系。起始段渐近线为水平线的系统对应于 0 型系统；起始段渐近线斜率为 -20dB/dec 的系统对应于 I 型系统；起始段渐近线斜率为 -40dB/dec 的系统对应于 II 型系统。

2) 对数幅频特性图各段渐近线的斜率变化取决于转折频率对应的典型环节类型。若为惯性环节，则在转折频率之后，渐近线斜率下降 20dB/dec；若为振荡环节，则在转折频率之后，渐近线斜率下降 40dB/dec；若为一阶微分环节，则在转折频率之后，渐近线斜率增加 20dB/dec；若为二阶微分环节，则在转折频率之后，渐近线斜率增加 40dB/dec。

例 4-7 某系统的开环传递函数为 $G_K(s) = \dfrac{1000}{s(s+10)}$，试绘制其伯德图。

解：1) 此系统为 I 型系统，系统的频率特性可以写成若干典型环节频率特性相乘的形式，即

$$G_K(j\omega) = \frac{1000}{j\omega(j\omega+10)} = \frac{100}{j\omega(j0.1\omega+1)}$$

由此可见，系统的开环传递函数是由一个比例环节、一个积分环节和一个惯性环节组成。

2) 确定各典型环节的转折频率。惯性环节的转折频率为 $\omega_{T1} = 10$。
3) 作出各典型环节的对数幅频特性的渐近线。
4) 根据误差修正值对渐近线进行修正，得出各环节的对数幅频特性的精确曲线（本例省略这一步）。
5) 将各环节的对数幅频特性叠加，得到系统的对数幅频特性，如图 4-30 所示。
6) 作出各典型环节的对数相频特性图，然后叠加得到系统的对数相频特性图，如图 4-30 所示。

例 4-8 某反馈控制系统的开环传递函数为 $G_K(s) = \dfrac{10(1+0.1s)}{s(1+0.5s)}$。试绘制该开环系统的伯德图。

It can be seen that the system consists of a proportion-integral element an inertia element and a first-order differential element. The logarithmic amplitude frequency characteristic and phase frequency characteristic of each element are analyzed as follows.

1) For the proportion-integral element, we have

$$L_1(\omega) = 20\lg \frac{10}{\omega} \quad (\omega > 0)$$

$$\varphi_1(\omega) = -90°$$

2) For the inertia element, we have

$$\omega_{T1} = \frac{1}{T_1} = 2$$

$$L_2(\omega) = -20\lg \sqrt{1+\left(\frac{\omega}{2}\right)^2} \approx \begin{cases} 0 & (0 < \omega \leq 2) \\ 20\lg \dfrac{2}{\omega} & (\omega > 2) \end{cases}$$

$$\varphi_2(\omega) = \arctan \frac{\omega}{2}$$

3) For the first-order differential element, we have

$$\omega_{T2} = \frac{1}{T_2} = 10$$

$$L_3(\omega) = 20\lg \sqrt{1+\left(\frac{\omega}{10}\right)^2} \approx \begin{cases} 0 & (0 < \omega \leq 10) \\ 20\lg \dfrac{\omega}{10} & (\omega > 10) \end{cases}$$

$$\varphi_3(\omega) = \arctan \frac{\omega}{10}$$

Thus, the total logarithmic amplitude frequency characteristic and phase frequency characteristic of the system are

$$L(\omega) = L_1(\omega) + L_2(\omega) + L_3(\omega) \approx \begin{cases} 20\lg \dfrac{10}{\omega} & (0 < \omega \leq 2) \\ 20\lg \dfrac{10 \times 2}{\omega \times \omega} & (2 < \omega \leq 10) \\ 20\lg \dfrac{10 \times 2 \times \omega}{\omega \times \omega \times 10} & (\omega > 10) \end{cases}$$

$$\varphi(\omega) = \varphi_1(\omega) + \varphi_2(\omega) + \varphi_3(\omega) = -90° - \arctan \frac{\omega}{2} + \arctan \frac{\omega}{10}$$

Finally, the logarithmic frequency characteristic of the system is plotted as Fig. 4-31.

2. Method to figure out open-loop transfer functions in terms of Bode plots of control systems

When an open-loop transfer function has no poles and zeros located in the right half of the s-plane, there is a one-to-one correspondence (Bode theorem) between the amplitude frequency characteristic and the phase frequency characteristic of the transfer function. Then, the open-loop transfer function of the system can be obtained according to the open-loop logarithmic amplitude fre-

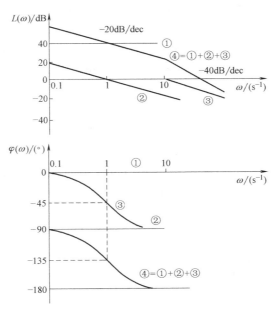

Fig. 4-30　Bode plot of example 4-7
（例 4-7 的伯德图）

解：系统的开环频率特性为

$$G_K(j\omega) = \frac{10\left(1+j\dfrac{\omega}{10}\right)}{j\omega\left(1+j\dfrac{\omega}{2}\right)}$$

可知该系统由比例积分环节、惯性环节以及一阶微分环节组成。各环节的对数幅频特性和相频特性分析如下。

1）对于比例积分环节，有

$$L_1(\omega) = 20\lg\frac{10}{\omega} \quad (\omega>0)$$

$$\varphi_1(\omega) = -90°$$

2）对于惯性环节，有

$$\omega_{T1} = \frac{1}{T_1} = 2$$

$$L_2(\omega) = -20\lg\sqrt{1+\left(\frac{\omega}{2}\right)^2} \approx \begin{cases} 0 & (0<\omega\leqslant 2) \\ 20\lg\dfrac{2}{\omega} & (\omega>2) \end{cases}$$

$$\varphi_2(\omega) = -\arctan\frac{\omega}{2}$$

3）对于一阶微分环节，有

$$\omega_{T2} = \frac{1}{T_2} = 10$$

quency characteristic plot of the system. The solving method is contrary to the above plotting process. Firstly, according to the change of curve slope at each corner frequency in the system open-loop logarithmic frequency characteristic plot, figure out whether there are inertia element, oscillation element, first-order differential element or second-order differential element. Then, determine the number of integral elements and differential elements from the slope of the low frequency band. Finally, determine the open loop gain K according to the height of the vertical axis at $\omega = 1$ in the starting segment (or its extension).

In addition, the magnitude of the open loop gain K can also be determined by the method shown in Fig. 4-32. As is shown in Fig. 4-32a, the initial segment slope is -20dB/dec, so the corresponding system is a Type-I system ($v=1$). Since the frequency at which the initial segment (or its extension) intersects with the 0dB line satisfies $20\lg\frac{K}{\omega} = 0\text{dB}$, it is equal to the value of K. As is shown in Fig. 4-32b, the initial segment slope is -40dB/dec, so the corresponding system is a Type-II system ($v=2$). Since the frequency at which the start segment (or its extension) intersects with the 0dB line satisfies $20\lg\frac{K}{\omega^2} = 0\text{dB}$, it is equal to the value of $\sqrt{K}$.

Example 4-9 A system open-loop logarithmic amplitude frequency characteristic plot is shown in Fig. 4-33. Figure out the open-loop transfer function $G_K(s)$ of the system.

Solution: According to Fig. 4-33, the slope change of the asymptotic curves of the system open-loop amplitude logarithmic frequency characteristic plot is $-20 \to -60 \to -40(\text{dB/dec})$, which indicates that the system contains at least three elements: an integral element, an oscillation element and a first-order differential element. Since the extension of the initial segment of the asymptotic curve does not pass the point of (1, 0), the system should also contain a proportion element. Therefore, the open-loop transfer function $G_K(s)$ of the system can be expressed as

$$G_K(s) = \frac{K\omega_n^2(1+Ts)}{s(s^2+2\xi\omega_n s+\omega_n^2)}$$

1) Calculate the value of K. The initial segment of the logarithmic amplitude frequency characteristic plot shows a proportional-integral element. When $\omega = 1$, the amplitude is $20\lg K$. According to $20\lg K = 15.6\text{dB}$, the value is obtained as $K = 6$.

2) Calculate the value of ω_n. In Fig. 4-33, the corner frequency from the integral element to the oscillation element is $\omega_{T1} = 0.2$. Thus, the natural frequency of the oscillation element is $\omega_n = \omega_{T1} = 0.2$.

3) Calculate the value of ξ. When $\omega = \omega_n$, the peak value of the oscillation element (i.e. the maximum error) is 10dB. We have $20\lg\frac{1}{2\xi} = 10$, $\xi = 0.158$.

4) Calculate the value of T. In Fig. 4-33, the corner frequency from the oscillation element to the first-order differential element is $\omega_{T2} = 4$, Thus the time constant of the differential element is $T = 0.25$.

$$L_3(\omega) = 20\lg\sqrt{1+\left(\frac{\omega}{10}\right)^2} \approx \begin{cases} 0 & (0<\omega\leq 10) \\ 20\lg\dfrac{\omega}{10} & (\omega>10) \end{cases}$$

$$\varphi_3(\omega) = \arctan\frac{\omega}{10}$$

因此，系统的对数幅频特性和相频特性分别为

$$L(\omega) = L_1(\omega)+L_2(\omega)+L_3(\omega) \approx \begin{cases} 20\lg\dfrac{10}{\omega} & (0<\omega\leq 2) \\ 20\lg\dfrac{10\times 2}{\omega\times\omega} & (2<\omega\leq 10) \\ 20\lg\dfrac{10\times 2\times\omega}{\omega\times\omega\times 10} & (\omega>10) \end{cases}$$

$$\varphi(\omega) = \varphi_1(\omega)+\varphi_2(\omega)+\varphi_3(\omega) = -90°-\arctan\frac{\omega}{2}+\arctan\frac{\omega}{10}$$

最后，绘制出系统的伯德图，如图 4-31 所示。

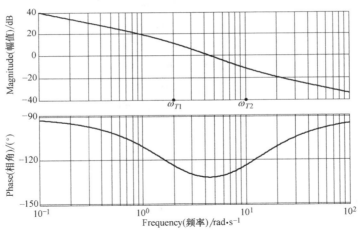

Fig. 4-31　Bode plot of example 4-8

（例 4-8 的伯德图）

2. 基于控制系统伯德图求系统开环传递函数的方法

当开环传递函数在复平面的右半面上无极点和零点时，传递函数的幅频特性和相频特性有一一对应的关系（伯德定理）。此时可以根据系统的开环对数幅频特性图求出系统的开环传递函数。求解方法与上述作图过程相反，首先根据系统开环对数幅频特性图上每个转折频率处曲线斜率的变化来确定惯性环节、振荡环节、一阶微分环节和二阶微分环节等；然后由低频段的斜率确定积分环节、微分环节的个数；最后由起始段（或其延长线）在 $\omega=1$ 处的纵坐标高度确定开环增益 K。

此外，开环增益 K 的大小还可以由图 4-32 所示的方法确定。图 4-32a 所示曲线起始段斜率为 -20dB/dec，故对应系统为 Ⅰ 型系统（$v=1$），起始段（或其延长线）与 0dB 线相交处存在 $20\lg\dfrac{K}{\omega}=0\text{dB}$ 相交处频率等于 K 值。对于 Ⅱ 型的情况下，图 4-32b 所示曲线起始段斜

Then, the above-mentioned parameter values are substituted into the expression of $G_K(s)$, we have

$$G_K(s) = \frac{0.06s + 0.24}{s(s^2 + 0.0632s + 0.04)}$$

4.4 Closed-loop Frequency Characteristics of Control Systems

4.4.1 Closed-loop frequency characteristics of control systems

For the control system with a unit feedback, the relation between the closed-loop transfer function and the open-loop transfer function is

$$\Phi(s) = \frac{G_K(s)}{1 + G_K(s)} \tag{4-39}$$

Substituting $s = j\omega$ into the above expression, we have

$$\Phi(j\omega) = \frac{G_K(j\omega)}{1 + G_K(j\omega)} \tag{4-40}$$

Since $\Phi(j\omega)$ and $G_K(j\omega)$ are both complex functions with respect to s, the amplitude frequency characteristic and phase frequency characteristic can be written as

$$|\Phi(j\omega)| = \frac{|G_K(j\omega)|}{|1 + G_K(j\omega)|} \tag{4-41}$$

$$\angle \Phi(j\omega) = \angle G_K(j\omega) - \angle [1 + G_K(j\omega)] \tag{4-42}$$

Therefore, when the open-loop frequency characteristic $G_K(j\omega)$ is known, the closed-loop frequency characteristic plot can be plotted point by point according to equation (4-41) and equation (4-42).

For the closed-loop control system with a non-unit feedback, the closed-loop frequency characteristic can be written as

$$\Phi(j\omega) = \frac{G(j\omega)}{1 + G(j\omega)H(j\omega)} = \frac{1}{H(j\omega)} \frac{G(j\omega)H(j\omega)}{1 + G(j\omega)H(j\omega)} \tag{4-43}$$

Obviously, the latter term on the right side of equation (4-43) is the frequency characteristic of the unit feedback system, and the transfer function of the forward channel is $G(s)H(s)$. After obtaining the term, it is multiplied by $1/H(j\omega)$, then we have $\Phi(j\omega)$. Therefore, we could consider replacing $G_K(j\omega)$ in the unit feedback system with $G(j\omega)H(j\omega)$ to study the non-unit feedback control system.

4.4.2 Estimation of closed-loop frequency characteristics using open-loop frequency characteristics

A general control system has a logarithmic amplitude frequency characteristic as is shown in Fig. 4-34. Its main characteristics can be described as: ①The amplitude of the open-loop frequency characteristic always decreases as the frequency increases; ②There is only one intersection point

率为-40dB/dec，故对应系统为Ⅱ型系统（$v=2$），起始段（或其延长线）与0dB线相交处存在 $20\lg\dfrac{K}{\omega^2}=0\mathrm{dB}$，相交处频率等于 $\sqrt{K}$ 值。

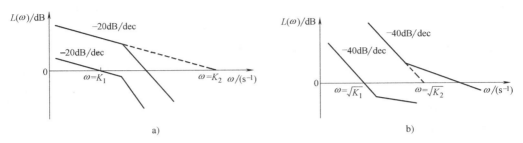

Fig. 4-32　The open-loop gain K determined by the open-loop logarithmic amplitude frequency characteristic plot
（由开环对数幅频特性图确定开环增益 K）

例 4-9　已知系统开环对数幅频特性图如图 4-33 所示，求系统的开环传递函数 $G_\mathrm{K}(s)$。

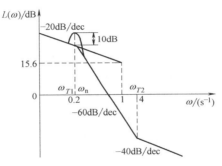

Fig. 4-33　System logarithmic amplitude frequency characteristic plot of example 4-9
（例 4-9 系统开环对数幅频特性图）

解：由图 4-33 可知，系统开环对数幅频特性图渐近线的斜率变化依次是 $-20\to-60\to-40(\mathrm{dB/dec})$，说明此系统至少包含三个环节：积分环节、振荡环节和一阶微分环节。又因渐近线起始段的延长线不通过（1，0）点，故系统还应包含比例环节。因此，系统的开环传递函数 $G_\mathrm{K}(s)$ 可表达为

$$G_\mathrm{K}(s)=\dfrac{K\omega_\mathrm{n}^2(1+Ts)}{s(s^2+2\xi\omega_\mathrm{n}s+\omega_\mathrm{n}^2)}$$

1）求 K。对数幅频特性图的起始段是比例积分环节，当 $\omega=1$ 时，幅值即为 $20\lg K$，根据 $20\lg K=15.6\mathrm{dB}$，可求得 K 值，即 $K=6$。

2）求 ω_n。在图 4-33 中，由积分环节到振荡环节的转折频率为 $\omega_{T1}=0.2$，此即振荡环节的固有频率，$\omega_\mathrm{n}=\omega_{T1}=0.2$。

3）求 ξ。当 $\omega=\omega_\mathrm{n}$ 时，振荡环节的峰值（即最大误差）为 10dB，即 $20\lg\dfrac{1}{2\xi}=10$，解得 $\xi=0.158$。

4）求 T。在图 4-33 中，由振荡环节到一阶微分环节的转折频率为 $\omega_{T2}=4$，所以微分环

between the open-loop amplitude frequency characteristic curve $L(\omega)$ and the 0dB line. The frequency of the intersection $[L(\omega)=0$ or $|G_K(j\omega)|=1]$ is called the open-loop amplitude crossing frequency or cutoff frequency ω_c.

According to the equation (4-43), a closed-loop frequency characteristic plot can be plotted. However, in the practical application of control engineering, the whole working frequency of the control system is divided into three frequency bands: the low frequency band, the medium frequency band and the high frequency band according to the open-loop cutoff frequency ω_c of the open-loop frequency characteristic, and then the closed-loop frequency characteristic is estimated according to the open-loop frequency characteristics.

1. Low frequency band ($\omega \ll \omega_c$)

The low frequency band refers to the frequency range of the open-loop logarithmic amplitude frequency characteristic before the first open-loop cutoff frequency. According to $|G_K(j\omega)| \gg 1$, we have

$$\Phi(j\omega) = \frac{G_K(j\omega)}{1+G_K(j\omega)} \approx 1$$

It can be seen that the amplitude frequency characteristic of the closed-loop system is approximate to 1 in the low frequency band. It means that the system can transmit the signal from input to output without attenuation for the low frequency signal input.

2. Medium frequency band

The medium band refers to the frequency range of the open-loop logarithmic amplitude frequency characteristic near the cut off frequency. At the cutoff frequency ω_c, we have $|G_K(j\omega_c)|=1$. As is shown in Fig. 4-35, the frequency of point A is the cutoff frequency. Thus, we have $|\overrightarrow{OA}|=|\overrightarrow{OP}|=1$. Then $\triangle OPA$ in Fig. 4-35 is an isosceles triangle, and thus the corresponding closed-loop amplitude frequency characteristic is

$$|G_B(j\omega_c)| = \frac{|\overrightarrow{OA}|}{|\overrightarrow{PA}|} = \frac{1}{|\overrightarrow{PA}|} = \frac{1}{2}\arcsin\frac{\angle AOP}{2}$$

After calculating $|G_B(j\omega_c)|$, according to the amplitude frequency characteristic curves of the low frequency band and the high frequency band regarding $|G_B(j\omega_c)|$, the amplitude frequency characteristic curve of the closed-loop system can be roughly plotted. In the medium frequency band, the amplitude frequency characteristic curve of the closed-loop system is quite different from the open-loop amplitude frequency characteristic curve.

3. High frequency band ($\omega \gg \omega_c$)

The high frequency band refers to the frequency range of the open-loop logarithmic amplitude frequency characteristic after the medium frequency band. According to $|G_K(j\omega)| \gg 1$, we have

$$\Phi(j\omega) = \frac{G_K(j\omega)}{1+G_K(j\omega)} \approx G_K(j\omega)$$

It can be seen that the amplitude frequency characteristic curve of the closed-loop system almost coincides with the open-loop amplitude frequency characteristic curve in the high frequency band,

节的时间常数 $T=0.25$。

将上述求得的各参数值代入到 $G_K(s)$ 的表达式，得

$$G_K(s) = \frac{0.06s+0.24}{s(s^2+0.0632s+0.04)}$$

4.4 控制系统的闭环频率特性

4.4.1 控制系统的闭环频率特性

具有单位反馈的控制系统，其闭环传递函数 $\Phi(s)$ 与开环传递函数 $G_K(s)$ 之间的关系为

$$\Phi(s) = \frac{G_K(s)}{1+G_K(s)} \tag{4-39}$$

将 $s=j\omega$ 代入上式中，则有

$$\Phi(j\omega) = \frac{G_K(j\omega)}{1+G_K(j\omega)} \tag{4-40}$$

由于 $\Phi(j\omega)$ 和 $G_K(j\omega)$ 均是 ω 的复变函数，因此 $\Phi(j\omega)$ 的幅频特性和相频特性可分别写为

$$|\Phi(j\omega)| = \frac{|G_K(j\omega)|}{|1+G_K(j\omega)|} \tag{4-41}$$

$$\angle \Phi(j\omega) = \angle G_K(j\omega) - \angle [1+G_K(j\omega)] \tag{4-42}$$

因此，当开环频率特性 $G_K(j\omega)$ 已知时，可根据式（4-41）和式（4-42）逐点绘制闭环频率特性图。

对于非单位反馈的闭环控制系统，其闭环频率特性为

$$\Phi(j\omega) = \frac{G(j\omega)}{1+G(j\omega)H(j\omega)} = \frac{1}{H(j\omega)} \frac{G(j\omega)H(j\omega)}{1+G(j\omega)H(j\omega)} \tag{4-43}$$

显然，式（4-43）右边的后一项是单位反馈系统的频率特性，其前向通道的传递函数为 $G(s)H(s)$，故求得此项后乘以 $1/H(j\omega)$ 即得 $\Phi(j\omega)$。因此，可以考虑将单位反馈系统的 $G_K(j\omega)$ 用 $G(j\omega)H(j\omega)$ 来替换，以便研究非单位反馈的控制系统。

4.4.2 利用开环频率特性估计闭环频率特性

一般的控制系统都具有如图 4-34 所示的对数幅频特性。它的主要特点可描述为：①开环频率特性的幅值总是随着频率的增加而降低；②开环幅频特性曲线与 0dB 线只有一个交点，将该交点处的频率 $[L(\omega)=0$ 或 $|G_K(j\omega)|=1]$ 称为开环幅值穿越频率或截止频率，用 ω_c 表示。

根据式（4-43），可以画出闭环频率特性图。但在控制工程的实际应用中，常常依据开环频率特性的开环截止频率 ω_c 的大小，将控制系统的整个工作频率分为低频、中频、高频三个频段，然后根据开环频率特性估计闭环频率特性。

which means $|\Phi(j\omega)|\ll 1$. It indicates that the system greatly attenuates the input signal with high frequency. Therefore, the amplitude of the system open-loop logarithmic amplitude frequency characteristic in the high frequency band directly reflects the ability to suppress the high frequency disturbance signal at the input end.

It should be noted that the division of the above three frequency bands is not absolute. The frequency range of 15dB to −10dB is usually considered as a medium frequency band in the open-loop amplitude frequency characteristic of the control system.

4.4.3 Closed-loop frequency-domain performance indicators of control systems

1. The closed-loop frequency domain performance indicators

Fig. 4-36 shows the closed-loop amplitude frequency characteristic of a closed-loop system. It can be seen that the amplitude-frequency characteristics change with frequency ω. These characteristics can be summarized by the characteristic valve, which constitutes the closed-loop frequency-domain performance indicators.

(1) Zero frequency amplitude $A(0)$ The zero frequency amplitude $A(0)$ represents the ratio of the output amplitude to the input magnitude when the frequency is near zero for the closed-loop system. At frequency $\omega \to 0$, if $A(0)=1$, then the output amplitude can fully accurately reflect the input amplitude. The closer $A(0)$ is to 1, the smaller the system steady-state error is.

(2) Recurrence frequency ω_M and recurrence frequency bandwidth $0 \sim \omega_M$ If Δ is specified in advance as an allowable error of system's response to the low-frequency input signal, then ω_M is the value that the difference between the amplitude of the amplitude frequency characteristic and the amplitude of the zero frequency reaches Δ for the first time. When $\omega > \omega_M$, the output cannot accurately "reproduce" the input. Thus the frequency range $0 \sim \omega_M$ is called the recurrence frequency bandwidth. If the value of ω_M is determined according to Δ, then the larger the value of ω_M is, the wider the frequency band in which the system can reproduce the input signal with a specified accuracy. Conversely, if the allowable error Δ is determined according to ω_M, the smaller the allowable error Δ is, the higher the accuracy of the system's response to the low frequency input signal is.

(3) Resonant frequency ω_r and resonant frequency peak M_r The frequency at which the amplitude frequency characteristic $A(\omega)$ shows the maximum value A_{max} is called the resonant frequency ω_r. The ratio $A_{max}/A(0)$ between the amplitude at $\omega=\omega_r$ and the zero frequency amplitude $A(0)$ at $\omega=0$ is called the resonant frequency peak M_r. M_r reflects the speed and relative stability of the system transient response. The larger the value is, the more severe the oscillation of the closed-loop system is and the worse the stability is.

(4) Cutoff frequency ω_b and cutoff frequency bandwidth $0 \sim \omega_b$ The cutoff frequency ω_b refers to the frequency that the amplitude of the closed-loop frequency characteristic of the system drops from $A(0)$ to $0.707A(0)$, that is, the frequency at which it drops by 3dB. The frequency range of $0 \sim \omega_b$ is called the cutoff frequency bandwidth of the system. The larger the value of ω_b is, the faster the closed-loop system responds to the input is, i.e. the shorter the transition time of the

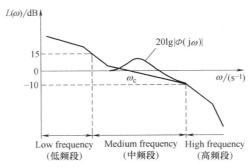

Fig. 4-34　Open-loop logarithmic amplitude frequency characteristic plot

(开环对数幅频特性图)

1. 低频段 ($\omega \ll \omega_c$)

低频段是指开环对数幅频特性在第一个开环截止频率以前的频率区段。根据 $|G_K(j\omega)| \gg 1$，有

$$\Phi(j\omega) = \frac{G_K(j\omega)}{1+G_K(j\omega)} \approx 1$$

可见在低频段内，闭环系统的幅频特性近似为 1。这表明对于低频信号输入，系统能无衰减地从输入传递信号到输出。

2. 中频段

中频段是指开环对数幅频特性在截止频率附近的频率区段。在截止频率 ω_c 处，有 $|G_K(j\omega_c)| = 1$。如图 4-35 所示，A 点的频率为截止频率 ω_c，则 $|\overrightarrow{OA}| = |\overrightarrow{OP}| = 1$，所以图 4-35 中 $\triangle OPA$ 是等腰三角形，从而对应的闭环幅频特性值为

$$|G_B(j\omega_c)| = \frac{|\overrightarrow{OA}|}{|\overrightarrow{PA}|} = \frac{1}{|\overrightarrow{PA}|} = \frac{1}{2}\arcsin\frac{\angle AOP}{2}$$

求出 $|G_B(j\omega_c)|$ 后，根据 $G_B(j\omega)$ 在低频段和高频段的幅频特性曲线，就可大致画出闭环系统的幅频特性曲线。在中频段内，闭环系统的幅频特性曲线与开环幅频特性曲线相差悬殊。

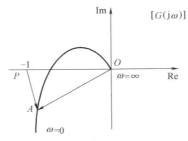

Fig. 4-35　Polar plot in the medium frequency band

(中频段的极坐标图)

3. 高频段 ($\omega \gg \omega_c$)

高频段是指开环对数幅频特性在中频段以后的频率区段。根据 $|G_K(j\omega)| \ll 1$，有

transient process is. It reflects the speed of the closed loop system response.

2. The closed-loop frequency domain performance analysis for typical systems

According to the frequency domain performance indicators of the closed-loop system described above, the performance analysis of the first-order system and the second-order system is as follows.

(1) First-order system For the first-order system with a unit feedback, the closed-loop frequency characteristic is

$$\Phi(j\omega) = \frac{1}{j\omega T + 1}$$

Let $|\Phi(j\omega_b)| = \frac{\sqrt{2}}{2}|\Phi(j0)| = \frac{\sqrt{2}}{2}$, we have

$$\frac{1}{\sqrt{1+T^2\omega_b^2}} = \frac{\sqrt{2}}{2}$$

The cutoff frequency of the first-order system is $\omega_b = \frac{1}{T}$.

It can be seen that the cutoff frequency is inversely proportional to the time constant of the first-order system. The wider the frequency band is, the smaller the inertia is, and the faster the response is.

(2) Second-order system The closed-loop transfer function of the second-order system is

$$\Phi(s) = \frac{\omega_n^2}{s^2 + 2\xi\omega_n s + \omega_n^2}$$

The closed-loop amplitude frequency characteristic is

$$A(\omega) = |\Phi(j\omega)| = \frac{1}{\sqrt{\left(1 - \frac{\omega^2}{\omega_n^2}\right)^2 + \left(2\xi\frac{\omega}{\omega_n}\right)^2}}$$

Let $\frac{dA(\omega)}{d\omega} = 0$, we have the resonant frequency ω_r and resonant frequency peak M_r of the system at $0 \leq \xi < 0.707$ as

$$\omega_r = \omega_n\sqrt{1 - 2\xi^2} \tag{4-44}$$

$$M_r = \frac{1}{2\xi\sqrt{1-\xi^2}} \tag{4-45}$$

It can be seen that when the damping ratio ξ is constant, the resonant frequency ω_r is proportional to the undamped natural frequency ω_n of the system. It indicates that the larger the value of ω_r is, the larger the value of ω_n is, and the faster the system response speed is. The resonant frequency peak M_r decreases as the damping ratio ξ increases. Its physical meaning is that when the closed-loop amplitude frequency characteristic has a resonant peak, the harmonic component of the input signal spectrum near $\omega = \omega_r$ is significantly enhanced after passing through the system, thus causing oscillation.

$$\Phi(j\omega) = \frac{G_K(j\omega)}{1+G_K(j\omega)} \approx G_K(j\omega)$$

在高频段内，闭环系统的幅频特性曲线几乎与开环幅频特性曲线重合，也就是说 $|\Phi(j\omega)| \ll 1$，这表明系统会极大衰减高频输入信号。因此，系统开环对数幅频特性在高频段的幅值直接反映对输入端高频干扰信号的抑制能力。

应当指出，以上三个频段的划分不是绝对的，通常将控制系统的开环幅频特性在 15dB 至 −10dB 的一段频率范围看作为中频段。

4.4.3 控制系统的闭环频域性能指标

1. 闭环频域性能指标

图 4-36 为闭环系统的闭环幅频特性图。可看出幅频特性随着频率变化的特征，这些特征可用特征量加以概述，也就构成闭环频域性能指标。

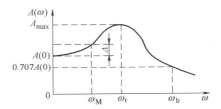

Fig. 4-36　The characteristic quantity of the closed-loop frequency characteristics
（闭环频率特性特征量）

（1）零频幅值 $A(0)$　零频幅值 $A(0)$ 表示当频率接近零时，闭环系统输出幅值与输入幅值之比。在 $\omega \to 0$ 时，若 $A(0)=1$，则输出幅值能完全准确地反映输入幅值。$A(0)$ 越接近于 1，系统稳态误差越小。

（2）复现频率 ω_M 与复现带宽 $0 \sim \omega_M$　若事先规定一个 Δ 作为系统对低频输入信号响应的允许误差，那么 ω_M 就是幅频特性与零频幅值之差第一次达到 Δ 时的频率值。当 $\omega > \omega_M$ 时，输出就不能准确"复现"输入。所以 $0 \sim \omega_M$ 这一频率范围称为复现带宽。由 Δ 确定的 ω_M 越大，表明系统能以规定精度复现输入信号的频带越宽。反之，由 ω_M 确定的允许误差 Δ 越小，说明系统对低频输入信号响应的精度越高。

（3）谐振频率 ω_r 及谐振频率峰值 M_r　幅频特性 $A(\omega)$ 出现最大值 A_{max} 时的频率称为谐振频率 ω_r。$\omega = \omega_r$ 时的幅值 $A(\omega_r) = A_{max}$ 与 $\omega = 0$ 时的零频幅值 $A(0)$ 之比 $[A_{max}/A(0)]$ 称为谐振频率峰值 M_r。M_r 反映系统瞬态响应的速度和相对稳定性。其值越大，则闭环系统振荡越严重，因而稳定性就越差。

（4）截止频率 ω_b 及截止频率带宽 $0 \sim \omega_b$　截止频率 ω_b 是指系统闭环频率特性的幅值由 $A(0)$ 下降到 $0.707A(0)$，即下降 3dB 时的频率。$0 \sim \omega_b$ 的频率范围称为系统的截止带宽。ω_b 值越大，闭环系统对输入的响应就越快，即瞬态过程的过渡过程时间越短。它反映闭环系统响应的快慢。

2. 典型系统闭环频域性能分析

下面根据上述闭环系统频域指标对一阶系统和二阶系统进行性能分析。

（1）一阶系统　对于具有单位反馈的一阶系统，其闭环频率特性为

Let $A(\omega) = 0.707$, the cutoff frequency ω_b of the second-order system can be obtained as

$$\omega_b = \omega_n \sqrt{1 - 2\xi^2 + \sqrt{2 - 4\xi^2 + 4\xi^4}}$$

It can be seen that when the damping ratio ξ is determined, the cutoff frequency ω_b is proportional to the undamped natural frequency ω_n, that is, the larger the ω_b is, the wider the bandwidth is, and the faster the response speed of the system is. However, if the bandwidth is too large, the system's ability to resist the high frequency disturbance will decrease.

$$\Phi(j\omega) = \frac{1}{j\omega T + 1}$$

令 $|\Phi(j\omega_b)| = \frac{\sqrt{2}}{2}|\Phi(j0)| = \frac{\sqrt{2}}{2}$，即

$$\frac{1}{\sqrt{1+T^2\omega_b^2}} = \frac{\sqrt{2}}{2}$$

解得一阶系统的截止频率 $\omega_b = \frac{1}{T}$。

可见，截止频率与一阶系统的时间常数成反比，既频带越宽，惯性越小，响应的快速性越好。

（2）二阶系统　二阶系统的闭环传递函数为

$$\Phi(s) = \frac{\omega_n^2}{s^2 + 2\xi\omega_n s + \omega_n^2}$$

闭环幅频特性为

$$A(\omega) = |\Phi(j\omega)| = \frac{1}{\sqrt{\left(1-\frac{\omega^2}{\omega_n^2}\right)^2 + \left(2\xi\frac{\omega}{\omega_n}\right)^2}}$$

令 $\frac{dA(\omega)}{d\omega} = 0$，可得当 $0 \leq \xi < 0.707$ 时，系统的谐振频率 ω_r 及谐振峰值 M_r 分别为

$$\omega_r = \omega_n\sqrt{1-2\xi^2} \tag{4-44}$$

$$M_r = \frac{1}{2\xi\sqrt{1-\xi^2}} \tag{4-45}$$

由此可见，当阻尼比 ξ 为常数时，谐振频率 ω_r 与系统的无阻尼固有频率 ω_n 成正比，表示 ω_r 值越大，ω_n 值也越大，系统响应速度越快。谐振峰值 M_r 随着阻尼比 ξ 的增大而减小。其物理意义在于：当闭环幅频特性有谐振峰值时，系统输入信号的频谱在 $\omega = \omega_r$ 附近的谐波分量通过系统后显著增强，从而引起振荡。

又令 $A(\omega) = 0.707$，可求得二阶系统的截止频率 ω_b 为

$$\omega_b = \omega_n\sqrt{1-2\xi^2 + \sqrt{2-4\xi^2+4\xi^4}}$$

由此可见，当阻尼比 ξ 确定后，截止频率 ω_b 与无阻尼固有频率 ω_n 成正比，即 ω_b 越大，带宽越宽，系统的响应速度越快。但带宽过大，系统抗高频干扰的能力就会下降。

Chapter 5   Performance Analysis of Control Systems

第5章   控制系统的性能分析

The performance of a control system mainly includes transient performance and steady-state performance, chapter 3 mainly analyzes the transient performance from the perspective of time domain, and chapter 4 introduces this frequency-domain analysis methods. This chapter will investigate the stability, steady-state error and rapidity of the system by using time-domain analysis and frequency-domain analysis methods.

5.1 Stability Analysis of Control Systems

5.1.1 Introduction

Stability is an important performance of a system, and is also a prerequisite for the system to work normally. Any system in actual work condition will be disturbed by internal and external factors, such as fluctuations of energy, changes of load, changes of environmental conditions, etc. If the system is unstable, it will deviate from the equilibrium state under the disturbance, and it will be divergent theoretically. Therefore, one of the basic tasks of this book is to study the stability of the system and propose measures to ensure the stability of the system.

1. Concept of stability

Assuming that a linear time-invariant system is in a certain equilibrium state firstly, and it deviates from the original equilibrium state under the influence of disturbance then, can the system return to the original equilibrium state when the disturbance disappears? In fact, it is the stability issue of the control system.

If the system can return to the original equilibrium state after the disturbance disappears, i.e. the zero input response of the system is convergent, the system is stable; on the contrary, if the system cannot return to the original equilibrium state or the zero input response of the system is divergent, the system is unstable.

As is shown in Fig. 5-1a, if the air damping and friction are ignored and an input $\delta(t)$ (unit impulse signal) is given in the equilibrium state, the mass m oscillates back and forth in a position centered on the equilibrium state. Thus, the system is not stable, the response curve is shown in Fig. 5-2a.

If a damper is added to the system, as is shown in Fig. 5-1b, the input $\delta(t)$ is given in the equilibrium state, the mass m attenuates and oscillates back and forth in a position centered on the equilibrium state, and finally returns to the initial equilibrium state. The system is stable and the response curve is as is shown in Fig. 5-2b or Fig. 5-2c. As is shown in Fig. 5-2b, although the system is stable, the transition process takes a long time and there is a large overshoot, thus the stability is not good enough. As is shown in Fig. 5-2c, the system is stable, meanwhile its transition time is short and its overshoot is also small, thus the stability is better.

Assuming that a simple pendulum shown in Fig. 5-1c is subjected to an external disturbance, whether it is in position A' or A'', it will eventually return to position A after several oscillations, when the external disturbance disappears. Thus it is a stable system. The ball shown in Fig. 5-1d will

第5章 控制系统的性能分析

控制系统的性能主要包括瞬态性能和稳态性能,第3章主要从时域的角度分析了瞬态性能,第4章介绍了频域分析法。本章将运用时域分析与频域分析方法研究系统的稳定性、稳态误差及快速性等。

5.1 控制系统的稳定性分析

5.1.1 概述

稳定性是系统的重要性能,也是系统能够正常工作的首要条件。任何系统在实际工作中都会受到内部和外部因素的扰动,如能源的波动、负载的变化、环境条件的改变等。系统若不稳定,则在扰动作用下将会偏离平衡状态,理论上呈发散状态。因此,本书的基本任务之一是研究系统的稳定性并提出保证系统稳定的措施。

1. 稳定的概念

设线性定常系统处于某一平衡状态,此系统在干扰作用下偏离了平衡状态,那么在干扰作用消失后,系统能否回到原来的平衡状态?事实上,这就是控制系统的稳定性问题。

如果系统在扰动作用消失后,能够恢复到原平衡状态,即系统的零输入响应是收敛的,则系统为稳定的;相反,若系统不能恢复到原平衡状态或系统的零输入响应是发散的,则系统为不稳定的。

对图 5-1a 所示系统,如果忽略空气阻尼和摩擦,在平衡状态下给定输入 $\delta(t)$(单位脉冲信号),则质量块 m 在以平衡状态为中心的位置来回等幅振荡,系统不稳定,其响应曲线如图 5-2a 所示。

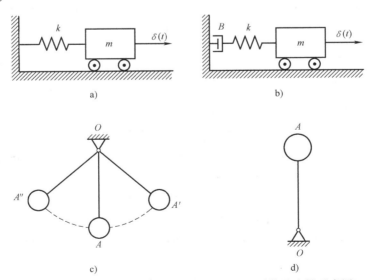

Fig. 5-1　Schematic diagram of system stability(系统稳定性示意图)

若给系统增加一个阻尼器,如图 5-1b 所示,在平衡状态下给定输入 $\delta(t)$,则质量块 m 在以平衡状态为中心的位置来回衰减振荡,最终回到最初的平衡状态,系统稳定,响应曲线如图 5-2b 或图 5-2c 所示。图 5-2b 所示系统虽是稳定的,但过渡过程时间较长并且超调量较

deviate from the original position A under the influence of disturbance, and cannot automatically return to the original equilibrium position A. Thus it is an unstable system. The response curve is shown in Fig. 5-2d, and the output amplitude increases gradually, which is a divergent oscillation.

The stability of the system is divided into the absolute stability and relative stability. The so-called absolute stability refers to whether the system is stable or unstable. For example, the systems corresponding to the response curves in Fig. 5-2b and Fig. 5-2c are stable; the so-called relative stability refers to the degree of stability. For example, the system transition process corresponding to the response curve in Fig. 5-2c is not only short in time but also small in overshoot, which means, it is not only fast but also smooth, and thus the system has a good degree of stability.

2. Necessary and sufficient conditions of stability for a linear system

According to the concept of stability, whether the system is stable depends mainly on whether the free motion response modal of the system gradually disappears with time (i.e. converges). If the initial condition of the system is zero, and the unit impulse signal $\delta(t)$ is applied on the system as the disturbance signal, the system will be in free motion model and its output response is

$$\omega(t) = x_o(t) = L^{-1}[G(s)] \tag{5-1}$$

If $\lim_{t \to \infty} \omega(t) = 0$, it means that the system can return to the original equilibrium state. Thus, the system is stable. It can be seen that the stability of the system is consistent with the convergence of the impulse response function. Otherwise, if $\lim_{t \to \infty} \omega(t) \neq 0$, the system is unstable.

According to equation (3-67) in Chapter 3, if the real parts of all real or complex poles of the system are less than zero, the modal amplitude of the system will converge and approach to zero as time tends to infinity. In other words, when the closed-loop poles of the system are all located in the left half of the s-plane, the characteristic roots (poles) are less than zero or have negative real parts, i.e. $-p_j < 0$ ($j = 1, 2, \cdots, n_1$) or $-\xi_k \omega_{nk} < 0$ ($k = 1, 2, \cdots, n_2$). $\sum_{j=1}^{n_1} A_j e^{-p_j t}$ and $\sum_{k=1}^{n_2} D_k e^{-\xi_k \omega_{nk} t} \sin(\omega_{dk} t + \beta_k)$ in the free motion response (modal) of the system will decay as time t increases, eventually approach to zero and disappear. i.e.

$$\lim_{t \to \infty} \omega(t) = \lim_{t \to \infty} x_o(t) = x_o(\infty) = 0 \tag{5-2}$$

Therefore, the necessary and sufficient condition for the stability of the linear time-invariant system is that all the roots of the system characteristic equation are smaller than zero or have negative real parts. In other words, the characteristic roots are all in the left half of the s-plane. If the system has a root in the right half of the s-plane (i.e. the root is greater than zero or its real part is positive), the system is unstable. If there is a pure virtual root in the characteristic roots and the other roots are in the left half of the s-plane, the system is in a critical stable state.

As long as there is a positive real root or a pair of complex roots with positive real parts in the characteristic roots of the system, i.e. they are located in the right half of the s-plane, the corresponding impulse response function of the system is divergent. It means that the system will never return to the original equilibrium state, which is an unstable system. Hence, the stability is an inherent characteristic of the system, which is determined by the structure and parameters of the system,

大，所以稳定性较差。如图 5-2c 所示的系统是稳定的同时其过渡过程时间很短，超调量也小，所以稳定性好。

图 5-1c 所示的单摆受到外界干扰作用时，不论处于 A′还是 A″位置，当外界扰动消失后，经过若干次振荡，最后一定恢复到 A 位置，是一个稳定的系统。图 5-1d 所示的球在平衡位置 A 受到干扰作用后，就会偏离平衡位置且不能自动恢复，所以是一个不稳定的系统。响应曲线如图 5-2d 所示，输出幅值逐渐增大，是发散振荡的。

系统的稳定性分为绝对稳定和相对稳定，绝对稳定指系统能否稳定。例如，图 5-2b 和图 5-2c 所示响应曲线所对应的系统都是稳定的。相对稳定指稳定程度的好坏。例如，图 5-2c 所示响应曲线所对应的系统过渡过程不仅时间短，并且超调量也小，即不仅响应速度快，而且平稳，所以该系统稳定的程度好。

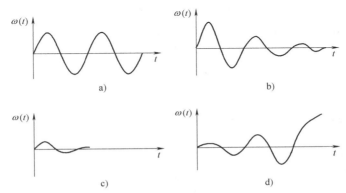

Fig. 5-2　Four possible scenarios of oscillating system impulse responses
（振荡系统脉冲响应可能出现的四种情况）

2. 线性系统稳定的充要条件

由稳定性概念可知，系统稳定与否主要取决于系统自由运动响应模态是否随时间增加而逐渐消失（收敛）。若系统［传递函数为 $G(s)$］的初始条件为零，用单位脉冲信号 $\delta(t)$ 作为干扰信号作用于系统之后，系统将处于自由运动状态，其输出响应为

$$\omega(t) = x_o(t) = L^{-1}[G(s)] \tag{5-1}$$

若 $\lim\limits_{t \to \infty} \omega(t) = 0$，表示系统仍能回到原有的平衡状态，因而系统是稳定的。由此可知，系统的稳定性是与其脉冲响应函数的收敛相一致的；反之若 $\lim\limits_{t \to \infty} \omega(t) \neq 0$，则系统不稳定。

由第 3 章式（3-67）可知，若系统的所有实数极点或复数极点的实部均小于零，则随着时间趋近于无穷大，系统的模态幅值将收敛并趋近于零。换言之，当系统的闭环极点全部在复平面的左半平面时，其特征根（极点）小于零或具有负实部，即 $-p_j < 0$（$j = 1, 2, \cdots, n_1$）或 $-\xi_k \omega_{nk} < 0$（$k = 1, 2, \cdots, n_2$）。系统的自由运动响应（模态）中，$\sum\limits_{j=1}^{n_1} A_j e^{-p_j t}$ 及 $\sum\limits_{k=1}^{n_2} D_k e^{-\xi_k \omega_{nk} t} \sin(\omega_{dk} t + \beta_k)$ 将随着时间 t 的增长而衰减，最终趋于零而消失。即

$$\lim\limits_{t \to \infty} \omega(t) = \lim\limits_{t \to \infty} x_o(t) = x_o(\infty) = 0 \tag{5-2}$$

因此，线性定常系统稳定的充分必要条件是系统特征方程的所有根小于零或具有负的实

and is independent of initial conditions and external input signals. Generally, the root on the right half of the s-plane is called the unstable root, and the root on the left half of the s-plane is called the stable root, as is shown in Fig. 5-3. If there is a pure virtual root in the system, and the other roots are in the left half of the s-plane, the system is critically stable. Due to some simplifications and assumptions in the process of establishing mathematical models of actual control systems to treat them as linearized systems, the estimation and measurement of system parameters may not be accurate enough. However, since the parameters of the system vary slightly during the actual working process, the original poles on the imaginary axis may move to the right half of the s-plane, which makes the system unstable. From the engineering point of view, it is generally considered that a critical stability system is an unstable system.

To sum up, the stability of the system can be figured out according to the roots of the characteristic equation. However, solving the roots of high-order systems is a very complex task. Generally, indirect methods are used to find out whether all the roots of the characteristic equation are distributed on the left half of the s-plane. The indirect methods often used are algebraic criterion and frequency criterion.

5.1.2 The algebraic stability criterion

The algebraic stability criterion can be described as: algebraic operations are carried out by using the coefficients of the characteristic equation to obtain the condition that all real parts of roots are negative, and then to determine whether the system is stable according to the condition. For example, the characteristic equation of a first-order system is

$$a_1 s + a_0 = 0$$

The root of the characteristic equation is $s_1 = -a_1/a_0$.

Obviously, the necessary and sufficient condition for the root of the characteristic equation to be negative is that a_1 and a_0 must be positive, i.e. $a_1 > 0$ and $a_0 > 0$.

1. Necessary conditions of the stability

Assume that the characteristic equation of the system is

$$D(s) = a_n s^n + a_{n-1} s^{n-1} + \cdots + a_1 s + a_0 = 0 \tag{5-3}$$

After dividing all terms in equation (5-3) by a_n, the polynomial factorization is

$$s^n + \frac{a_{n-1}}{a_n} s^{n-1} + \cdots + \frac{a_1}{a_n} s + \frac{a_0}{a_n} = (s-s_1)(s-s_2)\cdots(s-s_n) \tag{5-4}$$

where $s_1, s_2, \cdots, s_n$ are the characteristic roots of the system. The right hand side of the above equation is expanded as

$$(s-s_1)(s-s_2)\cdots(s-s_n) = (-1)^0 s^n + (-1)^1 \left(\sum_{i=1}^{n} s_i\right) s^{n-1} + (-1)^2 \left(\sum_{\substack{i,j=1 \\ i \neq j}}^{n} s_i s_j\right) s^{n-2} + \cdots + (-1)^n \prod_{i=1}^{n} s_i \tag{5-5}$$

By comparing equations (5-4) and equation (5-5), we have

部，换言之，特征根全部在复平面的左半平面。如果系统有一个根在复平面的右半平面（即根大于零或其实部为正），则系统不稳定。若特征根中有纯虚根，其余根均在复平面左半平面，则系统为临界稳定状态。

只要系统的特征根中有一个正实根或一对实部为正的复数根，即在复平面的右半平面，则系统对应的脉冲响应函数就是发散的。这意味着系统永远不会回到原平衡状态，这样的系统就是不稳定系统。可见稳定性是系统的固有特性，由系统的结构和参数决定，与初始条件及外界输入信号无关。一般把位于复平面右半平面的根称为不稳定根，把位于复平面左半平面的根称为稳定根，如图5-3所示。若系统有纯虚根，其余根均在复平面左半平面，则系统是临界稳定的。由于在对实际控制系统建立数学模型的过程中进行了一些简化和假设，所研究的系统都是线性化的系统，因此系统参数的估计和测量可能不够准确。然而系统在实际运行过程中，参数值处于微小变化之中，因此虚轴上的极点实际上可能移动到复平面的右半平面，致使系统不稳定。从工程实际来看，一般认为临界稳定系统属于不稳定系统。

总之，可以根据特征方程的根来判断系统稳定性。但求解高阶系统特征方程根是很复杂的工作，一般采用间接方法来判断特征方程的所有根是否都分布在复平面左半平面。经常采用的间接方法是代数判据和频率判据。

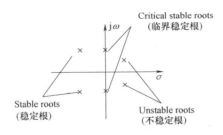

Fig. 5-3 The relationship between distribution of characteristic roots and stability
（特征根分布与稳定性的关系）

5.1.2 代数稳定性判据

代数稳定性判据可以描述为：利用特征方程的系数进行代数运算，得出全部根为负实部的条件，进而根据此条件确定系统是否稳定。例如，一阶系统的特征方程为

$$a_1 s + a_0 = 0$$

特征方程的根为 $s_1 = -a_1/a_0$。

显然，特征方程的根为负的充分必要条件是 a_1 和 a_0 必须为正值。即 $a_1 > 0$ 且 $a_0 > 0$。

1. 稳定的必要条件

设系统的特征方程为

$$D(s) = a_n s^n + a_{n-1} s^{n-1} + \cdots + a_1 s + a_0 = 0 \tag{5-3}$$

将式（5-3）中的各项同时除以 a_n 后，进行多项式因式分解，可得

$$s^n + \frac{a_{n-1}}{a_n} s^{n-1} + \cdots + \frac{a_1}{a_n} s + \frac{a_0}{a_n} = (s - s_1)(s - s_2) \cdots (s - s_n) \tag{5-4}$$

$$\begin{cases} \dfrac{a_{n-1}}{a_n} = -\sum_{i=1}^{n} s_i \\ \dfrac{a_{n-2}}{a_n} = \sum_{\substack{i,j=1 \\ i \neq j}}^{n} s_i s_j \\ \dfrac{a_{n-3}}{a_n} = -\sum_{\substack{i,j,k=1 \\ i \neq j \neq k}}^{n} s_i s_j s_k \\ \vdots \\ \dfrac{a_0}{a_n} = (-1)^n \prod_{i=1}^{n} s_i \end{cases} \tag{5-6}$$

According to the above equation, in order to make all the characteristic roots s_1, s_2, $\cdots$, s_n have negative real parts, the following two conditions must be satisfied.

1) The coefficients a_i ($i=0, 1, 2, \cdots, n$) of the characteristic equations are not equal to zero. Intuitively, the characteristic equations are arranged in descending order and have no missing items.

2) All signs of the coefficients a_i in the characteristic equation are the same.

Generally, a_i is taken as a positive value, thus the necessary condition for system stability is

$$a_n, a_{n-1}, \cdots, a_1, a_0 > 0 \tag{5-7}$$

The condition is not sufficient, and the characteristic equations with positive coefficients are likely to have positive real part roots. The reason is that no matter whether the characteristic roots are positive or negative, their combination could still satisfy all equations in equation (5-6).

If the system does not meet the necessary conditions of stability, it will be unstable. If the system meets the necessary conditions of stability, it is necessary to further judge whether it satisfies sufficient conditions of the stability.

2. The Routh criterion

The Routh criterion was proposed by the mathematics professor Routh at Cambridge University in 1876. He solved the problem of determining the distribution of the characteristic roots on the s-plane without calculating the characteristic roots, so as to determine whether a system is stable. If the system is unstable, it can be further determined that how many unstable roots i. e. roots with positive real parts) there are. Due to the complex mathematical principle of the Routh criterion, here we directly give the conclusion.

(1) Arrangement of the Routh table The first two rows of the Routh table are composed of the coefficients of the system characteristic equation shown in equation (5-3). The first row consists of the coefficients of odd terms in the characteristic equation (such as the first term, third term, fifth term); the second row consists of the coefficients of even terms in the characteristic equation (such as the second term, fourth term, sixth term). The values of the remaining rows need to be calculated line by line. The arrangement continues until the nth row, forming the following Routh table.

式中，s_1，s_2，$\cdots$，s_n 为系统的特征根。将上式的右边展开得

$$(s-s_1)(s-s_2)\cdots(s-s_n) = (-1)^0 s^n + (-1)^1 \left(\sum_{i=1}^n s_i\right) s^{n-1} + $$

$$(-1)^2 \left(\sum_{\substack{i,j=1 \\ i \neq j}}^n s_i s_j\right) s^{n-2} + \cdots + (-1)^n \prod_{i=1}^n s_i \tag{5-5}$$

比较式（5-4）与式（5-5）可得

$$\begin{cases} \dfrac{a_{n-1}}{a_n} = -\sum_{i=1}^n s_i \\ \dfrac{a_{n-2}}{a_n} = \sum_{\substack{i,j=1 \\ i \neq j}}^n s_i s_j \\ \dfrac{a_{n-3}}{a_n} = -\sum_{\substack{i,j,k=1 \\ i \neq j \neq k}}^n s_i s_j s_k \\ \vdots \\ \dfrac{a_0}{a_n} = (-1)^n \prod_{i=1}^n s_i \end{cases} \tag{5-6}$$

从上式可知，要使全部特征根 s_1、s_2、s_3、$\cdots$、s_n 均具有负实部，就必须满足以下两个条件。

1）特征方程的各项系数 $a_i(i=0,1,2,\cdots,n)$ 都不等于零，直观上看即特征方程按降幂排列且没有缺项。

2）特征方程中各项系数 a_i 符号相同。

按习惯，一般取 a_i 为正值，因此系统稳定的必要条件是

$$a_n, a_{n-1}, \cdots, a_1, a_0 > 0 \tag{5-7}$$

这一条件并不充分，对各项系数均为正的特征方程，还有可能存在具有正实部的根。原因在于不论特征根是正还是负，它们的组合仍可能满足式（5-6）中各式。

若系统不满足稳定的必要条件，则系统必不稳定。若系统满足稳定的必要条件，还要进一步判断其是否满足稳定的充分条件。

2. 劳斯判据

劳斯判据由剑桥大学数学教授劳斯于 1876 年提出，该方法无需求解特征根就能确定特征根在复平面上的分布情况，从而判断一个系统是否稳定；如果系统不稳定，还能进一步确定有几个不稳定的根（不稳定的根指实部为正的根）。由于劳斯判据的数学原理比较复杂，此处我们直接给出结论。

（1）劳斯表的排列　劳斯表的前两行由式（5-3）中的系统特征方程系数构成。其中，第一行由特征方程的奇数项系数（如第 1、3、5 项等）组成；第二行由特征方程的偶数项系数（如第 2、4、6 项等）组成。剩余各行的数值需逐行计算，这种排列一直进行到第 n 行，形成如下劳斯表。

s^n	a_n	a_{n-2}	a_{n-4}	a_{n-6}	$\cdots$
s^{n-1}	a_{n-1}	a_{n-3}	a_{n-5}	a_{n-7}	$\cdots$
s^{n-2}	b_1	b_2	b_3	b_4	$\cdots$
s^{n-3}	c_1	c_2	c_3	c_4	$\cdots$
$\vdots$	$\vdots$	$\vdots$	$\vdots$	$\vdots$	$\vdots$
s^2	e_1	e_2	e_3		
s^1	f_1				
s^0	g_1				

where

$$\begin{cases} b_1 = \dfrac{a_{n-1}a_{n-2} - a_n a_{n-3}}{a_{n-1}} \\ b_2 = \dfrac{a_{n-1}a_{n-4} - a_n a_{n-5}}{a_{n-1}} \\ b_3 = \dfrac{a_{n-1}a_{n-6} - a_n a_{n-7}}{a_{n-1}} \\ \vdots \end{cases} \tag{5-8}$$

$$\begin{cases} c_1 = \dfrac{b_1 a_{n-3} - a_{n-1} b_2}{b_1} \\ c_2 = \dfrac{b_1 a_{n-5} - a_{n-1} b_3}{b_1} \\ c_3 = \dfrac{b_1 a_{n-7} - a_{n-1} b_4}{b_1} \\ \vdots \end{cases} \tag{5-9}$$

The process is calculated to the nth row. In addition, to simplify numerical operations, a positive integer can be used to multiply or divide by each item in a row, which does not change the stability conclusion.

(2) The Routh stability criterion The Routh stability criterion points out that the necessary and sufficient condition for the stability of the system is that the signs of the first column elements in the Routh table are positive. The number of positive real roots in the characteristic equation is equal to the time of sign changes of the first column element in the Routh table. Note that the number of negative elements in the first column is not included.

Example 5-1 The characteristic equation of a system is
$$s^5 + 6s^4 + 14s^3 + 17s^2 + 10s + 2 = 0$$
Determine whether the system is stable according to the Routh criterion.

s^n	a_n	a_{n-2}	a_{n-4}	a_{n-6}	$\cdots$
s^{n-1}	a_{n-1}	a_{n-3}	a_{n-5}	a_{n-7}	$\cdots$
s^{n-2}	b_1	b_2	b_3	b_4	$\cdots$
s^{n-3}	c_1	c_2	c_3	c_4	$\cdots$
$\vdots$	$\vdots$	$\vdots$	$\vdots$	$\vdots$	$\vdots$
s^2	e_1	e_2	e_3	s^2	
s^1	f_1				
s^0	g_1				

表中，

$$\begin{cases} b_1 = \dfrac{a_{n-1}a_{n-2}-a_n a_{n-3}}{a_{n-1}} \\ \\ b_2 = \dfrac{a_{n-1}a_{n-4}-a_n a_{n-5}}{a_{n-1}} \\ \\ b_3 = \dfrac{a_{n-1}a_{n-6}-a_n a_{n-7}}{a_{n-1}} \\ \\ \quad\vdots \end{cases} \tag{5-8}$$

$$\begin{cases} c_1 = \dfrac{b_1 a_{n-3}-a_{n-1}b_2}{b_1} \\ \\ c_2 = \dfrac{b_1 a_{n-5}-a_{n-1}b_3}{b_1} \\ \\ c_3 = \dfrac{b_1 a_{n-7}-a_{n-1}b_4}{b_1} \\ \\ \quad\vdots \end{cases} \tag{5-9}$$

这一过程一直计算到第 n 行为止。此外，为简化数值运算，可用一个正整数乘或除某一行的各项，这并不改变稳定性的结论。

（2）劳斯稳定判据　劳斯判据指出系统稳定的充分必要条件是劳斯表中第一列各元素的符号均为正，特征方程中具有正实部根的个数等于劳斯表中第一列元素符号改变的次数，注意不包括第一列元素为负的元素个数。

例 5-1　系统的特征方程为

$$s^5 + 6s^4 + 14s^3 + 17s^2 + 10s + 2 = 0$$

根据劳斯判据确定系统是否稳定。

Solution: All coefficients of the characteristic equation are positive real numbers, and the Routh table is listed as

s^5	1	14	10
s^4	6	17	2
s^3	67/6	58/6	
s^2	791/67	2	
s^1	36900/791		
s^0	2		

It can be seen that the first column element values of the Routh table are all positive, thus the system is stable.

Example 5-2   The characteristic equation of a system is
$$s^5+3s^4+2s^3+s^2+5s+6=0$$
Determine whether the system is stable according to the Routh criterion.

Solution: All coefficients of the characteristic equation are positive real numbers, and the Routh table is listed as

s^5	1	2	5
s^4	3	1	6
s^3	5	9	
s^2	−11	15	
s^1	174/11		
s^0	15		

First of all, the first column element values of the Routh table are not all positive, so the system is definitely an unstable system. Further, the changes of sign in the first column element values can be observed: from 5 to−11, then from−11 to 174/11. Therefore, the signs of the first column element of the Routh table change twice, and there are two unstable roots in the system.

(3) Special cases in the Routh criterion

1) The first element in a row of the Routh table is zero, and the rest elements are not equal to zero. Since the first element is zero, the elements of the next row will approach to infinity, and the Routh table cannot be arranged. Thus, a small positive number ε can be used instead of zero to continue to list the Routh table, and then the time of sign changes in the first column of the Routh table is studied when $\varepsilon \rightarrow 0$.

Example 5-3   The characteristic equation of a system is
$$s^4+2s^3+s^2+2s+1=0$$
Determine whether the system is stable according to the Routh criterion.

解：特征方程的所有系数均为正实数，列出劳斯表，有

s^5	1	14	10
s^4	6	17	2
s^3	67/6	58/6	
s^2	791/67	2	
s^1	36900/791		
s^0	2		

可以看出劳斯表的第一列数值全部为正，所以系统是稳定的。

例 5-2 系统的特征方程为
$$s^5+3s^4+2s^3+s^2+5s+6=0$$
根据劳斯判据确定系统是否稳定。

解：特征方程的所有系数均为正实数，列出劳斯表，有

s^5	1	2	5
s^4	3	1	6
s^3	5	9	
s^2	−11	15	
s^1	174/11		
s^0	15		

首先劳斯表的第一列不全为正，系统一定是不稳定的系统。再考察第一列元素符号的改变：由 5 变成−11，再由−11 变成 174/11。所以劳斯表的第一列元素有两次符号变化，系统中有两个不稳定根。

(3) 劳斯判据中的特殊情况

1) 劳斯表中任一行的第一个元素为零，其余元素不全为零。此时由于第一个元素为零，因此下一行的各元素将趋于无穷大，劳斯表无法排列。这时可用一个很小的正数 ε 代替零，继续完成劳斯表，然后令 $\varepsilon\to 0$ 研究劳斯表中第一列的符号。

例 5-3 系统的特征方程为
$$s^4+2s^3+s^2+2s+1=0$$
根据劳斯判据确定系统是否稳定。

解：根据特征方程列劳斯表，得

Solution: According to the characteristic equation, the Routh table is listed as

s^4	1	1	1
s^3	2	2	0
s^2	$\varepsilon \approx 0$	1	0
s^1	$2-2/\varepsilon$	0	
s^0	1	0	

The system is unstable due to the changes of sign in the first column elements ($\varepsilon \to 2-2/\varepsilon \to 1$). The sign of the first column elements of the Routh table change twice, thus the characteristic equation has two roots with positive real parts.

Example 5-4 The characteristic equation of a system is
$$s^3+2s^2+s+2=0$$
Determine whether the system is stable according to the Routh criterion.

Solution: According to the characteristic equation, the Routh table is listed as

s^3	1	1
s^2	2	2
s^1	ε	0
s^0	2	0

It can be seen that the sign of the first column elements never change, so there is a pair of virtual roots. The characteristic equation can be decomposed as
$$(s^2+1)(s+2)=0$$
The roots of the equation are
$$s_{1,2}=\pm j1, \quad s_3=-2$$

The case where the first element of a row is zero is summarized as follows: if the signs of the first column elements change, the time of sign changes is equal to the number of unstable roots; if the signs of the first column elements never change, the system is in a critical state.

2) All elements of a row in the Routh table are zero, which indicates that there are symmetrically distributed roots in the s-plane. At this time, an auxiliary polynomial can be formed by using the elements of the preceding row, and the coefficients of the first derivative of the auxiliary polynomial are taken instead of the all-zero row in the Routh table to continue the arrangement of the Routh table. Therefore, the symmetrically distributed roots in the characteristic equation can be obtained by solving the auxiliary polynomial.

Example 5-5 The characteristic equation of a system is
$$s^6+2s^5+8s^4+12s^3+20s^2+16s+16=0$$
Determine whether the system is stable according to the Routh criterion.

s^4	1	1	1
s^3	2	2	0
s^2	$\varepsilon \approx 0$	1	0
s^1	$2-2/\varepsilon$	0	
s^0	1	0	

由于第一列元素符号有改变（$\varepsilon \to 2-2/\varepsilon \to 1$），因此系统不稳定。劳斯表中第一列元素符号改变两次，因此特征方程有两个具有正实部的根。

例 5-4 系统的特征方程为

$$s^3 + 2s^2 + s + 2 = 0$$

根据劳斯判据确定系统是否稳定。

解：根据特征方程列劳斯表，得

s^3	1	1
s^2	2	2
s^1	ε	0
s^0	2	0

可以看出，第一列元素符号不改变，故有一对虚根。将特征方程分解，有

$$(s^2+1)(s+2) = 0$$

解得的根为

$$s_{1,2} = \pm j1, \quad s_3 = -2$$

某行第一个元素为零的情况总结如下：如第一列元素有符号的改变，则符号改变的次数等于不稳定根的个数；如第一列元素符号没有改变，则该系统处于临界稳定状态。

2) 劳斯表中任一行的所有元素均为零，表明在复平面内有对称分布的根。此时可用该行上一行的元素构成一个辅助多项式，取辅助多项式的一阶导数的系数代替劳斯表中的零行，继续劳斯表的排列。因此，可通过求解辅助多项式来获取特征方程中对称分布的根。

例 5-5 系统的特征方程为

$$s^6 + 2s^5 + 8s^4 + 12s^3 + 20s^2 + 16s + 16 = 0$$

根据劳斯判据确定系统是否稳定。

解：根据特征方程列出劳斯表，得

s^6	1	8	20	16	
s^5	1	6	8	0	（为简化运算，各项同除了2）
s^4	1	6	8	0	
s^3	0	0	0		

Solution: According to the characteristic equation, the Routh table is listed as

s^6	1	8	20	16	
s^5	1	6	8	0	(To simplify calculation, the
s^4	1	6	8	0	items are all divided by 2.)
s^3	0	0	0		

According to elements in the third row, an auxiliary polynomial is constructed as

$$F(s) = s^4 + 6s^2 + 8 = 0$$

By analyzing the above equation, it can be seen that there are two pairs of roots with identical size and opposite sign in the characteristic equation. Further, the derivative of the auxiliary polynomial with respect to s is obtained as

$$\frac{d}{ds}F(s) = 4s^3 + 12s$$

Replace the zero elements in the s^3 row with 4 and 12, and list the new Routh table as

s^6	1	8	20	
s^5	1	6	8	
s^4	1	6	8	
s^3	1	3		(To simplify calculation, the
s^2	3	8		items are all divided by 2.)
s^1	1/3			
s^0	8			

Since the first column elements in the new arrangement have no sign changes. It can be concluded that there isn't a root with positive real parts in the characteristic equation. In addition, it is calculated that the auxiliary polynomial is constructed as

$$F(s) = s^4 + 6s^2 + 8 = 0$$

$$s_{1,2} = \pm j\sqrt{2}, \quad s_{3,4} = \pm j2$$

It can be seen that the system is critically stable.

The case where elements of a row are all zero is summarized as follows: there must be symmetric distribution roots in the system. If the sign of the first column elements changes, the time of sign changes is equal to the number of unstable roots; if the sign of first column elements never changes, the system is in a critical state.

(4) Application of the Routh criterion

1) Determine the stability of a system. If it is unstable, one can know the distribution of the system poles in the s-plane.

2) Obtain the range of parameters when the system is stable. These parameters can be the open-loop gain of a system, and can be also the time constant, and so on.

Example 5-6 The open-loop transfer function of a unit feedback system is

由第三行各元素构造一个辅助多项式，得
$$F(s) = s^4 + 6s^2 + 8 = 0$$

分析上式可知该特征方程有两对大小相等、符号相反的根存在。进一步求得辅助多项式对 s 的导数为
$$\frac{d}{ds}F(s) = 4s^3 + 12s$$

用 4 和 12 代替 s^3 行中的零元素，列出新的劳斯表，得

s^6	1	8	20
s^5	1	6	8
s^4	1	6	8
s^3	1	3	（为简化运算，各项同除了2）
s^2	3	8	
s^1	1/3		
s^0	8		

由于在新排列的第一列中符号没有改变，因此可以断定，特征方程没有一个具有正实部的根。此外，经计算，辅助多项式根为
$$s_{1,2} = \pm j\sqrt{2}, \quad s_{3,4} = \pm j2$$
可见系统是临界稳定的。

某行元素全为零的情况总结如下：系统肯定有对称分布的根。如第一列元素有符号的改变，则符号改变的次数就是不稳定根的个数；如第一列元素没有符号的改变，则该系统处于临界稳定状态。

(4) 劳斯判据的应用

1) 用来判断系统的稳定性。如不稳定，则可以了解系统极点在复平面的分布情况。

2) 求取使系统稳定的参数取值范围，这些参数可以是系统的开环增益，也可以是时间常数等。

例 5-6 单位反馈系统的开环传递函数为
$$G(s) = \frac{K}{s(s+1)(0.25s+1)}$$
试求使系统稳定的 K 值取值范围。

解：系统的闭环传递函数：
$$G(s) = \frac{K}{s(s+1)(0.25s+1)+K}$$

系统的特征方程为
$$D(s) = 0.25s^3 + 1.25s^2 + s + K = 0$$

$$G(s) = \frac{K}{s(s+1)(0.25s+1)}$$

Try to find the value range of K when the system is stable.

Solution: The closed-loop transfer function of the system is

$$G(s) = \frac{K}{s(s+1)(0.25s+1)+K}$$

The characteristic equation of the system is

$$D(s) = 0.25s^3 + 1.25s^2 + s + K = 0$$

So the Routh table is listed as

s^3	0.25	1
s^2	1.25	K
s^1	$\dfrac{1.25-0.25K}{1.25}$	
s^0	K	

To make the system stable, the first column elements in the Routh table must be all greater than zero, i.e.

$$\begin{cases} K>0 \\ \dfrac{1.25-0.25K}{1.25}>0 \end{cases}$$

Thus the value range of K is

$$0<K<5$$

5.1.3 The stability criterion of control systems in the frequency domain

The method to judge the stability of control systems. can not only be established based on the relationship between the coefficients of the closed-loop system characteristic equation and the distribution of its characteristic roots, but also a domain criterion of system stability in terms of the frequency characteristic plot.

Since the frequency characteristic plot of the system is divided into the polar plot (also called amplitude and phase frequency characteristic plot or Nyquist plot) and the Bode plot (also called logarithmic amplitude frequency characteristic plot), the corresponding frequency domain criteria are divided into the Nyquist criterion and Bode criterion.

Compared to the algebraic criterion, the frequency domain criterion has the following characteristics.

1) It is a graphical method based on the polar plot or Bode plot to determine whether the system is stable without calculating the characteristic roots.

2) The frequency domain criterion judges whether the closed loop system is stable by using the open-loop frequency characteristics.

3) According to the frequency domain criterion, it can not only judge whether the system is

因此劳斯表为

s^3	0.25	1
s^2	1.25	K
s^1	$\dfrac{1.25-0.25K}{1.25}$	
s^0	K	

要保证系统稳定，劳斯表中第一列元素必须全大于零，即

$$\begin{cases} K>0 \\ \dfrac{1.25-0.25K}{1.25}>0 \end{cases}$$

故 K 的取值范围为

$$0<K<5$$

5.1.3 控制系统的频域稳定性判据

除了可以根据闭环系统特征方程系数与其特征根分布的关系判断控制系统稳定性，也可以根据频率特性图确定系统稳定性。

系统频率特性图分为极坐标图（又称为幅相频率特性图或奈奎斯特图）和伯德图（又称为对数幅频特性图），相应的频域判据分别为奈奎斯特判据和伯德判据。

频域判据相对于代数判据有以下几个特点。

1）频域判据根据极坐标图或伯德图来判断系统是否稳定而不需计算特征根，是一种图解方法。

2）频域判据是利用开环频率特性来判断闭环系统是否稳定的。

3）应用频域判据不仅能判断系统是否稳定，还可确定系统稳定程度如何，即系统的相对稳定性。

4）应用频域判据易于确定组成系统的不同环节如何影响系统稳定性。

1. 控制系统闭环特征函数及其特点

对于如图 5-4 所示的闭环系统，其闭环传递函数为

$$\varPhi(s)=\dfrac{G(s)}{1+G(s)H(s)}$$

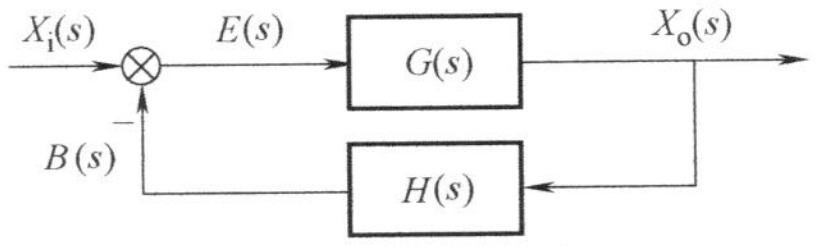

Fig. 5-4 Block diagram of the closed-loop system
（闭环系统的框图）

开环传递函数为

stable, but also know how the system stability is, i.e. the relative stability of the system.

4) According to the frequency domain criterion, it is easy to know how different elements of the system affect the stability of the system.

1. The closed-loop characteristic function of a control system and its characteristics

For a closed-loop system shown in Fig. 5-4, the closed-loop transfer function is

$$\Phi(s) = \frac{G(s)}{1 + G(s)H(s)}$$

The open-loop transfer function is

$$G_K(s) = G(s)H(s) = \frac{M(s)}{N(s)} \quad (n>m) \tag{5-10}$$

where n is the highest order of the denominator of $G_K(s)$; m is the highest order of the numerator of $G_K(s)$.

Substituting equation (5-10) into the closed-loop transfer function, we have

$$\Phi(s) = \frac{G(s)}{[N(s) + M(s)]/N(s)} \tag{5-11}$$

The system closed-loop characteristic auxiliary function is defined as

$$F(s) = 1 + G_K(s) = 1 + \frac{M(s)}{N(s)} = \frac{N(s) + M(s)}{N(s)} = \frac{D_B(s)}{D_K(s)} \tag{5-12}$$

It can be seen that the characteristic function $F(s)$ is the ratio of the closed-loop characteristic polynomial to the open-loop characteristic polynomial, and its numerator $D_B(s) = N(s) + M(s)$ and denominator $D_K(s) = N(s)$ are the system closed-loop and open-loop characteristic polynomial, respectively. The characteristic function $F(s)$ reflects the relationship between closed-loop poles and open-loop poles of the system, and has the following characteristics.

1) The zeros of $F(s)$ are the closed-loop poles of the system.

2) The poles of $F(s)$ are the open-loop poles of the system.

3) The order of the numerator and the denominator of $F(s)$ is the same, i.e. the number of closed-loop poles is equal to the number of open-loop poles and is equal to n.

4) The difference value between the characteristic function $F(s)$ and open-loop transfer function $G_K(s)$ is 1.

The above characteristic function is a complex function, and has all the characteristics of the complex function. When $s = j\omega$, the characteristic function becomes the characteristic frequency function $F(j\omega)$ with the input signal frequency of the system as the variable. According to equation (5-12), we have

$$F(j\omega) = 1 + G_K(j\omega) = \frac{D_B(j\omega)}{D_K(j\omega)} \tag{5-13}$$

According to the characteristics of $F(s)$, the polar plot of $F(j\omega)$ [or $G_K(j\omega)$] can be used to judge the stability of the closed-loop system.

2. The Nyquist stability criterion based on polar plot

For a linear control system with nth-order, according to equation (5-13), the characteristic

$$G_K(s) = G(s)H(s) = \frac{M(s)}{N(s)} \quad (n > m) \tag{5-10}$$

式中，n 为 $G_K(s)$ 分母的最高阶次；m 为 $G_K(s)$ 分子的最高阶次。

将式（5-10）代入闭环传递函数式可得

$$\Phi(s) = \frac{G(s)}{[N(s)+M(s)]/N(s)} \tag{5-11}$$

定义系统闭环特征辅助函数为

$$F(s) = 1 + G_K(s) = 1 + \frac{M(s)}{N(s)} = \frac{N(s)+M(s)}{N(s)} = \frac{D_B(s)}{D_K(s)} \tag{5-12}$$

可见，特征函数 $F(s)$ 是闭环特征多项式与开环特征多项式之比，其分子 $D_B(s) = N(s) + M(s)$ 和分母 $D_K(s) = N(s)$ 分别是系统闭环和开环特征多项式。特征函数 $F(s)$ 反映了系统的闭环极点与开环极点之间的关系，其有如下特点。

1）$F(s)$ 的零点就是系统的闭环极点。
2）$F(s)$ 的极点就是系统的开环极点。
3）$F(s)$ 的分子与分母的阶次相同，即说明闭环极点数与开环极点数相等且等于 n。
4）特征函数 $F(s)$ 与开环传递函数 $G_K(s)$ 之差为 1。

上述特征函数是复变函数，具有复变函数的所有特性。当 $s = j\omega$ 时，特征函数是以系统输入信号频率为变量的特征频率函数 $F(j\omega)$。由式（5-12）可得

$$F(j\omega) = 1 + G_K(j\omega) = \frac{D_B(j\omega)}{D_K(j\omega)} \tag{5-13}$$

依据 $F(s)$ 的特点，可以利用 $F(j\omega)$ ［或 $G_K(j\omega)$］的极坐标图来判断闭环系统的稳定性。

2. 奈奎斯特稳定性判据

对于 n 阶线性控制系统，由式（5-13）可将特征频率函数表示为

$$F(j\omega) = A(\omega)e^{j\varphi(\omega)} = \frac{D_B(j\omega)}{D_K(j\omega)}$$

$$= \frac{K(j\omega+z_1)(j\omega+z_2)\cdots(j\omega+z_n)}{(j\omega+p_1)(j\omega+p_2)\cdots(j\omega+p_n)} \tag{5-14}$$

式中，$-z_i = -\sigma_{zi} + j\omega_{zi}(i=1,2,\cdots,n)$ 为函数 $F(s)$ 的零点；$-p_k = -\sigma_{pk} + j\omega_{pk}(k=1,2,\cdots,n)$ 为函数 $F(s)$ 的极点。

$F(j\omega)$ 在 ［$F(j\omega)$］平面上相角随频率 ω 变化的关系为

$$\varphi(\omega) = \varphi_B(\omega) - \varphi_K(\omega)$$

$$= \sum_{i=1}^{n} \arctan \frac{\omega + \omega_{zi}}{\sigma_{zi}} - \sum_{k=1}^{n} \arctan \frac{\omega + \omega_{pk}}{\sigma_{pk}} \tag{5-15}$$

由此可知，$F(j\omega)$ 在 ［$F(j\omega)$］平面上的极坐标曲线的形状及绕向与 $F(s)$ 的零点和极点分布密切相关。若 $F(s)$ 的零点位于复平面的左半平面（$-\sigma_{zi} < 0$）［或右半平面（$-\sigma_{zi} > 0$）］，则频率 $\omega \to \pm \infty$ 时有

frequency function can be expressed as

$$F(j\omega) = A(\omega) e^{j\varphi(\omega)} = \frac{D_B(j\omega)}{D_K(j\omega)}$$

$$= \frac{K(j\omega+z_1)(j\omega+z_2)\cdots(j\omega+z_n)}{(j\omega+p_1)(j\omega+p_2)\cdots(j\omega+p_n)} \qquad (5\text{-}14)$$

where $-z_i = -\sigma_{zi} + j\omega_{zi}$ ($i = 1, 2, \cdots, n$) is the zero of function $F(s)$; $-p_k = -\sigma_{pk} + j\omega_{pk}$ ($k = 1, 2, \cdots, n$) is the pole of function $F(s)$.

The relationship of the phase angle of $F(j\omega)$ with frequency ω in the plane $[F(j\omega)]$ is

$$\varphi(\omega) = \varphi_B(\omega) - \varphi_K(\omega)$$

$$= \sum_{i=1}^{n} \arctan \frac{\omega + \omega_{zi}}{\sigma_{zi}} - \sum_{k=1}^{n} \arctan^{-1} \frac{\omega + \omega_{pk}}{\sigma_{pk}} \qquad (5\text{-}15)$$

It can be seen that the shape and the direction of the polar plot with respect to $F(j\omega)$ in the plane $[F(j\omega)]$ are closely related to the zeros and poles distribution of $F(s)$. If the zero of $F(j\omega)$ is located in the left half ($-\sigma_{zi}<0$) [or the right half ($-\sigma_{zi}>0$)] of the s-plane, then when $\omega \to \pm\infty$, we have

$$\lim_{\omega \to +\infty} \arctan \frac{\omega+\omega_{zi}}{\sigma_{zi}} = \begin{cases} \dfrac{\pi}{2} & (-\sigma_{zi}<0) \\ -\dfrac{\pi}{2} & (-\sigma_{zi}>0) \end{cases} \qquad (5\text{-}16)$$

or

$$\lim_{\omega \to -\infty} \arctan \frac{\omega+\omega_{zi}}{\sigma_{zi}} = \begin{cases} -\dfrac{\pi}{2} & (-\sigma_{zi}<0) \\ \dfrac{\pi}{2} & (-\sigma_{zi}>0) \end{cases} \qquad (5\text{-}17)$$

If $F(s)$ has N_z zeros in the right half of the s-plane, and the remaining $n-N_z$ zeros are located in the left half of the s-plane, the following expressions can be obtained according to equations (5-16) and equation (5-17) as

$$\begin{cases} \varphi_B(+\infty) = \sum_{i=1}^{N_z} \lim_{\omega \to +\infty} \left(\arctan \frac{\omega+\omega_{zi}}{\sigma_{zi}}\right)_{-\sigma_{zi}>0} + \sum_{i=N_z+1}^{n} \lim_{\omega \to +\infty} \left(\arctan \frac{\omega+\omega_{zi}}{\sigma_{zi}}\right)_{-\sigma_{zi}<0} \\ \qquad = N_z \times \left(-\dfrac{\pi}{2}\right) + (n - N_z) \times \dfrac{\pi}{2} = (n - 2N_z) \times \dfrac{\pi}{2} \\ \varphi_B(-\infty) = \sum_{i=1}^{N_z} \lim_{\omega \to -\infty} \left(\arctan \frac{\omega+\omega_{zi}}{\sigma_{zi}}\right)_{-\sigma_{zi}>0} + \sum_{i=N_z+1}^{n} \lim_{\omega \to -\infty} \left(\arctan \frac{\omega+\omega_{zi}}{\sigma_{zi}}\right)_{-\sigma_{zi}<0} \\ \qquad = N_z \times \dfrac{\pi}{2} + (n - N_z) \times \left(-\dfrac{\pi}{2}\right) = -(n - 2N_z) \times \dfrac{\pi}{2} \end{cases}$$

$$(5\text{-}18)$$

If $F(s)$ has N_p poles in the right half of the s-plane, and the remaining $n-N_p$ poles are located in the left half of the s-plane, the expressions can be obtained according to equations (5-16) and

$$\lim_{\omega \to +\infty} \arctan \frac{\omega+\omega_{zi}}{\sigma_{zi}} = \begin{cases} \dfrac{\pi}{2} & (-\sigma_{zi}<0) \\ -\dfrac{\pi}{2} & (-\sigma_{zi}>0) \end{cases} \tag{5-16}$$

或

$$\lim_{\omega \to -\infty} \arctan \frac{\omega+\omega_{zi}}{\sigma_{zi}} = \begin{cases} -\dfrac{\pi}{2} & (-\sigma_{zi}<0) \\ \dfrac{\pi}{2} & (-\sigma_{zi}>0) \end{cases} \tag{5-17}$$

若 $F(s)$ 有 N_z 个零点位于复平面的右半平面，其余 $n-N_z$ 个零点位于复平面的左半平面，由式（5-16）和式（5-17）可得

$$\begin{cases} \varphi_B(+\infty) = \sum_{i=1}^{N_z} \lim_{\omega \to +\infty}\left(\arctan\dfrac{\omega+\omega_{zi}}{\sigma_{zi}}\right)_{-\sigma_{zi}>0} + \sum_{i=N_z+1}^{n} \lim_{\omega \to +\infty}\left(\arctan\dfrac{\omega+\omega_{zi}}{\sigma_{zi}}\right)_{-\sigma_{zi}<0} \\ \qquad = N_z \times \left(-\dfrac{\pi}{2}\right) + (n-N_z) \times \dfrac{\pi}{2} = (n-2N_z) \times \dfrac{\pi}{2} \\ \varphi_B(-\infty) = \sum_{i=1}^{N_z} \lim_{\omega \to -\infty}\left(\arctan\dfrac{\omega+\omega_{zi}}{\sigma_{zi}}\right)_{-\sigma_{zi}>0} + \sum_{i=N_z+1}^{n} \lim_{\omega \to -\infty}\left(\arctan\dfrac{\omega+\omega_{zi}}{\sigma_{zi}}\right)_{-\sigma_{zi}<0} \\ \qquad = N_z \times \dfrac{\pi}{2} + (n-N_z) \times \left(-\dfrac{\pi}{2}\right) = -(n-2N_z) \times \dfrac{\pi}{2} \end{cases}$$
$$\tag{5-18}$$

若 $F(s)$ 有 N_p 个极点位于复平面的右半平面，其余 $n-N_p$ 个极点位于复平面的左半平面，由式（5-16）和式（5-17）可得

$$\begin{cases} \varphi_K(+\infty) = \sum_{k=1}^{N_p} \lim_{\omega \to +\infty}\left(\arctan\dfrac{\omega+\omega_{pk}}{\sigma_{pk}}\right)_{-\sigma_{pk}>0} + \sum_{i=N_p+1}^{n} \lim_{\omega \to +\infty}\left(\arctan\dfrac{\omega+\omega_{pk}}{\sigma_{pk}}\right)_{-\sigma_{pk}<0} \\ \qquad = N_p \times \left(-\dfrac{\pi}{2}\right) + (n-N_p) \times \dfrac{\pi}{2} = (n-2N_p) \times \dfrac{\pi}{2} \\ \varphi_K(-\infty) = \sum_{k=1}^{N_p} \lim_{\omega \to -\infty}\left(\arctan\dfrac{\omega+\omega_{pk}}{\sigma_{pk}}\right)_{-\sigma_{pk}>0} + \sum_{k=N_p+1}^{n} \lim_{\omega \to -\infty}\left(\arctan\dfrac{\omega+\omega_{pk}}{\sigma_{pk}}\right)_{-\sigma_{pk}<0} \\ \qquad = N_p \times \dfrac{\pi}{2} + (n-N_p) \times \left(-\dfrac{\pi}{2}\right) = -(n-2N_p) \times \dfrac{\pi}{2} \end{cases}$$
$$\tag{5-19}$$

当频率 $\omega = -\infty \to +\infty$ 时，函数 $F(j\omega)$ 在 $[F(j\omega)]$ 平面上的相角变化为闭环特征矢量相角变化与开环特征矢量相角变化之差，即

$$\Delta \arg F(j\omega) = \Delta \arg D_B(j\omega) - \Delta \arg D_K(j\omega) \tag{5-20}$$

设系统在复平面的右半平面存在 N_z 个闭环极点和 N_p 个开环极点，当频率 $\omega = -\infty \to +\infty$ 时，系统的闭环特征矢量的相角变化为

$$\Delta \arg D_B(j\omega) = \varphi_B(+\infty) - \varphi_B(-\infty) = (n-2N_z)\pi \tag{5-21}$$

开环特征矢量的相角变化为

(5-17) as

$$\begin{cases} \varphi_K(+\infty) = \sum_{k=1}^{N_p} \lim_{\omega \to +\infty} \left(\arctan \dfrac{\omega+\omega_{pk}}{\sigma_{pk}}\right)_{-\sigma_{pk}>0} + \sum_{i=N_p+1}^{n} \lim_{\omega \to +\infty} \left(\arctan \dfrac{\omega+\omega_{pk}}{\sigma_{pk}}\right)_{-\sigma_{pk}<0} \\ \qquad = N_p \times \left(-\dfrac{\pi}{2}\right) + (n-N_p) \times \dfrac{\pi}{2} = (n-2N_p) \times \dfrac{\pi}{2} \\ \varphi_K(-\infty) = \sum_{k=1}^{N_p} \lim_{\omega \to -\infty} \left(\arctan \dfrac{\omega+\omega_{pk}}{\sigma_{pk}}\right)_{-\sigma_{pk}>0} + \sum_{k=N_p+1}^{n} \lim_{\omega \to -\infty} \left(\arctan \dfrac{\omega+\omega_{pk}}{\sigma_{pk}}\right)_{-\sigma_{pk}<0} \\ \qquad = N_p \times \dfrac{\pi}{2} + (n-N_p) \times \left(-\dfrac{\pi}{2}\right) = -(n-2N_p) \times \dfrac{\pi}{2} \end{cases}$$

(5-19)

When $\omega = -\infty \to +\infty$, the phase angle change of function $F(j\omega)$ in the $[F(j\omega)]$ plane is the difference between the phase angle change of the closed-loop characteristic vector and the phase angle change of the open-loop characteristic vector, i. e.

$$\Delta\arg F(j\omega) = \Delta\arg D_B(j\omega) - \Delta\arg D_K(j\omega) \qquad (5\text{-}20)$$

Assume that the system has N_z closed-loop poles and N_p open-loop poles in the right half of the s-plane. When $\omega = -\infty \to +\infty$, the phase angle change of the closed-loop characteristic vector of the system is

$$\Delta\arg D_B(j\omega) = \varphi_B(+\infty) - \varphi_B(-\infty) = (n-2N_z)\pi \qquad (5\text{-}21)$$

The phase angle change of the open-loop characteristic vector of the system is

$$\Delta\arg D_K(j\omega) = \varphi_K(+\infty) - \varphi_K(-\infty) = (n-2N_p)\pi \qquad (5\text{-}22)$$

Thus, when $\omega = -\infty \to +\infty$, the phase angle change of function $F(j\omega)$ in the $[F(j\omega)]$ plane is

$$\Delta\arg F(j\omega) = (N_p - N_z) \times 2\pi = 2\pi N \qquad (5\text{-}23)$$

where

$$N = N_p - N_z \qquad (5\text{-}24)$$

Equation (5-24) shows that when $\omega = -\infty \to +\infty$, the polar plot $F(j\omega)$ in the $[F(j\omega)]$ plane encircles the coordinate origin $N = N_p - N_z$ times in the anticlockwise direction. Since the difference value between $F(j\omega)$ and the system open-loop transfer function $G_K(j\omega)$ is only one constant "1", the coordinate origin of the $[F(j\omega)]$ plane that is mapped to the complex $[G_K(j\omega)]$ plane is the point of $(-1, j0)$ as is shown in Fig. 5-5.

Therefore, the meaning of equation (5-24) is expressed as: when $\omega = -\infty \to +\infty$, the open-loop polar plot $G_K(j\omega)$ encircles the point of $(-1,j0)$ $N = N_p - N_z$ times in the $[G_K(j\omega)]$ plane.

When the system is stable, there is no closed-loop pole $(N_z = 0)$ in the right half of the s-plane, i. e. $N = N_p$. Thus, the Nyquist stability criterion based on the polar plot can be expressed as follows.

Nyquist stability criterion: If the open-loop system has N_p poles in the right half of the s-plane, and the polar plot $G_K(j\omega)$ of the open-loop frequency characteristic encircles the point of $(-1, j0)$ N times when $\omega = -\infty \to +\infty$, then the necessary and sufficient condition for the stability of the closed-loop system is

$$\Delta \arg D_K(j\omega) = \varphi_K(+\infty) - \varphi_K(-\infty) = (n - 2N_p)\pi \qquad (5\text{-}22)$$

所以，当频率 $\omega = -\infty \to +\infty$ 时，函数 $F(j\omega)$ 在 $[F(j\omega)]$ 平面上的相角变化量为

$$\Delta \arg F(j\omega) = (N_p - N_z) \times 2\pi = 2\pi N \qquad (5\text{-}23)$$

式中，

$$N = N_p - N_z \qquad (5\text{-}24)$$

式（5-24）表明，当频率 $\omega = -\infty \to +\infty$ 时，极坐标曲线 $F(j\omega)$ 在 $[F(j\omega)]$ 平面上按逆时针方向绕坐标原点 $N = N_p - N_z$ 圈。由于 $F(j\omega)$ 与系统开环传递函数 $G_K(j\omega)$ 只差一个常数"1"，$[F(j\omega)]$ 平面的坐标原点映射到 $[G_K(j\omega)]$ 平面上就是点 $(-1, j0)$，如图 5-5 所示。

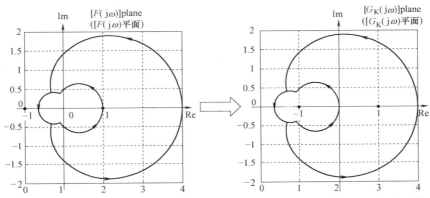

Fig. 5-5 The relationship between $[F(j\omega)]$ plane and $[G_K(j\omega)]$ plane（$[F(j\omega)]$ 平面与 $[G_K(j\omega)]$ 平面之间的关系）

因此，式（5-24）表示为：当 $\omega = -\infty \to +\infty$ 时，开环极坐标曲线 $G_K(j\omega)$ 在 $[G_K(j\omega)]$ 平面上绕点 $(-1, j0)$ $N = N_p - N_z$ 圈。

当系统稳定时，在复平面的右半平面不存在闭环极点（$N_z = 0$），即 $N = N_p$。于是，奈奎斯特稳定判据可表述如下。

奈奎斯特稳定判据：如果开环系统在复平面的右半平面具有 N_p 个极点，由 $\omega = -\infty \to +\infty$ 所对应的开环频率特性的极坐标曲线 $G_K(j\omega)$ 围绕点 $(-1, j0)$ 的圈数为 N，那么闭环系统稳定的充分必要条件是

$$N = N_p \qquad (5\text{-}25)$$

显然，若闭环系统是稳定的，则其开环极坐标曲线 $G_K(j\omega)$ 围绕点 $(-1, j0)$ 的圈数 $N \geq 0$，即 $G_K(j\omega)$ 一定是按逆时针方向围绕点 $(-1, j0)$ N 圈；若开环极坐标曲线 $G_K(j\omega)$ 按顺时针方向围绕点 $(-1, j0)$（$N<0$），则系统一定不稳定，不论其开环系统在复平面的右半平面是否具有极点。

例 5-7 设某开环系统的传递函数为

$$G_K(s) = \frac{K}{Ts - 1}$$

判断此系统的稳定性，并讨论稳定性与系数 K 的关系。

解：这个开环系统是一个非最小相位系统，在复平面的右半平面有一个极点，即 $N_p = 1$。因此，其幅频特性和相频特性分别为

$$N = N_p \qquad (5\text{-}25)$$

Obviously, if the closed-loop system is stable, the time of encircling the point of $(-1, j0)$ in the open-loop polar plot $G_K(j\omega)$ is $N \geqslant 0$, i.e. $G_K(j\omega)$ must encircle the point of $(-1, j0)$ N times in the anticlockwise direction. If the open-loop polar plot $G_K(j\omega)$ encircles the point of $(-1, j0)$ ($N<0$) in the clockwise direction, the system must be unstable regardless of whether the open-loop system has a pole in the right half of the s-plane.

Example 5-7 The transfer function of an open-loop system is

$$G_K(s) = \frac{K}{Ts-1}$$

Judge the stability of the system and discuss the relationship between stability and coefficient K.

Solution: The open-loop system is a non-minimum phase system, which has a pole in the right half of s-plane, and $N_p = 1$. Thus, its amplitude-frequency characteristic and phase-frequency characteristic are respectively

$$A(\omega) = \frac{K}{\sqrt{(\omega T)^2 + 1}}, \quad \varphi(\omega) = -\arctan\frac{\omega T}{-1}$$

When $\omega = 0$ and $\omega = \infty$, the start and end points of the polar plot are respectively

$$\omega = 0, \quad A(\omega) = K, \quad \varphi(\omega) = -180°$$
$$\omega = \infty, \quad A(\omega) = 0, \quad \varphi(\omega) = -90°$$

As is shown in Fig. 5-6, when $K>1$, the open-loop polar plot encircles the point of $(-1, j0)$ once in the clockwise direction, i.e. $N = N_p = 1$, so the closed-loop system is stable; when $K<1$, the open-loop polar plot does not encircle the point of $(-1, j0)$, i.e. $N = 0 \neq N_p = 1$, thus the closed-loop system is unstable.

Example 5-8 The open-loop transfer function of a system is

$$G_K(s) = \frac{K(s+3)}{s(s-1)}$$

Judge the stability of the closed-loop system.

Solution: The open-loop system has a pole in the right half of the s-plane, i.e. $N_p = 1$. When $\omega = -\infty \rightarrow +\infty$, its polar plot is shown in Fig. 5-7. When $K>1$, the open-loop polar plot encircles the point of $(-1, j0)$ once in the anticlockwise direction, i.e. $N = N_p = 1$, so the closed-loop system is stable; when $K<1$, the open-loop polar plot encircles the point of $(-1, j0)$ once in the clockwise direction, i.e. $N = -1 \neq N_p = 1$, so the closed-loop system is unstable.

When the shape of the open-loop polar plot $G_K(j\omega)$ is more complicated, it becomes difficult to distinguish the number of times of encircling the point of $(-1, j0)$. In this case, it is convenient to use a "crossover" in the polar plot for calculating the number N of times of encircling the point of $(-1, j0)$.

As is shown in Fig. 5-8, when the Nyquist stability criterion is used to analyze the number of times of the polar plot $G_K(j\omega)$ encircling the point of $(-1, j0)$, it can be determined by calculating the intersection points of polar plot $G_K(j\omega)$ and $(-\infty, -1)$ segment on the negative real axis in the range of $\omega = 0 \rightarrow +\infty$ to obtain the number of times regarding positive and negative cross-

$$A(\omega) = \frac{K}{\sqrt{(\omega T)^2 + 1}}, \quad \varphi(\omega) = -\arctan\frac{\omega T}{-1}$$

当 $\omega = 0$ 和 $\omega = \infty$ 时，极坐标图的起点和终点分别为

$$\omega = 0, \quad A(\omega) = K, \quad \varphi(\omega) = -180°$$
$$\omega = \infty, \quad A(\omega) = 0, \quad \varphi(\omega) = -90°$$

如图 5-6 所示，当 $K > 1$ 时，开环极坐标曲线按逆时针方向包围点（-1，j0）1 圈，即 $N = N_p = 1$，所以闭环系统稳定；当 $K < 1$ 时，开环极坐标曲线不包围点（-1，j0），即 $N = 0 \neq N_p = 1$，故闭环系统不稳定。

例 5-8 设系统开环传递函数为

$$G_K(s) = \frac{K(s+3)}{s(s-1)}$$

判别闭环系统的稳定性。

解： 该开环系统在复平面右半平面有一个极点，即 $N_p = 1$。当 $\omega = -\infty \to +\infty$ 时，其极坐标曲线如图 5-7 所示。当 $K > 1$ 时，开环极坐标曲线逆时针包围点（-1，j0）1 圈，即 $N = N_p = 1$，所以闭环系统稳定；当 $K < 1$ 时，开环极坐标曲线顺时针包围点（-1，j0）1 圈，即 $N = -1 \neq N_p = 1$，故闭环系统不稳定。

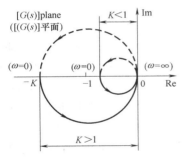

Fig. 5-6 The polar plot of example 5-8

（例 5-8 的极坐标图）

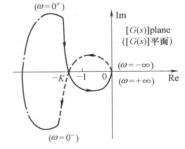

Fig. 5-7 The polar plot of example 5-9

（例 5-9 的极坐标图）

当开环极坐标曲线 $G_K(j\omega)$ 图形状较复杂时，便不易分辨它对点（-1，j0）包围的次数了，这时采用极坐标曲线中"穿越"来计算绕点（-1，j0）的圈数 N 比较方便。

如图 5-8 所示，利用奈奎斯特稳定判据分析极坐标曲线 $G_K(j\omega)$ 绕点（-1，j0）的圈数时，可以通过计数极坐标曲线 $G_K(j\omega)$ 在 $\omega = 0 \to +\infty$ 范围内与负实轴上（-∞，-1）段相交的交点，以获得正、负穿越的次数来确定。正穿越是指在 $\omega = 0 \to +\infty$ 范围内，极坐标曲线 $G_K(j\omega)$ 按相位增大方向穿过负实轴（-∞，-1）段，即 $G_K(j\omega)$ 曲线绕坐标点（-1，j0）由上向下按逆时针方向穿过负实轴（-∞，-1）段，正穿越的次数记为 N_+，为正数。负穿越是指在 $\omega = 0 \to +\infty$ 范围内，极坐标曲线 $G_K(j\omega)$ 按相位减小方向穿过负实轴（-∞，-1）段，即 $G_K(j\omega)$ 曲线绕坐标点（-1，j0）由下向上顺时针穿越负实轴（-∞，-1）段，负穿越的次数记为 N_-，为负数。若极坐标曲线 $G_K(j\omega)$ 以负实轴（-∞，-1）段上的点为起点，则由上离开实轴的次数为 $N_-/2$，由下离开实轴的次数为 $N_+/2$；若极坐标曲线 $G_K(j\omega)$ 以负实轴（-∞，-1）段上的点为终点，则由上进入实轴的次数为 $N_-/2$，由下进入实轴的次数为 $N_+/2$。于是，对应于封闭的极坐标曲线包围（-1，j0）点的圈数 N 就为

overs. The positive crossover means that in the range of $\omega = 0 \rightarrow +\infty$, the polar plot passes through $(-\infty, -1)$ segment on the negative real axis in the direction of increasing the phase, that is, the curve $G_K(j\omega)$ passes through $(-\infty, -1)$ segment on the negative real axis from the top to the bottom in the anticlockwise direction. The number of times of the positive crossover is denoted as N_+, and it is a positive number. The negative crossover means that in the range of $\omega = 0 \rightarrow +\infty$, the polar plot $G_K(j\omega)$ passes through $(-\infty, -1)$ segment on the negative real axis in the phase decreasing direction, that is, the curve $G_K(j\omega)$ passes through $(-\infty, -1)$ segment on the negative real axis from the bottom to the top in the clockwise direction. The number of times of the negative crossover is denoted as N_-, and it is a negative number. If the polar plot $G_K(j\omega)$ starts from the point on $(-\infty, -1)$ segment on the negative real axis, the number of times of leaving the real axis from above is $N_-/2$, and the number of times of leaving the real axis from below is $N_+/2$; if the polar plot $G_K(j\omega)$ ends at the point on $(-\infty, -1)$ segment on the negative real axis, the number of times of entering the real axis from above is $N_-/2$, and the number of times of entering the real axis from below is $N_+/2$. Hence, the number of times N corresponding to the closed polar plot encircling the point of $(-1, j0)$ is

$$N = 2(N_+ - N_-) \quad (5\text{-}26)$$

Fig. 5-9 shows a complex encircling case, $N_+ = 2$, $N_- = 1$, thus we have $N = 2$.

3. The Nyquist stability criterion based on Bode plot

If an open-loop polar plot is redrawn as an open-loop logarithmic plot (i.e. Bode plot), the stability of the system can also be judged by the Nyquist criterion. In this case, the number of times N can be determined according to the relationship between the open-loop logarithmic amplitude-frequency characteristic and the logarithmic phase-frequency characteristic. This method is called the logarithmic frequency domain criterion or the Bode criterion. It is essentially the same as the Nyquist criterion.

As is shown in Fig. 5-10, according to the correspondence between the open-loop frequency characteristic polar plot and the Bode plot in Chapter 4, the Nyquist stability criterion based on the polar plot can be converted into the Nyquist stability criterion based on the Bode plot.

According to the concept of "crossover" introduced above, the number of times N_+ of the positive crossover in the polar plot indicates that the phase increases after the curve passes through the $-180°$ phase line in the logarithmic phase frequency characteristic plot, i.e. the crossover is from down to up; the negative crossing number of the polar graph N_- of the positive crossover in the polar plot indicates that the phase decreases after the curve passes through the $-180°$ phase line in the logarithmic phase frequency characteristic plot, i.e. the crossover is from up to down. Therefore, according to the number of crossover of the polar plot defined above, the total number N of phase-frequency characteristic curve $\varphi(\omega)$ crossing the $-180°$ phase line is

$$N = 2(N_+ - N_-) \quad (5\text{-}27)$$

According to above analysis, if the open-loop transfer function $G_K(s)$ of the closed-loop system has N_p poles in the right half of the s-plane, and the total number N of phase-frequency characteristic curve $\varphi(\omega)$ crossing the $-180°$ phase line is N within the range of frequency ω regarding the logarithmic amplitude-frequency characteristic curve $L(\omega) > 0$, then the necessary and sufficient

$$N = 2(N_+ - N_-) \tag{5-26}$$

对如图 5-9 所示的复杂包围情况，$N_+ = 2$，$N_- = 1$，所以 $N = 2$。

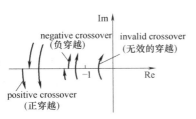

Fig. 5-8 Crossover cases of a curve
（曲线的穿越情况）

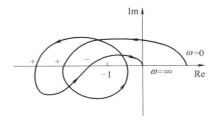

Fig. 5-9 Complex encircling case
（复杂包围情况）

3. 基于 Bode 图的奈奎斯特稳定判据

如果将开环极坐标图改画为开环对数坐标图（即伯德图），同样可以用奈奎斯特判据来判断系统的稳定性，这时要按开环对数幅频特性和对数相频特性的对应关系来确定 N。这种方法称为对数频域判据或伯德判据，它与奈奎斯特判据本质是相同的。

如图 5-10 所示，根据第 4 章中开环频率特性的极坐标图与伯德图的对应关系，可以将基于极坐标图的奈奎斯特稳定判据转化为基于伯德图的奈奎斯特稳定判据。

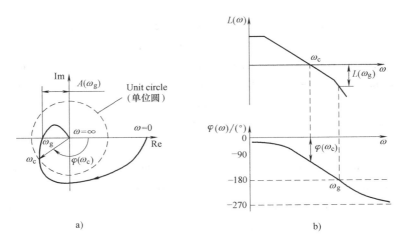

Fig. 5-10 The relationship between the polar plot and Bode plot
（极坐标图与伯德图的关系）

根据前面介绍的"穿越"概念，极坐标图的正穿越次数 N_+ 意味着曲线穿过对数相频特性图的 $-180°$ 相位线后相位增大，即穿越由下向上；极坐标图的负穿越次数 N_- 意味着曲线穿过对数相频特性图的 $-180°$ 相位线后相位减小，即穿越由上向下。于是，按照前面定义的极坐标曲线穿越次数，相频特性曲线 $\varphi(\omega)$ 穿越 $-180°$ 相位线的总次数 N 为

$$N = 2(N_+ - N_-) \tag{5-27}$$

由上述分析可知，如果闭环系统的开环传递函数 $G_K(s)$ 在复平面的右半平面有 N_p 个极点，且在对数幅频特性曲线 $L(\omega) > 0$ 的频率 ω 范围内，其相频特性曲线 $\varphi(\omega)$ 穿越 $-180°$ 相位线的总次数为 N。那么，闭环系统稳定的充分必要条件就是

condition for the stability of the closed-loop system is

$$N = N_p \tag{5-28}$$

Example 5-9 The open-loop Bode plots of two systems are depicted in Fig. 5-11, where N_p is the number of poles of the open-loop systems in the right-hand s-plane. Evaluate the stability of the corresponding closed-loop systems.

Solution: For the system shown in Fig. 5-11a: within the range of $L(\omega) > 0$, the number of the positive crossover of the phase-frequency characteristic $\varphi(\omega)$ is $N_+ = 1$ and the number of the negative crossover of the phase-frequency characteristic $\varphi(\omega)$ is $N_- = 2$, and then $N = 2(N_+ - N_-) = -2 \neq N_p = 2$. Thus the closed-loop system is unstable.

For the system shown in Fig. 5-11b: within the range of $L(\omega) > 0$, the number of the positive crossover of the phase-frequency characteristic $\varphi(\omega)$ is $N_+ = 2$ and the number of the negative crossover of the phase-frequency characteristic $\varphi(\omega)$ is $N_- = 1$ and then $N = 2(N_+ - N_-) = N_p = 2$. Thus the closed-loop system is stable.

5.1.4 Relative stability analysis of control systems

When designing a control system, it is required not only to be absolutely stable, but also to have a certain degree of stability. Only in this way can the performance index of the system be satisfied. Thus the concept of relative stability will be introduced in the section.

The so-called relative stability refers to the degree to which the stable state of the stable system is away from the unstable state. When discussing the relative stability of the system, it is generally assumed that the open-loop system is stable and it is a minimum phase system, that is, the zeros and poles of the open-loop system are located in the left half of the s-plane. At this time, if the polar plot of the open-loop frequency characteristic encircles the point of $(-1, j0)$, the system is unstable; if it doesn't encircle the point of $(-1, j0)$, the system is stable; if it passes through the point of $(-1, j0)$, then the system is in a critical steady state. Therefore, for a stable system, the relative stability of the system can be judged by the degree to which the polar plot is close to the point of $(-1, j0)$. The closer it is to the point of $(-1, j0)$, the worse the relative stability is. The indicator that reflects the relative stability is the stability margin. The quantitative representation of the stability margin mainly includes phase margin and amplitude (gain) margin.

1. The stability margin based on open-loop frequency characteristics

(1) The phase margin Let the system be a minimum phase system, its frequency characteristics are shown in Fig. 5-12, and its cutoff frequency (or the amplitude crossing frequency) corresponds to $A(\omega_c) = |G_K(j\omega_c)| = 1$ or $L(\omega_c) = 0$. Therefore, the phase margin is defined as the phase difference γ between the phase angle $\varphi(\omega_c)$ of the polar plot at the cutoff frequency and $-180°$, i.e.

$$\gamma = 180° + \varphi(\omega_c) \tag{5-29}$$

The meaning of the phase margin is that, for a closed-loop stable system, the larger the value of γ is, the farther the open-loop phase frequency characteristic curve of the system is from the $-180°$ phase line, and the higher the stability of the closed-loop system is; in other words, if the open-loop phase frequency characteristic further lags γ, the system will become critically stable. In

$$N = N_p \qquad (5-28)$$

例 5-9 已知两个系统的开环伯德图如图 5-11 所示,其中 N_p 为其开环系统在复平面右半平面的极点数,评估对应闭环系统的稳定性。

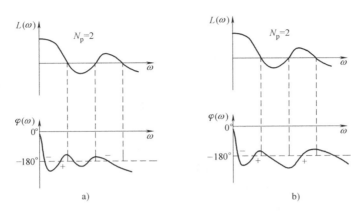

Fig. 5-11 Bode diagram of example 5-9(例 5-9 的伯德图)

解:对于图 5-11a 所示的系统,在 $L(\omega)>0$ 的范围内,相频特性 $\varphi(\omega)$ 正穿越次数 $N_+ = 1$,相频特性 $\varphi(\omega)$ 负穿越次数 $N_- = 2$,所以 $N = 2(N_+ - N_-) = -2 \neq N_p = 2$,因此闭环系统不稳定。

对于图 5-11b 所示的系统,在 $L(\omega)>0$ 的范围内,相频特性 $\varphi(\omega)$ 正穿越次数 $N_+ = 2$,相频特性 $\varphi(\omega)$ 负穿越次数 $N_- = 1$,所以 $N = 2(N_+ - N_-) = 2 = N_p$,因此闭环系统稳定。

5.1.4 控制系统的相对稳定性分析

在设计控制系统时,不仅需要绝对稳定性,还需要有一定的稳定程度,只有这样才能满足系统的性能指标。因此,下面将介绍相对稳定性的概念。

所谓相对稳定性是指稳定系统的稳定状态偏离不稳定状态的程度。在讨论系统的相对稳定性时,一般假定开环系统是稳定的,是最小相位系统,即开环系统的零、极点均位于复平面的左半平面。这时开环频率特性的极坐标曲线若包围点 (-1, j0),则系统不稳定;若不包围点 (-1, j0),则系统稳定;若穿过点 (-1, j0),则系统处于临界稳定状态。因此对于稳定的系统,可利用极坐标图靠近点 (-1, j0) 的程度来判断系统的相对稳定性,越靠近点 (-1, j0),相对稳定性越差。反映相对稳定性的指标就是稳定裕度,稳定裕度的定量表示主要有相位裕度和幅值(增益)裕度。

1. 基于开环频率特性的稳定裕度

(1) 相位裕度 设系统为最小相位系统,其频率特性如图 5-12 所示,其截止频率(或幅值穿越频率)所对应的 $A(\omega_c) = |G_K(j\omega_c)| = 1$ 或 $L(\omega_c) = 0$。定义系统的极坐标曲线在截止频率处的相位角 $\varphi(\omega_c)$ 距离 $-180°$ 的相位差 γ 为相位裕度,即为

$$\gamma = 180° + \varphi(\omega_c) \qquad (5-29)$$

相位裕度的意义是,对于闭环稳定系统,γ 值越大,系统的开环相频特性曲线距离 $-180°$ 相位线越远,闭环系统的稳定程度就越高;换言之,如果开环相频特性再滞后 γ,则系统将变为临界稳定的。为了使最小相位系统稳定,相位裕度必须为正。

order to obtain a stable minimum phase system, the phase margin must be positive.

(2) The amplitude margin The amplitude margin is also called the gain margin. If the phase crossing frequency ω_g of the system corresponds phase angle $\varphi(\omega_g) = -180°$, the reciprocal of the open-loop amplitude-frequency characteristic $A(\omega_g)$ at the phase-crossing frequency is defined as the amplitude margin, i. e.

$$K_g = \frac{1}{A(\omega_g)} \quad (5\text{-}30)$$

The meaning of the amplitude margin K_g is that, for the closed-loop stable system, if the open-loop amplitude-frequency characteristic of the system is further increased by K_g times, the system will become critical stable. The larger K_g is, the higher the stability of the closed-loop system is.

The amplitude margin on the Bode plot is expressed as

$$L_g = 20\lg K_g = -20\lg A(\omega_g) = -L(\omega_g) \quad (\text{dB}) \quad (5\text{-}31)$$

If $K_g > 1$, then $L_g > 0$, the magnitude margin is positive; if $K_g < 1$, then $L_g < 0$, the magnitude margin is negative. The positive amplitude margin (in dB) indicates that the system is stable; the negative amplitude margin (in dB) indicates that the system is unstable.

For a stable minimum phase system, the amplitude margin indicates how much the gain can be increased before the system becomes unstable. For an unstable system, the amplitude margin indicates how much the gain should be reduced to stabilize the system.

(3) Some explanations about phase margin and amplitude margin

1) For a minimum phase system, only if the conditions $\gamma > 0$ and $L_g > 0$ are satisfied, the system is stable. In order to make the closed-loop system have good dynamic performance, it is usually required for $\gamma = 30° \sim 60°$ and $K_g > 2$ (or $L_g > 6$dB).

2) Generally it is not enough to illustrate the relative stability of the system by only using the phase margin or amplitude margin. It is necessary to consider both the phase margin and the amplitude margin to explain the relative stability of the system. As is shown in Fig. 5-13a, the phase margin γ of the system is large enough, but the amplitude margin K_g is very small, so the stability of the system is not very high. This may occur in systems with oscillation elements in engineering practice. Therefore, the curve $L(\omega)$ should be corrected when analyzing the relative stability of the closed-loop system. As is shown in Fig. 5-13b, the amplitude margin K_g of the system is large, but the phase margin γ is small, so the stability of the system is not satisfactory.

3) For a minimum phase system, there is a certain correspondence between the open-loop amplitude-frequency characteristics and the phase-frequency characteristics. The phase margin $\gamma = 30° \sim 60°$ indicates that the slope of the open-loop logarithmic amplitude-frequency characteristic at frequency ω_c should be greater than or equal to -40dB/dec. It is often taken -20dB/dec in practice. If the slope at frequency ω_c is equal to -40dB/dec, the closed-loop system may be unstable. Even if it is stable, its stability is very poor. Therefore, as long as the open-loop logarithmic amplitude-frequency characteristic is studied, the stability can be roughly determined.

4) For complex control systems, when there are multiple crossing frequencies ω_c and (or) ω_g, the corresponding stability margin should be investigated for each crossing frequency.

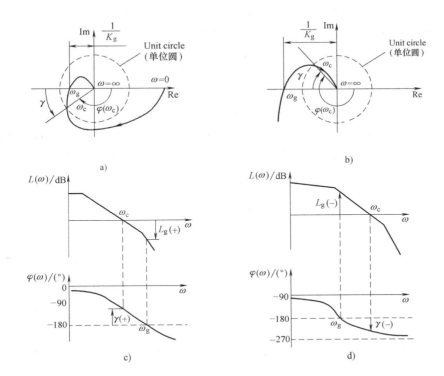

Fig. 5-12 Phase margin and amplitude margin（相位裕度与幅值裕度）
a) Polar plot with positive phase margin and amplitude margin（具有正相位裕度与幅值裕度的极坐标图）
b) Polar plot with negative phase margin and amplitude margin（具有负相位裕度与幅值裕度的极坐标图）
c) Bode diagram with positive phase margin and amplitude margin（具有正相位裕度与幅值裕度的伯德图）
d) Bode diagram with negative phase margin and amplitude margin（具有负相位裕度与幅值裕度的伯德图）

（2）幅值裕度 幅值裕度也称为增益裕度。若系统的相频穿越频率为 ω_g，对应有相位角 $\varphi(\omega_g) = -180°$。于是，定义相频穿越频率处的开环幅频特性 $A(\omega_g)$ 的倒数为幅值裕度，即

$$K_g = \frac{1}{A(\omega_g)} \tag{5-30}$$

幅值裕度 K_g 的意义是，对于闭环稳定系统，如果系统开环幅频特性再增大 K_g 倍，则系统将变为临界稳定状态。K_g 越大，闭环系统的稳定程度越高。

在伯德图上幅值裕度表示为

$$L_g = 20\lg K_g = -20\lg A(\omega_g) = -L(\omega_g) \quad (\mathrm{dB}) \tag{5-31}$$

如果 $K_g > 1$，则 $L_g > 0$，幅值裕度为正值；如果 $K_g < 1$，则 $L_g < 0$，幅值裕度为负值。正幅值裕度（以 dB 表示）表示系统是稳定的，负幅值裕度（以 dB 表示）表示系统不稳定。

对于稳定的最小相位系统，幅值裕度指出系统在变得不稳定之前，增益能够增大多少。对于不稳定系统，幅值裕度指出为使系统稳定，增益应当减少多少。

（3）关于相位裕度和幅值裕度的几点说明

1) 对于最小相位系统，只有满足 $\gamma > 0$ 且 $L_g > 0$，系统才能稳定。为了使闭环系统具有良

2. Stability analysis of typical elements

Assume that the open-loop transfer function of the unit feedback system is

$$G_K(s) = \frac{K\prod_{i=1}^{m}(\tau_i s + 1)}{s^v \prod_{j=1}^{n}(T_j s + 1)} e^{-\tau s} \tag{5-32}$$

It consists of several typical elements, such as proportional element with K, v integral elements, n inertia elements, m first-order differential elements, etc. Its frequency characteristic is

$$G_K(j\omega) = \frac{K\prod_{i=1}^{m}(j\omega\tau_i + 1)}{(j\omega)^v \prod_{j=1}^{n}(j\omega T_j + 1)} e^{-j\omega\tau} \tag{5-33}$$

If the open-loop crossing frequencies ω_c and ω_g are calculated, then the amplitude margin and phase margin of the system are respectively

$$L_g = -20\lg K + 20v\lg\omega_g - \sum_{i=1}^{m}20\lg\sqrt{1+(\omega_g\tau_i)^2} + \sum_{j=1}^{n}20\lg\sqrt{1+(\omega_g T_j)^2} \tag{5-34}$$

$$\gamma = 180° + \varphi(\omega_c) = 180° + \sum_{i=1}^{m}\arctan\omega_c\tau_i - v90° - \sum_{j=1}^{n}\arctan\omega_c T_j - \omega\tau \tag{5-35}$$

where T_j is the time constant of the inertia element; τ_i is the first-order differential time constant; τ is the delay time constant.

It can be seen from equation (5-34) or equation (5-35) that an integral element can reduce the phase margin γ by 90°, and the two integral elements reduce the phase margin γ by 180°. Therefore, if the system contains integral elements in the forward path, the stability of the system will be seriously deteriorated; the inertia element of the system will also make the stability of the system worse, the greater the time constant of the inertia element is, the more significant the effect is; the greater the amplification factor of the proportional element (or the open-loop gain) is, the worse the stability of the system is; the time delay element is the non-minimum phase element, which will delay the phase of the system and reduce the phase margin γ of the system. The larger the time constant τ is, the worse the stability is; the differential element is to increase the phase margin γ, so the differential element can be added to the forward path to improve the stability of the system. But the differential element is a high-pass filter, which is easy to introduce disturbance.

For the first-order system, its open-loop frequency characteristic is

$$G_K(j\omega) = \frac{1}{j\omega T}$$

Its stability margin is

$$\begin{cases} L_g = 20\lg\omega_g T = +\infty \\ \gamma = 180° - 90° = 90° > 0 \end{cases} \tag{5-36}$$

It shows that the first-order system is a stable system, but if the time constant becomes smaller, the stability of the system will deteriorate.

好的动态性能，通常要求 $\gamma = 30° \sim 60°$，$K_g > 2$（或 $L_g > 6\text{dB}$）。

2）一般情况下，仅用相位裕度或幅值裕度是不足以说明系统的相对稳定性的，必须同时采用相位裕度和幅值裕度两个指标才能说明系统的相对稳定性。如图 5-13a 所示，系统的相位裕度 γ 足够大，但幅值裕度 K_g 很小，因而系统的稳定程度不是很高。工程实践中含有振荡环节的系统可能出现这种情况。因而分析闭环系统的相对稳定性时需要对 $L(\omega)$ 曲线加以修正。如图 5-13b 所示，系统的幅值裕度 K_g 很大，但相位裕度 γ 小，因而系统的稳定性也不能令人满意。

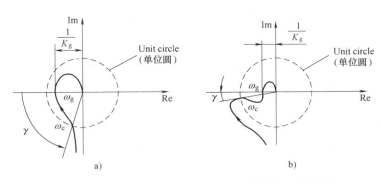

Fig. 5-13　Relative stability of a system（系统的相对稳定性）

3）对于最小相位系统，其开环幅频特性和相频特性之间具有一定的对应关系。相位裕度 $\gamma = 30° \sim 60°$ 表明开环对数幅频特性在 ω_c 处的斜率应大于或等于 -40dB/dec。在实际中常取 -20dB/dec。如果在 ω_c 处的斜率等于 -40dB/dec，则闭环系统可能不稳定，即使稳定，其稳定性也是很差。因此，只要研究开环对数幅频特性就可以大致判别其稳定性。

4）对于复杂的控制系统，当存在多个穿越频率 ω_c 和（或）ω_g 时，则需对每一个穿越频率考查相应的稳定裕度。

2. 典型环节的稳定性分析

设单位反馈系统的开环传递函数形式为

$$G_K(s) = \frac{K \prod_{i=1}^{m}(\tau_i s + 1)}{s^v \prod_{j=1}^{n}(T_j s + 1)} e^{-\tau s} \tag{5-32}$$

即为由比例环节 K、v 个积分环节、n 个惯性环节和 m 个一阶微分环节等若干典型环节组成的系统。其频率特性为

$$G_K(j\omega) = \frac{K \prod_{i=1}^{m}(j\omega \tau_i + 1)}{(j\omega)^v \prod_{j=1}^{n}(j\omega T_j + 1)} e^{-j\omega \tau} \tag{5-33}$$

若计算得开环穿越频率为 ω_c 和 ω_g，则系统的幅值裕度与相位裕度分别为

$$L_g = -20\lg K + 20v\lg\omega_g - \sum_{i=1}^{m} 20\lg\sqrt{1+(\omega_g\tau_i)^2} + \sum_{j=1}^{n} 20\lg\sqrt{1+(\omega_g T_j)^2} \tag{5-34}$$

For the second-order system, the open-loop frequency characteristic can be expressed as

$$G_K(j\omega) = \frac{K}{j\omega(j\omega T+1)}$$

The crossing frequency of the system is approximately

$$\omega_c = K \text{ or } \omega_c = \sqrt{\frac{K}{T}}, \quad \omega_g = +\infty$$

Therefore, its stability margin is

$$\begin{cases} L_g = -20\lg K + 20\lg\omega_g + 20\lg\sqrt{1+(\omega_g T)^2} = +\infty \\ \gamma = 90° - \arctan\omega_c T > 0 \end{cases} \quad (5\text{-}37)$$

It can be seen that the second-order system is always a stable system. Increasing the system gain K will cause an increase on ω_c, and will reduce the phase margin of the system and deteriorate the system stability.

Note that the magnitude margin of a first-order or second-order system is infinite because the polar plots of such systems do not intersect the negative real axis. Therefore, the first-order or second-order systems cannot be unstable in theory. Of course, a system can only be approximated as a first-order or second-order system in a certain sense. Because some small time lags are neglected when establishing system equations, it is not really a first-order or second-order system. If these small lags are taken into account, the so-called first-order or second-order systems may be unstable.

Example 5-10 The open-loop transfer function of a system is

$$G_K(s) = \frac{K}{s(s+1)(0.2s+1)}$$

Figure out the phase margin γ and the amplitude margin K_g when K is equal to 2 and 20, respectively.

Solution: The open-loop system is a minimum phase system, $N_p = 0$. Plot the Bode plot, as is shown in Fig. 5-14.

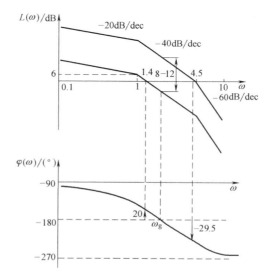

Fig. 5-14 Bode diagram of example 5-10 (例 5-10 的伯德图)

$$\gamma = 180° + \varphi(\omega_c) = 180° + \sum_{i=1}^{m} \arctan\omega_c\tau_i - v90° - \sum_{j=1}^{n} \arctan\omega_c T_j - \omega\tau \quad (5\text{-}35)$$

式中，T_j 为惯性环节的时间常数；τ_i 为一阶微分时间常数；τ 为延迟时间。

由式（5-34）或式（5-35）可见，一个积分环节可使相位裕度 γ 减小 $90°$，两个积分环节可使相位裕度 γ 减小 $180°$，因此，若系统在前向通路中含有积分环节，将使系统的稳定性严重变差；系统含惯性环节也会使系统的稳定性变差，其惯性环节的时间常数越大，这种影响越显著；比例环节（或开环增益）放大系数越大，系统的稳定性越差；延迟环节是非最小相位环节，将使系统的相位滞后，减小系统的相位裕量 γ，时间常数 τ 越大，稳定性越差；而微分环节会增大相位裕度 γ，因此可以在前向通路中增加微分环节改善系统的稳定性，但微分环节是高通滤波器，易引入干扰。

对于一阶系统，其开环频率特性为

$$G_K(j\omega) = \frac{1}{j\omega T}$$

其稳定裕度为

$$\begin{cases} L_g = 20\lg\omega_g T = +\infty \\ \gamma = 180° - 90° = 90° > 0 \end{cases} \quad (5\text{-}36)$$

由此可见，一阶系统是稳定的系统，但若时间常数 T 变小，系统的稳定性将变差。

对于二阶系统，其开环频率特性可表示为

$$G_K(j\omega) = \frac{K}{j\omega(j\omega T+1)}$$

系统的穿越频率近似为

$$\omega_c = K \text{ 或 } \omega_c = \sqrt{\frac{K}{T}}, \quad \omega_g = +\infty$$

故其稳定裕度为

$$\begin{cases} L_g = -20\lg K + 20\lg\omega_g + 20\lg\sqrt{1+(\omega_g T)^2} = +\infty \\ \gamma = 90° - \arctan\omega_c T > 0 \end{cases} \quad (5\text{-}37)$$

由此可见，二阶系统总是稳定的，增大系统增益 K，将引起 ω_c 增大，并将使系统的相位裕度减小，系统稳定性变差。

注意，一阶或二阶系统的幅值裕度为无穷大，因为这类系统的极坐标图与负实轴不相交。因此，理论上一阶或二阶系统不可能是不稳定的。当然，一个系统在一定意义上说只能近似是一阶或二阶系统，因为在建立系统方程时，忽略了一些小的时间滞后，因此其不是真正的一阶或二阶系统。如果考虑这些小的滞后，则所谓的一阶或二阶系统可能是不稳定的。

例 5-10 系统的开环传递函数

$$G_K(s) = \frac{K}{s(s+1)(0.2s+1)}$$

试分别求取 $K=2$ 及 $K=20$ 时的相位裕度 γ 和幅值裕度 K_g。

解： 此开环系统为最小相位系统，$N_p = 0$。画出其伯德图，如图 5-14 所示。

（1）求相位穿越频率 ω_g　因为 K 值的变化不会影响对数相频特性，所以无论 $K=2$，还

(1) Calculating the phase crossing frequency ω_g the change of value of K does not affect the logarithmic phase-frequency characteristic, so the phase crossing frequency is the same regardless of whether $K = 2$ or $= 20$. The open-loop frequency characteristic and phase-frequency characteristic are

$$G_K(j\omega) = \frac{K}{j\omega(j\omega+1)(0.2j\omega+1)}$$

$$\varphi(\omega_g) = -90° - \arctan\omega_g - \arctan 0.2\omega_g = -180°$$

Thus we have $\omega_g = \sqrt{5}\,\text{s}^{-1}$.

(2) When $K = 2$

1) Calculating the amplitude crossing frequency ω_c, we have

$$A(\omega_c) = \frac{2}{\omega_c\sqrt{\omega_c^2+1}\sqrt{(0.2\omega_c)^2+1}} = 1$$

Obviously, the calculation of the above formula is very complicated. We generally calculate it according to the Bode plot of approximate asymptotic curves. According to the transfer function, the corner frequencies of the two inertia elements are $\omega_{T1} = 1$ and $\omega_{T2} = \frac{1}{0.2} = 5$ respectively, thus the piecewise function equation of the asymptotic curve can be expressed as

$$L(\omega) = \begin{cases} 20\lg\dfrac{K}{\omega} & (0 < \omega \leq 1) \\ 20\lg\dfrac{K}{\omega \cdot \omega} & (1 < \omega \leq 5) \\ 20\lg\dfrac{K}{\omega \cdot \omega \cdot 0.2\omega} & (\omega > 5) \end{cases}$$

When $K = 2$, let $L(\omega)$ equals zero, we have $\omega_c = \sqrt{2} \approx 1.4$.

2) Calculating the phase margin, we have

$$\gamma = 180° + \varphi(\omega_c) = 180° + (-90° - \arctan 1.4 - \arctan 0.28) = 20°$$

3) Calculating the amplitude margin, we have

$$A(\omega_g) = \frac{2}{\sqrt{5}\sqrt{5+1}\sqrt{0.04\times 5+1}} = \frac{1}{3}$$

$$L_g = -20\lg A(\omega_g) = 20\lg 3 = 9.5\,(\text{dB})$$

Therefore, the system is stable when $K = 2$.

(3) When $K = 20$

1) Calculating the amplitude crossing frequency ω_c, we have

$$A(\omega_c) = \frac{20}{\omega_c\sqrt{\omega_c^2+1}\sqrt{0.04\omega_c^2+1}} = 1$$

Let $K = 20$, we have $\omega_c \approx 4.5$.

2) Calculating the phase margin, we have

$$\gamma = 180° + \varphi(\omega_c) = 180° + (-90° - \arctan 4.5 - \arctan 0.9) = -29.5°$$

是 $K=20$，其相位穿越频率 ω_g 都是相同的。系统的开环频率特性和相频特性为

$$G_K(j\omega) = \frac{K}{j\omega(j\omega+1)(0.2j\omega+1)}$$

$$\phi(\omega_g) = -90° - \arctan\omega_g - \arctan0.2\omega_g = -180°$$

解得 $\omega_g = \sqrt{5}\,\text{s}^{-1}$。

(2) 当 $K=2$ 时

1) 求幅值穿越频率 ω_c，有

$$A(\omega_c) = \frac{2}{\omega_c\sqrt{\omega_c^2+1}\sqrt{(0.2\omega_c)^2+1}} = 1$$

显然采用上式计算比较复杂，我们一般根据近似渐进线的伯德图来计算，由传递函数可知两个惯性环节的转折频率分别为 $\omega_{T1}=1$，$\omega_{T2}=\frac{1}{0.2}=5$，故渐近线的分段函数方程可表示为

$$L(\omega) = \begin{cases} 20\lg\dfrac{K}{\omega} & (0<\omega\leq 1) \\ 20\lg\dfrac{K}{\omega\cdot\omega} & (1<\omega\leq 5) \\ 20\lg\dfrac{K}{\omega\cdot\omega\cdot 0.2\omega} & (\omega>5) \end{cases}$$

当 $K=2$ 时，令 $L(\omega)$ 为零可解出 $\omega_c = \sqrt{2} \approx 1.4$。

2) 求相位裕度，有

$$\gamma = 180° + \varphi(\omega_c) = 180° + (-90° - \arctan1.4 - \arctan0.28) = 20°$$

3) 求幅值裕度，有

$$A(\omega_g) = \frac{2}{\sqrt{5}\sqrt{5+1}\sqrt{0.04\times 5+1}} = \frac{1}{3}$$

$$L_g = -20\lg A(\omega_g) = 20\lg 3 = 9.5(\text{dB})$$

可见，当 $K=2$ 时系统是稳定的。

(3) 当 $K=20$ 时

1) 求幅值穿越频率 ω_c，有

$$A(\omega_c) = \frac{20}{\omega_c\sqrt{\omega_c^2+1}\sqrt{0.04\omega_c^2+1}} = 1$$

对前面的分段函数，令 $K=20$ 解得

$$\omega_c \approx 4.5$$

2) 求相位裕度，有

$$\gamma = 180° + \varphi(\omega_c) = 180° + (-90° - \arctan4.5 - \arctan0.9) = -29.5°$$

3) 求幅值裕度，有

$$A(\omega_g) = \frac{20}{\sqrt{5}\sqrt{5+1}\sqrt{0.04\times 5+1}} = \frac{10}{3}$$

$$L_g = -20\lg A(\omega_g) = -20\lg\frac{10}{3} = -10.5(\text{dB})$$

3) Calculating the amplitude margin, we have

$$A(\omega_g) = \frac{20}{\sqrt{5}\sqrt{5+1}\sqrt{0.04 \times 5 + 1}} = \frac{10}{3}$$

$$L_g = -20\lg A(\omega_g) = -20\lg \frac{10}{3} = -10.5 (\text{dB})$$

Therefore, the system is unstable when $K = 20$.

Example 5-11. The open-loop logarithmic amplitude-frequency characteristic of a system is shown in Fig. 5-15. Do the following solution and drawing: 1) figure out the open-loop transfer function $G_K(s)$; 2) calculate the amplitude crossing frequency ω_c; 3) calculate the phase margin γ; 4) draw the open-loop logarithmic phase-frequency characteristic curve $\varphi(\omega)$.

Solution: 1) Figure out the open-loop transfer function of the system. According to each segment slope of $L(\omega)$, we have

$$G_K(s) = \frac{K(T_1 s + 1)}{s^2(T_2 s + 1)} \quad (T_1 > T_2)$$

The initial segment slope of $L(\omega)$ is -40dB/dec, and it intersects the 0dB line at $\omega_0 = 20$, thus we have

$$K = \omega_0^2 = 20^2 = 400$$

Based on the logarithmic amplitude-frequency characteristic of the frequency between ω_1 and ω_0, the slope of an extension line of the low-frequency asymptotic curve is -40dB/dec, we have

$$\frac{0 - 20\lg 4}{\lg 20 - \lg \omega_1} = -40 \Rightarrow \omega_1 = 10$$

so

$$T_1 = \frac{1}{\omega_1} = 0.1$$

According to Fig 5-15, we have $\omega_2 = 100$, $T_2 = \dfrac{1}{\omega_2} = 0.01$

$$G_K(s) = \frac{400(0.1s + 1)}{s^2(0.01s + 1)}$$

2) Calculate the amplitude crossing frequency ω_c. Based on the logarithmic amplitude-frequency characteristic of the frequency between ω_1 and ω_0, the slope of an extension line of the low-frequency asymptotic curve is -20dB/dec, then we have

$$\frac{0 - 20\lg 4}{\lg \omega_c - \lg 10} = -20 \Rightarrow \omega_c = 40$$

3) Calculate the phase margin, we have

$$\gamma = 180° + \varphi(\omega_c) = 180° + (-180° + \arctan 0.1\omega_c - \arctan 0.01\omega_c) = 54.2°$$

4) Plot the curve $\varphi(\omega)$, as is shown in Fig. 5-15. According to the figure, when $\omega \to \infty$, $\varphi(\omega) \to -180°$, we have $L_g = \infty$. Therefore, the system has good relative stability.

5.2 Error Analysis of Control Systems

With an input signal, the response output of the control system is divided into two stages: the

可见，当 $K=20$ 时系统不稳定。

例 5-11 某系统的开环对数幅频特性如图 5-15 所示。完成如下求解及作图：1) 求开环传递函数 $G_K(s)$；2) 求幅值穿越频率 ω_c；3) 求相位裕度 γ；4) 概略绘出开环对数相频特性曲线 $\varphi(\omega)$。

解：1) 求系统的开环传递函数。由 $L(\omega)$ 的各段斜率可知

$$G_K(s) = \frac{K(T_1 s+1)}{s^2(T_2 s+1)} \quad (T_1 > T_2)$$

$L(\omega)$ 的起始段斜率为 -40dB/dec，且与 0dB 线相交于 $\omega_0 = 20$ 处，故

$$K = \omega_0^2 = 20^2 = 400$$

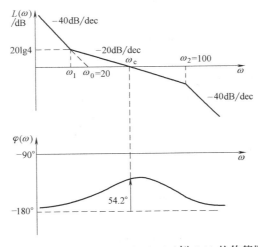

Fig. 5-15 Bode plot of example 5-11（例 5-11 的伯德图）

根据频率在 ω_1 与 ω_0 之间的对数幅频特性，可知低频段渐近线延长线的斜率为 -40dB/dec。可得

$$\frac{0 - 20\lg 4}{\lg 20 - \lg \omega_1} = -40 \Rightarrow \omega_1 = 10$$

故

$$T_1 = \frac{1}{\omega_1} = 0.1$$

由图 5-15 可知

$$\omega_2 = 100, \quad T_2 = \frac{1}{\omega_2} = 0.01$$

$$G_K(s) = \frac{400(0.1s+1)}{s^2(0.01s+1)}$$

2) 计算幅值穿越频率 ω_c。根据频率在 ω_1 与 ω_c 之间的对数幅频特性，可知低频段渐近线延长线的斜率为 -20dB/dec。可得

$$\frac{0 - 20\lg 4}{\lg \omega_c - \lg 10} = -20 \Rightarrow \omega_c = 40$$

3) 计算相位裕度 γ，有

$$\gamma = 180° + \varphi(\omega_c) = 180° + (-180° + \arctan 0.1\omega_c - \arctan 0.01\omega_c) = 54.2°$$

transient process and steady-state process. The transient process reflects the dynamic response performance of the control system, which is mainly reflected in the response speed and the response stability of the system to the input signal. For a stable system, the transient process will gradually disappear over time. The steady-state process reflects the steady-state response performance of the control system, which is mainly reflected in the accuracy of the system tracking the input signal and the capacity of suppressing the disturbance signal. Therefore, the steady-state performance of the control system mainly reflects the variation law and size of the error under the influence of the input signal or the disturbance signal after the system reaches a steady state.

5.2.1 System error and error transfer function

1. System error

The so-called error is the difference between the desired output (or the reference input) and the actual output. When elements' performance is imperfect or there are nonlinear factors such as dry friction, clearance, dead zones and so on, there will be errors. when there is no random disturbance in the system and the elements are also ideally linear, there may still be errors. This section will focus on the errors in the latter case. The output response process of the system is divided into the transient response and steady-state response, and the corresponding errors are also divided into the transient error and steady-state error. i. e.

$$e(t) = e_t(t) + e_s(t) \tag{5-38}$$

where $e(t)$ is the system error; $e_t(t)$ is the error of the system transient output, which is called the transient error, and for a stable system, it is generally considered $e_t(t) = 0$ after the adjustment process ends; $e_s(t)$ is the error of the system steady state output, which is called the steady-state error. Different from the steady-state error, the dynamic error is a function of time, which can provide the rule of control error changing with time when the system is in steady state. The final value of the steady-state error is called the static error, i. e $e_{ss} = \lim_{t \to \infty} e(t)$.

For the control system shown in Fig. 5-16, the error is the difference between the desired output $x_{or}(t)$ and its actual output $x_o(t)$, i. e.

$$e(t) = x_{or}(t) - x_o(t) \tag{5-39}$$

Taking the Laplace transformation, we have

$$E(s) = X_{or}(s) - X_o(s) \tag{5-40}$$

In the system, when the feedback transfer function $H(s) = 1$, the reference input $x_i(t)$ of the system is the expected output $x_{or}(t)$. Thus, the output error $e(t)$ of the system is the deviation $\varepsilon(t)$ of the system, i. e.

$$E(s) = \varepsilon(s) = X_i(s) - X_o(s) \tag{5-41}$$

When the feedback transfer function $H(s) \neq 1$, the output error $e(t)$ of the system is not the same as the deviation $\varepsilon(t)$ of the system. Thus, it can be seen from Fig. 5-16 that the deviation of the system is

$$\varepsilon(s) = X_i(s) - H(s) X_o(s) \tag{5-42}$$

There is internal relation between the system deviation $\varepsilon(s)$ and error $E(s)$. If the deviation is

4) 作 $\varphi(\omega)$ 曲线，如图 5-15 所示，由图可知当 $\omega \to \infty$，$\varphi(\omega) \to -180°$，故 $L_g = \infty$，该系统具有良好的相对稳定性。

5.2 控制系统的误差分析

控制系统在输入信号作用下，响应输出分为瞬态过程和稳态过程两个阶段。瞬态过程反映控制系统的动态响应性能，主要体现在系统对输入信号的响应速度和响应的平稳性这两个方面。对于稳定系统，瞬态过程将随着时间的推移而逐渐消失。稳态过程反映控制系统的稳态响应性能，它主要表现在系统跟踪输入信号的准确性和抑制干扰信号的能力。因此，控制系统的稳态性能主要反映系统到达稳定状态后，在输入信号或干扰信号作用下的误差变化规律和大小等情况。

5.2.1 系统误差和误差传递函数

1. 系统误差

所谓误差是指期望输出（或参考输入）与实际输出的差。当元件性能有缺陷，或者存在干摩擦、间隙、死区等非线性因素时，系统会存在误差；没有随机干扰作用且元件满足理想线性条件时，系统仍然可能存在的误差。本节将重点讨论后一种情况下的误差。系统的输出响应过程分为瞬态响应和稳态响应，相应的误差也分为瞬态误差和稳态误差，即

$$e(t) = e_t(t) + e_s(t) \tag{5-38}$$

式中，$e(t)$ 为系统的误差；$e_t(t)$ 是系统的瞬态输出误差，称为瞬态误差，对于稳定系统，一般可认为调整过程结束后 $e_t(t) = 0$；$e_s(t)$ 是系统的稳态输出误差，称为稳态误差。与稳态误差不同，动态误差是以时间为变量的函数，能提供系统为稳态时控制误差随时间变化的规律。稳态误差的终值称为静态误差，即 $e_{ss} = \lim_{t \to \infty} e(t)$。

对于图 5-16 所示控制系统，误差即为所期望的输出 $x_{or}(t)$ 与其实际输出 $x_o(t)$ 之差，即

$$e(t) = x_{or}(t) - x_o(t) \tag{5-39}$$

对上式进行拉氏变换，有

$$E(s) = X_{or}(s) - X_o(s) \tag{5-40}$$

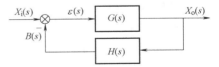

Fig. 5-16 The block diagram of a control system
（控制系统框图）

在该系统中，当反馈传递函数 $H(s) = 1$ 时，系统的参考输入 $x_i(t)$ 就是期望输出 $x_{or}(t)$。因此，系统的输出误差 $e(t)$ 就是系统的偏差 $\varepsilon(t)$，即

$$E(s) = \varepsilon(s) = X_i(s) - X_o(s) \tag{5-41}$$

当反馈传递函数 $H(s) \neq 1$ 时，系统的输出误差 $e(t)$ 与系统的偏差 $\varepsilon(t)$ 并不相同。此时，由图 5-16 可得系统的偏差为

$$\varepsilon(s) = X_i(s) - H(s)X_o(s) \tag{5-42}$$

系统偏差 $\varepsilon(s)$ 与误差 $E(s)$ 之间存在内在联系。偏差为零则误差也为零，偏差越大则误差越大。由前面章节的分析可知，闭环控制系统之所以对输出 $X_o(s)$ 起自动控制作用，就在于运用偏差 $\varepsilon(s)$ 进行控制。当 $X_o(s) \neq X_{or}(s)$ 时，由于误差 $E(s)$ 或偏差 $\varepsilon(s)$ 不等于零，故系统应利用偏差 $\varepsilon(s)$ 起调节控制作用，力图将 $X_o(s)$ 调节到 $X_{or}(s)$ 值，使误差

zero, the error will be zero. The larger the deviation is, the larger the error is. According to the analysis in the previous chapters, the reason why the closed-loop control system automatically controls the output $X_o(s)$ is that the system uses the deviation $\varepsilon(s)$ for control. When $X_o(s) \neq X_{or}(s)$, since the error or deviation is not equal to zero, the system uses the deviation $\varepsilon(s)$ as an adjustment and control role, aiming at adjusting $X_o(s)$ to $X_{or}(s)$, so that the error $E(s)$ or deviation $\varepsilon(s)$ becomes zero. Conversely, when $X_o(s) = X_{or}(s)$, and the error $E(s)$ or deviation $\varepsilon(s)$ is equal to zero, the adjustment of $\varepsilon(s)$ to $X_o(s)$ is no longer performed. According to such deviation adjustment principle, when the output $X_o(s)$ is equal to the desired output $X_{or}(s)$, the deviation $\varepsilon(s)$ of the system is zero. According to equation (5-42), we have

$$0 = X_i(s) - H(s)X_{or}(s) \Rightarrow X_{or}(s) = \frac{X_i(s)}{H(s)}$$

According to equation (5-40), we have

$$E(s) = X_{or}(s) - X_o(s) = \frac{X_i(s)}{H(s)} - X_o(s)$$

$$= \frac{1}{H(s)}[X_i(s) - H(s)X_o(s)] = \frac{\varepsilon(s)}{H(s)} \tag{5-43}$$

Thus, the relation between the deviation and the error of the system is

$$\varepsilon(s) = H(s)E(s) \tag{5-44}$$

For a unit feedback system, according to $H(s) = 1$, we have $\varepsilon(s) = E(s)$, then $\varepsilon(t) = e(t)$; for a non-unit feedback system, we have $\varepsilon(t) \neq e(t)$ in general.

According to above analysis, the system error can be obtained by calculating the system deviation. Therefore, if it is not specified in the following contents, the error analysis of the system is carried out by using the deviation instead of the error for the convenience.

2. The transfer function of the system error

(As is shown in Fig. 5-17), for a control system with both a given input and disturbance input the Laplace transform of the system error can be obtained as

$$E(s) = \frac{1}{1 + G_1(s)G_2(s)H(s)}X_i(s) + \frac{-G_2(s)H(s)}{1 + G_1(s)G_2(s)H(s)}N(s)$$

$$= \Phi_{X_i}(s)X_i(s) + \Phi_N(s)N(s) \tag{5-45}$$

where $G_1(s)$, $G_2(s)$ and $H(s)$ depends on structural parameters of the system; $\Phi_{X_i}(s)$ is the transfer function of an error signal to the reference input signal when there is no disturbance signal; $\Phi_N(s)$ is the transfer function of an error signal to the disturbance signal when there is no input signal, i. e.

$$\begin{cases} \Phi_{X_i}(s) = \dfrac{1}{1 + G_1(s)G_2(s)H(s)} \\ \Phi_N(s) = \dfrac{-G_2(s)H(s)}{1 + G_1(s)G_2(s)H(s)} \end{cases} \tag{5-46}$$

The errors caused by the reference input signal and the disturbance input signal are respectively

$E(s)$ 或偏差 $\varepsilon(s)$ 变为零。反之，当 $X_o(s) = X_{or}(s)$ 时，应有误差 $E(s)$ 或偏差 $\varepsilon(s)$ 等于零，而使 $\varepsilon(s)$ 不再对 $X_o(s)$ 有调节作用。根据这种偏差调节原理，当输出量 $X_o(s)$ 等于期望输出量 $X_{or}(s)$ 时，系统的偏差 $\varepsilon(s)$ 即为零。由式（5-42）得

$$0 = X_i(s) - H(s)X_{or}(s) \Rightarrow X_{or}(s) = \frac{X_i(s)}{H(s)}$$

接着由式（5-40）得

$$E(s) = X_{or}(s) - X_o(s) = \frac{X_i(s)}{H(s)} - X_o(s)$$

$$= \frac{1}{H(s)}[X_i(s) - H(s)X_o(s)] = \frac{\varepsilon(s)}{H(s)} \tag{5-43}$$

因此，系统的偏差与误差之间的关系为

$$\varepsilon(s) = H(s)E(s) \tag{5-44}$$

对单位反馈系统来说，$H(s) = 1$，有 $\varepsilon(s) = E(s)$，则 $\varepsilon(t) = e(t)$，对非单位反馈系统，一般来说 $\varepsilon(t) \neq e(t)$。

由以上分析可知，求得系统偏差即可求出系统误差，因此，后文如不特别说明，为分析方便，均用偏差代替误差而对系统进行误差分析。

2. 系统误差的传递函数

如图 5-17 所示，对于给定输入和干扰输入同时作用的控制系统，可求得系统误差的拉氏变换，即

$$E(s) = \frac{1}{1 + G_1(s)G_2(s)H(s)} X_i(s) + \frac{-G_2(s)H(s)}{1 + G_1(s)G_2(s)H(s)} N(s)$$

$$= \Phi_{X_i}(s)X_i(s) + \Phi_N(s)N(s) \tag{5-45}$$

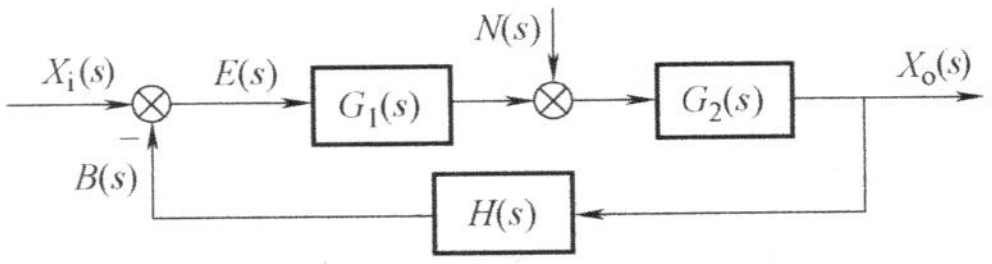

Fig. 5-17 The control system under disturbance（干扰作用下的控制系统）

式中，$G_1(s)$、$G_2(s)$、$H(s)$ 取决于系统的结构参数；$\Phi_{X_i}(s)$ 为无干扰信号时误差信号对于参考输入信号的传递函数；$\Phi_N(s)$ 为无输入信号时误差信号对于干扰输入信号的传递函数。即

$$\begin{cases} \Phi_{X_i}(s) = \dfrac{1}{1 + G_1(s)G_2(s)H(s)} \\ \Phi_N(s) = \dfrac{-G_2(s)H(s)}{1 + G_1(s)G_2(s)H(s)} \end{cases} \tag{5-46}$$

由此得参考输入信号和干扰输入信号引起的误差分别为

$$\begin{cases} E_{X_i}(s) = \dfrac{1}{1 + G_1(s)G_2(s)H(s)} X_i(s) \\ E_N(s) = \dfrac{-G_2(s)H(s)}{1 + G_1(s)G_2(s)H(s)} N(s) \end{cases} \tag{5-47}$$

$$\begin{cases} E_{X_i}(s) = \dfrac{1}{1+G_1(s)G_2(s)H(s)} X_i(s) \\ E_N(s) = \dfrac{-G_2(s)H(s)}{1+G_1(s)G_2(s)H(s)} N(s) \end{cases} \quad (5\text{-}47)$$

Using the final value theorem of Laplace transform, the static errors are respectively

$$\begin{cases} \varepsilon_{si} = \lim\limits_{s\to 0} s E_{X_i}(s) = \lim\limits_{s\to 0} \dfrac{sX_i(s)}{1+G_1(s)G_2(s)H(s)} \\ \varepsilon_{sn} = \lim\limits_{s\to 0} s E_N(s) = \lim\limits_{s\to 0} \dfrac{-sG_2(s)H(s)N(s)}{1+G_1(s)G_2(s)H(s)} \end{cases} \quad (5\text{-}48)$$

It can be seen that the system error $E(s)$ is the sum of the error $X_i(s)$ caused by the reference input signal and the error $N(s)$ caused by the disturbance input signal, i.e.

$$\varepsilon_s = \lim_{t\to\infty} \varepsilon(t) = \lim_{t\to 0} sE(s) = \varepsilon_{si} + \varepsilon_{sn} \quad (5\text{-}49)$$

According to above analysis, the steady-state error of the system consists of two parts: the following error and the disturbance error. They are related not only to the structure and parameters of the system, but also to the magnitude, variation and action point of variables (like the input and disturbance).

If the general form of the system open loop transfer function is

$$\begin{aligned} G_K(s) &= G(s)H(s) = G_1(s)G_2(s)H(s) \\ &= \dfrac{K\prod\limits_{k=1}^{p}(T_k s + 1)\prod\limits_{l=1}^{q}(T_l^2 s^2 + 2\xi_l T_l s + 1)}{s^v \prod\limits_{i=1}^{g}(T_i s + 1)\prod\limits_{j=1}^{h}(T_j^2 s^2 + 2\xi_j T_j s + 1)} e^{-T_d s} \\ &= \dfrac{K}{s^v} G_0(s) \end{aligned} \quad (5\text{-}50)$$

where K is an open-loop transfer coefficient (or open-loop amplification coefficient) of the system; v is the number of integrals in the open-loop transfer function; m is the order of the system's molecular polynomial and $m = p+2q$; n is the order of the system's denominator polynomial and $n = v+g+2h$; for an actual system, $n \geqslant m$;

$$G_0(s) = \dfrac{\prod\limits_{k=1}^{p}(T_k s + 1)\prod\limits_{l=1}^{q}(T_l^2 s^2 + 2\xi_l T_l s + 1)}{\prod\limits_{i=1}^{g}(T_i s + 1)\prod\limits_{j=1}^{h}(T_j^2 s^2 + 2\xi_j T_j s + 1)} e^{-T_d s}$$

$G_0(s)$ is composed of p first-order differential elements, q second-order differential elements, g inertial elements, h oscillation elements and time delay element, where T_k, T_l, T_i, T_j, T_d represent the time constant of the corresponding element respectively. Obviously, when $s \to 0$, we have $G_0(s) = G_0(0) = 1$. According to equation (5-47) and equation (5-48), the static error of the system is obtained as

$$\varepsilon_{si} = \lim_{s\to 0} s E_{X_i}(s) = \lim_{s\to 0} \dfrac{sX_i(s)}{1+K/s^v} \quad (5\text{-}51)$$

利用拉氏变换的终值定理，其静态误差分别为

$$\begin{cases} \varepsilon_{si} = \lim_{s\to 0} sE_{X_i}(s) = \lim_{s\to 0} \dfrac{sX_i(s)}{1+G_1(s)G_2(s)H(s)} \\ \varepsilon_{sn} = \lim_{s\to 0} sE_N(s) = \lim_{s\to 0} \dfrac{-sG_2(s)H(s)N(s)}{1+G_1(s)G_2(s)H(s)} \end{cases} \quad (5\text{-}48)$$

可见，系统的误差 $E(s)$ 就是参考输入信号 $X_i(s)$ 引起的误差和干扰输入信号 $N(s)$ 引起的误差之和，即

$$\varepsilon_s = \lim_{t\to\infty} \varepsilon(t) = \lim_{s\to 0} sE(s) = \varepsilon_{si} + \varepsilon_{sn} \quad (5\text{-}49)$$

由以上分析可知，系统的稳态误差由跟随误差和扰动误差两部分组成。它们不仅与系统的结构、参数有关，而且还与变量（如输入量和扰动量）的大小、变化规律和作用点有关。

若系统开环传递函数的一般形式为

$$G_K(s) = G(s)H(s) = G_1(s)G_2(s)H(s)$$

$$= \dfrac{K\prod_{k=1}^{p}(T_k s + 1)\prod_{l=1}^{q}(T_l^2 s^2 + 2\xi_l T_l s + 1)}{s^v \prod_{i=1}^{g}(T_i s + 1)\prod_{j=1}^{h}(T_j^2 s^2 + 2\xi_j T_j s + 1)} e^{-T_d s}$$

$$= \dfrac{K}{s^v} G_0(s) \quad (5\text{-}50)$$

式中，K 为系统开环传递系数（或开环放大系数）；v 为开环传递函数所包含积分环节的个数；$p+2q=m$ 为系统分子多项式的阶数；$v+g+2h=n$ 为系统分母多项式的阶数。对于实际的系统有 $n \geqslant m$；

$$G_0(s) = \dfrac{\prod_{k=1}^{p}(T_k s + 1)\prod_{l=1}^{q}(T_l^2 s^2 + 2\xi_l T_l s + 1)}{\prod_{i=1}^{g}(T_i s + 1)\prod_{j=1}^{h}(T_j^2 s^2 + 2\xi_j T_j s + 1)} e^{-T_d s}$$

$G_0(s)$ 即由 p 个一阶微分环节、q 个二阶微分环节、g 个惯性环节、h 个二阶振荡环节和延迟环节等若干典型环节组成，其中 T_k、T_l、T_i、T_j、T_d 分别代表对应环节的时间常数。显然，当 $s\to 0$，$G_0(s) = G_0(0) = 1$。由式（5-47）和式（5-48）得系统静态误差的计算公式为

$$\varepsilon_{si} = \lim_{s\to 0} sE_{X_i}(s) = \lim_{s\to 0} \dfrac{sX_i(s)}{1+K/s^v} \quad (5\text{-}51)$$

$$\varepsilon_{sn} = \lim_{s\to 0} sE_N(s) = \lim_{s\to 0} \dfrac{-sG_2(s)H(s)N(s)}{1+K/s^v} \quad (5\text{-}52)$$

可见，系统静态误差只与系统的类型（积分环节的个数）、系统开环传递系数和给定（或干扰）输入信号有关，而与开环传递函数的结构参数无关。

5.2.2 系统静态误差分析与计算

工程实际中，往往要求系统静态误差必须在某一给定范围之内。只有静态误差处于这个给定范围之内，分析和研究系统的动态误差才有实际意义。下面分别就给定输入信号和干扰

$$\varepsilon_{sn} = \lim_{s \to 0} sE_N(s) = \lim_{s \to 0} \frac{-sG_2(s)H(s)N(s)}{1+K/s^v} \tag{5-52}$$

It can be seen that the static error of the system is only related to the type of the system (number of integral elements), open-loop transfer coefficient and given input signal (or disturbance signal), but has nothing to do with the structural parameters of the open-loop transfer function.

5.2.2 Analysis and calculation of system static errors

In engineering practice, it is often required that the static error of the system must be within a given value range. It is of practical significance to analyze the dynamic error of the system only if the static error is within the given range. The static errors of the system with the given input signal and the disturbance input signal are further discussed below.

1. Calculation of static errors with a given signal

1) When the input signal is a unit step signal, the static error of the system is

$$\varepsilon_{si} = \lim_{t \to 0} sE_{X_i}(s) = \lim_{s \to 0} s\frac{X_i(s)}{1+G(s)H(s)} = \lim_{s \to 0} \frac{1}{1+G(s)H(s)} = \frac{1}{1+K_p}$$

where K_p is the static error coefficient on position and we have

$$K_p = \lim_{s \to 0} G(s)H(s) = \lim_{s \to 0} \frac{KG_0(s)}{s^v} = \lim_{s \to 0} \frac{K}{s^v} = \begin{cases} K & (v=0) \\ \infty & (v=1,2,\cdots) \end{cases} \tag{5-53}$$

Thus, the static error of the system with the unit step signal is

$$\varepsilon_{si} = \frac{1}{1+K_p} = \begin{cases} \dfrac{1}{1+K} & (v=0) \\ 0 & (v=1,2,\cdots) \end{cases} \tag{5-54}$$

According to above equation, when the given signal is a step signal, a fixed error value $\dfrac{1}{1+K}$ will be generated when the system is a type-0 system; it will not generate the static error when the system is a type-I, type-II or above system. That is, the output accurately reflects the input. It can be seen that when the given signal is a step signal, the steady-state value of the system step response will be error-free if there is an integral element in the system's open-loop transfer function, but the steady-state value will have an error when there is no integral element. In order to reduce the error, the amplification factor should be appropriately increased. But if the value of K is too large, it will affect the relative stability of the system.

2) When the input signal is a unit ramp signal, i.e. $X_i(s) = \dfrac{1}{s^2}$, the static error of the system is

$$\varepsilon_{si} = \lim_{s \to 0} sE_{X_i}(s) = \lim_{s \to 0} s\frac{X_i(s)}{1+G(s)H(s)}$$

$$= \lim_{s \to 0} \frac{1}{s+sG(s)H(s)} = \frac{1}{\lim_{s \to 0} sG(s)H(s)} = \frac{1}{K_v}$$

输入信号作用下系统的静态误差进行进一步讨论。

1. 给定信号下静态误差的计算

1) 当输入信号为单位阶跃信号时，即 $X_i(s) = \dfrac{1}{s}$，则系统的静态误差为

$$\varepsilon_{si} = \lim_{t \to 0} s E_{X_i}(s) = \lim_{s \to 0} s \frac{X_i(s)}{1+G(s)H(s)} = \lim_{s \to 0} \frac{1}{1+G(s)H(s)} = \frac{1}{1+K_p}$$

式中，K_p 称为位置静态误差系数，且有

$$K_p = \lim_{s \to 0} G(s)H(s) = \lim_{s \to 0} \frac{KG_0(s)}{s^v} = \lim_{s \to 0} \frac{K}{s^v} = \begin{cases} K & (v=0) \\ \infty & (v=1,2,\cdots) \end{cases} \tag{5-53}$$

所以，系统在单位阶跃信号作用下的静态误差为

$$\varepsilon_{si} = \frac{1}{1+K_p} = \begin{cases} \dfrac{1}{1+K} & (v=0) \\ 0 & (v=1,2,\cdots) \end{cases} \tag{5-54}$$

由上式可知，在给定信号为阶跃信号且系统为 0 型系统时，将产生固定的误差值 $\dfrac{1}{1+K}$；当系统为 Ⅰ 型、Ⅱ 型及以上系统时，将不会产生静态误差，也就是说输出可以准确地反映输入。可见当给定信号为阶跃信号且系统开环传递函数中有积分环节时，系统阶跃响应的稳态值将是有误差的，而没有积分环节时，稳态值是有误差的，为了减小误差，应适当提高放大倍数。但过大的 K 值将影响系统的相对稳定性。

2) 当输入信号为单位斜坡信号时，即 $X_i(s) = \dfrac{1}{s^2}$，系统的静态误差为

$$\begin{aligned}\varepsilon_{si} &= \lim_{s \to 0} s E_{X_i}(s) = \lim_{s \to 0} s \frac{X_i(s)}{1+G(s)H(s)} \\ &= \lim_{s \to 0} \frac{1}{s+sG(s)H(s)} = \frac{1}{\lim_{s \to 0} sG(s)H(s)} = \frac{1}{K_v}\end{aligned}$$

式中，K_v 称为速度静态误差系数，且有

$$\begin{aligned}K_v &= \lim_{s \to 0} sG(s)H(s) = \lim_{s \to 0} \frac{sKG_0(s)}{s^v} = \lim_{s \to 0} \frac{K}{s^{v-1}} \\ &= \begin{cases} 0 & (v=0) \\ K & (v=1) \\ \infty & (v=2,3,\cdots) \end{cases}\end{aligned} \tag{5-55}$$

系统在单位斜坡信号作用下的静态误差为

$$\varepsilon_{si} = \frac{1}{K_v} = \begin{cases} \infty & (v=0) \\ \dfrac{1}{K} & (v=1) \\ 0 & (v=2,3,\cdots) \end{cases} \tag{5-56}$$

由此可知，若给定信号为单位斜坡信号且系统为 0 型系统时，将产生无穷大的静态误差；当系统为 Ⅰ 型系统时，将产生固定的静态误差值 $1/K$；当系统为 Ⅱ 型及以上系统时，其静态误差为零。

where K_v is the static error coefficient on speed and we have

$$K_v = \lim_{s \to 0} sG(s)H(s) = \lim_{s \to 0} \frac{sKG_0(s)}{s^v} = \lim_{s \to 0} \frac{K}{s^{v-1}}$$

$$= \begin{cases} 0 & (v=0) \\ K & (v=1) \\ \infty & (v=2,3,\cdots) \end{cases} \tag{5-55}$$

The static error of the system with the unit ramp signal is

$$\varepsilon_{si} = \frac{1}{K_v} = \begin{cases} \infty & (v=0) \\ \dfrac{1}{K} & (v=1) \\ 0 & (v=2,3,\cdots) \end{cases} \tag{5-56}$$

Therefore, if the given signal is a unit ramp signal, it will generate an infinite static error when the system is a type-0 system; it will generate a static error of fixed value $1/K$ when the system is a type-I system; the static error will be zero when the system is a type-II or above system.

3) When the input signal is a unit parabolic signal, i.e. $X_i(s) = \dfrac{1}{s^3}$ the static error of the system is

$$\varepsilon_{si} = \lim_{s \to 0} sE_{X_i}(s) = \lim_{s \to 0} s\frac{X_i(s)}{1+G(s)H(s)} = \lim_{s \to 0} \frac{1}{s^2 + s^2 G(s)H(s)}$$

$$= \frac{1}{\lim_{s \to 0} s^2 G(s)H(s)} = \frac{1}{K_a}$$

where K_a is the static error coefficient on acceleration and we have

$$K_a = \lim_{s \to 0} s^2 G(s)H(s) = \lim_{s \to 0} \frac{K}{s^{v-2}} = \begin{cases} 0 & (v=0,1) \\ K & (v=2) \\ \infty & (v=3,4,\cdots) \end{cases} \tag{5-57}$$

The static error of the system with the unit parabolic signal is

$$\varepsilon_{si} = \frac{1}{K_a} = \begin{cases} \infty & (v=0,1) \\ \dfrac{1}{K} & (v=2) \\ 0 & (v=3,4,\cdots) \end{cases} \tag{5-58}$$

Therefore, if the given signal is a unit parabolic signal, it will generate infinite static error when the system is a type-0 or type-I system; it will generate a static error of fixed value $1/K$ when the system is a type-II system; the static error will be zero when the system is a type-III or above system.

The static errors of different types of systems with different input signals are shown in Fig. 5-18 ~ Fig. 5-20.

In summary, the static errors and their static error coefficients of different types of systems with different inputs are listed in Table 5-1. According to above analysis, the static error (following er-

3) 当输入信号为单位抛物线信号时，有 $X_i(s) = \dfrac{1}{s^3}$，系统的静态误差为

$$\varepsilon_{si} = \lim_{s \to 0} s E_{X_i}(s) = \lim_{s \to 0} s \dfrac{X_i(s)}{1+G(s)H(s)} = \lim_{s \to 0} \dfrac{1}{s^2 + s^2 G(s)H(s)}$$

$$= \dfrac{1}{\lim\limits_{s \to 0} s^2 G(s)H(s)} = \dfrac{1}{K_a}$$

式中，K_a 称为加速度静态误差系数，且有

$$K_a = \lim_{s \to 0} s^2 G(s)H(s) = \lim_{s \to 0} \dfrac{K}{s^{v-2}} = \begin{cases} 0 & (v = 0,1) \\ K & (v = 2) \\ \infty & (v = 3,4,\cdots) \end{cases} \tag{5-57}$$

系统在单位抛物线信号作用下的静态误差为

$$\varepsilon_{si} = \dfrac{1}{K_a} = \begin{cases} \infty & (v = 0,1) \\ \dfrac{1}{K} & (v = 2) \\ 0 & (v = 3,4,\cdots) \end{cases} \tag{5-58}$$

由此可知，当给定信号为单位抛物线信号且系统为 0 型或 I 型系统时，将产生无穷大的静态误差；当系统为 II 型系统时，将产生固定的静态误差值 $1/K$；当系统为 III 型及以上系统时，其静态误差为零。

不同类型系统在不同输入信号作用下的静态误差情况如图 5-18 ~ 图 5-20 所示。

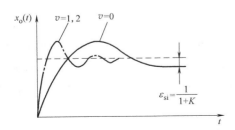

Fig. 5-18 The deviation with unit step signal of different types of systems

（不同类型系统对单位阶跃信号产生的偏差）

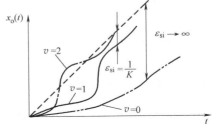

Fig. 5-19 The deviation with unit ramp signal of different types of systems

（不同类型系统对单位斜坡信号产生的偏差）

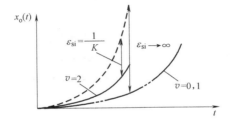

Fig. 5-20 The deviation with unit parabolic signal of different types of systems

（不同类型系统对单位抛物线信号产生的偏差）

ror) of the system is related to the type of system (the number of integral elements) v and the system open-loop transfer (amplification) coefficient K with the given input signal $X_i(s)$. The more the value v is and the larger the value K is, the higher the steady-state accuracy is, that is, the better the steady-state performance of the system is. When the type of the system is increased, the static error of the system can be reduced or eliminated. However, when the system adopts the method of increasing the number of integral elements in the open-loop transfer function to increase the type of the system, the stability of the system will be deteriorated. The reason is that when the open-loop transfer function contains more than two integral elements, it is difficult to ensure the stability of the system. Therefore, it is hard to implement for a type-Ⅲ or higher system, and it is rarely used. Increasing the open-loop gain K of the system can also effectively reduce the static error, but it cannot eliminate the error, and it will also make the stability of the system worse.

Example 5-12 The block diagram of a system is shown in Fig. 5-21. The input signal is $x_i(t) = 1+t+t^2$. Calculate the system static error ε_{si}.

Solution: The open loop transfer function of the system is

$$G_K(s) = \frac{k_1 k_2 (\tau s+1)}{s^2 (Ts+1)}$$

It can be seen that the system is a type-Ⅱ system, the open loop gain $K = k_1 k_2$. From the composition of input signals, three input signals, including the unit step signal, unit ramp signal and parabolic signal, simultaneously are applied on the system. According to the principle of linear superposition, the static error of the system is

$$\varepsilon_{si} = \varepsilon_{si1} + \varepsilon_{si2} + \varepsilon_{si3}$$

where ε_{si1} is the static deviation generated by the input of unit step signal, whose value is 0 according to Table 5-1; ε_{si2} is the static deviation generated by the input of unit ramp signal, whose value is also 0 according to Table 5-1; ε_{si3} is the static deviation generated by the input of unit parabolic signal, whose value is $\dfrac{2}{k_1 k_2}$ according to Table 5-1.

Therefore, the static error of the system is

$$\varepsilon_{si} = \varepsilon_{si1} + \varepsilon_{si2} + \varepsilon_{si3} = 0+0+\frac{2}{k_1 k_2} = \frac{2}{k_1 k_2}$$

Example 5-13 A unit negative feedback system whose forward channel transfer function is

$$G_K(s) = \frac{5}{s(s+2)(s+4)}$$

When the input is $x_i(t) = 1+2t$, calculate the static error of the system.

Solution: Transforming the given function to the standard form, we have

$$G_K(s) = \frac{\dfrac{5}{8}}{s\left(\dfrac{1}{2}s+1\right)\left(\dfrac{1}{4}s+1\right)}$$

According to the standard form of the open-loop transfer function, its open-loop gain is $K =$

第5章 控制系统的性能分析

综上所述，在不同输入信号作用下不同类型系统中的静态误差及相应的静态误差系数可以归纳为表 5-1。由以上分析可知，在给定输入信号 $X_i(s)$ 作用下，系统的静态误差（跟随误差）与系统的类型（积分环节的个数）v、系统开环传递（放大）系数 K 有关。若 v 愈大，K 愈大，则跟随稳态精度愈高，即系统的稳态性愈好。提高系统的型别可以减小或消除系统的静态误差，然而当系统采用增加开环传递函数中积分环节数目的办法来提高系统的型别时，系统的稳定性将变差。由于系统的开环传递函数包含两个以上积分环节时，要保证系统的稳定性比较困难，因此Ⅲ型或更高型的系统较难实现，实际上也极少采用。增大系统的开环增益 K，也可以有效减小静态误差，但不能消除误差，而且会使系统的稳定性变差。

Table 5-1　The static errors of different types of systems with different inputs
（在不同输入作用下不同类型系统中的静态误差）

System type（系统类型）	Static error coefficient on position K_p（位置静态误差系数 K_p）	Static error coefficient on speed K_v（速度静态误差系数 K_v）	Static error coefficient on acceleration K_a（加速度静态误差系数 K_a）	Static error ε_{si} with different inputs（不同输入时的静态误差 ε_{si}）		
				Step input（阶跃输入）	Ramp input（斜坡输入）	Parabolic input（抛物线输入）
0	K	0	0	$1/(1+K)$	∞	∞
Ⅰ	∞	K	0	0	$1/K$	∞
Ⅱ	∞	∞	K	0	0	$1/K$
Ⅲ and above	∞	∞	∞	0	0	0

例 5-12　某系统的框图如图 5-21 所示。已知输入信号 $x_i(t) = 1+t+t^2$，求系统的静态误差 ε_{si}。

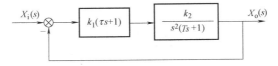

Fig. 5-21　The block diagram of example 5-14（例 5-12 框图）

解：求出系统的开环传递函数为

$$G_K(s) = \frac{k_1 k_2 (\tau s+1)}{s^2 (Ts+1)}$$

可见，系统是Ⅱ型系统，开环增益 $K=k_1 k_2$。由输入信号的组成可知，单位阶跃信号、单位斜坡信号和抛物线信号三种输入信号同时作用于系统。根据线性叠加原理，系统的静态误差为

$$\varepsilon_{si} = \varepsilon_{si1} + \varepsilon_{si2} + \varepsilon_{si3}$$

式中，ε_{si1} 是由单位阶跃输入信号产生的静态偏差，根据表 5-1 可知其值为 0；ε_{si2} 是由单位斜坡输入信号产生的静态偏差，根据表 5-1 可知其值也为 0；ε_{si3} 是由 2 倍的单位加速度输入信号产生的静态偏差，根据表 5-1 可知其值为 $\dfrac{2}{k_1 k_2}$。

5/8. Thus

$$e_s = \varepsilon_{si} = 0 + 2 \times \frac{8}{5} = \frac{16}{5}$$

Example 5-14 The block diagram of a system is shown in Fig. 5-22. When the input signal is $X_i(s) = \dfrac{1}{s^2}$, calculate the static error.

Solution: First, transforming the open-loop transfer function into the standard form, we have

$$G_K(s) = \frac{\dfrac{7}{8}(s+1)}{s\left(\dfrac{1}{4}s+1\right)\left(\dfrac{1}{2}s+s+1\right)}$$

According to the standard form, the open-loop gain is $K = 7/8$. Since the open-loop transfer function contains an integral element, it is a type-I system. According to Table 5-1, we have

$$\varepsilon_{si} = \frac{1}{K} = \frac{8}{7}$$

Example 5-15 The Closed-loop transfer function of a unit negative feedback system is

$$\Phi(s) = \frac{a_1 s + a_0}{s^3 + a_2 s^2 + a_1 s + a_0}$$

When the input is $x_i(t) = t^2$, calculate the static error of the system

Solution: First, figure out the open-loop transfer function. According to $x_i(t) = t^2$, we have $G_K(s) = \dfrac{\Phi(s)}{1 - \Phi(s)}$. Then

$$G_K(s) = \frac{\dfrac{a_1 s + a_0}{s^3 + a_2 s^2 + a_1 s + a_0}}{1 - \dfrac{a_1 s + a_0}{s^3 + a_2 s^2 + a_1 s + a_0}} = \frac{a_1 s + a_0}{s^3 + a_2 s^2} = \frac{a_1 s + a_0}{s^2(s + a_2)}$$

Then, the standard form can be obtained as

$$G_K(s) = \frac{a_0\left(\dfrac{a_1}{a_0}s + 1\right)}{a_2 s^2 \left(\dfrac{1}{a_2}s + 1\right)} = \frac{\dfrac{a_0}{a_2}\left(\dfrac{a_1}{a_0}s + 1\right)}{s^2 \left(\dfrac{1}{a_2}s + 1\right)}$$

According to the standard form of the open-loop transfer function, we have

$$K = \frac{a_0}{a_2}, \quad v = 2$$

$$\varepsilon_{si} = \frac{1}{K} = \frac{a_2}{a_0}$$

因此，系统的静态误差为

$$\varepsilon_{si} = \varepsilon_{si1} + \varepsilon_{si2} + \varepsilon_{si3} = 0 + 0 + \frac{2}{k_1 k_2} = \frac{2}{k_1 k_2}$$

例 5-13 某单位负反馈系统，其前向通道传递函数为

$$G_K(s) = \frac{5}{s(s+2)(s+4)}$$

当输入 $x_i(t) = 1 + 2t$ 时，求其静态误差。

解：将给定的传递函数化为标准形式，即

$$G_K(s) = \frac{\frac{5}{8}}{s\left(\frac{1}{2}s+1\right)\left(\frac{1}{4}s+1\right)}$$

由开环传递函数的标准形式可知，其开环增益 $K = 5/8$。于是，

$$e_s = \varepsilon_{si} = 0 + 2 \times \frac{8}{5} = \frac{16}{5}$$

例 5-14 某系统的框图如图 5-22 所示，当输入信号 $X_i(s) = \frac{1}{s^2}$ 时，求静态误差。

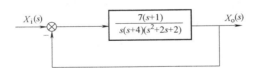

Fig. 5-22　The block diagram of example 5-14（例 5-14 框图）

解：首先将开环传递函数转化为标准形式，得

$$G_K(s) = \frac{\frac{7}{8}(s+1)}{s\left(\frac{1}{4}s+1\right)\left(\frac{1}{2}s+1\right)}$$

由标准形式可知其开环增益 $K = 7/8$，又因为开环传递函数含有一个积分环节，属于 Ⅰ 型系统，根据表 5-1 可知

$$\varepsilon_{si} = \frac{1}{K} = \frac{8}{7}$$

例 5-15 某单位负反馈系统的闭环传递函数为

$$\Phi(s) = \frac{a_1 s + a_0}{s^3 + a_2 s^2 + a_1 s + a_0}$$

当 $x_i(t) = t^2$ 时，求系统的静态误差。

解：首先求开环传递函数。由 $x_i(t) = t^2$，推出 $G_K(s) = \frac{\Phi(s)}{1 - \Phi(s)}$

2. Calculation of the error caused by the disturbance signal

When calculating the steady-state error of the system in above part, only the static error caused by the given input (control variable) is considered, and the static error caused by the disturbance input is regarded as zero. In following discussion, we regard the given input as zero, and calculate the static error caused by the disturbance input applying on the system.

As is shown in Fig. 5-23, the static error caused by the disturbance signal $N(s)$ is

$$\varepsilon_{sn} = \lim_{s \to 0} sE_N(s) = \lim_{s \to 0} \frac{-sG_2(s)H(s)N(s)}{1+G_1(s)G_2(s)H(s)}$$

It can be seen that the static error caused by the disturbance signal is not only related to the characteristics of the open-loop transfer function $G_1(s)G_2(s)H(s)$ of the system and the form of the disturbance input signal $N(s)$, but also related to the input point of the disturbance signal, that is, it is also related to $G_2(s)H(s)$. In other words, it is related to the number of integral elements and gain of the forward path before the operating point of variables. The more the integral elements are and the greater the gain is, the higher the steady-state precision of anti-perturbations will be.

Example 5-16 A servo system is shown in Fig. 5-24. The input signal $x_i(t) = t$ and the disturbance signal $n(t) = 1$ are known. Find the static error of the system.

Solution: 1) Assuming that the disturbance signal is zero, the static error with the given input signal is calculated. Since the system is a type-I system, we can obtain the open-loop gain $K = 5$ and the deviation $\varepsilon(s) = 1/5$. According to $H(s) = 1/2$, we have

$$\varepsilon_{si} = \frac{1}{5} \Big/ \frac{1}{2} = \frac{2}{5}$$

2) Let the input signal be zero, the static error with the disturbance signal is calculated as

$$X_{oN}(s) = \frac{\dfrac{2}{s(s+1)}}{1+\dfrac{5}{s(s+1)(0.2s+1)}} \cdot \frac{1}{s} = \frac{2(0.2s+1)}{s^2(s+1)(0.2s+1)+5s}$$

$$\begin{aligned} e_{sn} &= \lim_{s \to 0} sE_N(s) = -\lim_{s \to 0} sX_{oN}(s) \\ &= -\lim_{s \to 0} s \frac{2(0.2s+1)}{s^2(s+1)(0.2s+1)+5s} = -\lim_{s \to 0} \frac{2(0.2s+1)}{s(s+1)(0.2s+1)+5} \\ &= -\frac{2}{5} \end{aligned}$$

3) Calculating the total error of the system, we have

$$e_s = e_{si} + \varepsilon_{sn} = 0$$

5.2.3 Analysis and calculation of system dynamic errors

The above equations for calculating the static error of the system is derived according to the final value theorem. The advantages of this method are that it is convenient and fast to solve the differential equation and not necessary to calculate the derivative. However, since the result of the solution method is either a finite non-zero value, or the infinite value or zero, it cannot reflect the

$$G_K(s) = \frac{\dfrac{a_1s+a_0}{s^3+a_2s^2+a_1s+a_0}}{1-\dfrac{a_1s+a_0}{s^3+a_2s^2+a_1s+a_0}} = \frac{a_1s+a_0}{s^3+a_2s^2} = \frac{a_1s+a_0}{s^2(s+a_2)}$$

然后，就可以得到标准形式为

$$G_K(s) = \frac{a_0\left(\dfrac{a_1}{a_0}s+1\right)}{a_2s^2\left(\dfrac{1}{a_2}s+1\right)} = \frac{\dfrac{a_0}{a_2}\left(\dfrac{a_1}{a_0}s+1\right)}{s^2\left(\dfrac{1}{a_2}s+1\right)}$$

由开环传递函数的标准形式可知

$$K = \frac{a_0}{a_2}, \quad v = 2$$

$$\varepsilon_{si} = \frac{1}{K} = \frac{a_2}{a_0}$$

2. 干扰信号产生的误差的计算

上面在计算系统稳态误差时，只考虑给定输入量（控制量）引起的静态误差，而将干扰输入引起的静态误差视为零。下面令给定输入量为零，计算干扰输入作用于系统所产生的静态误差。

如图 5-23 所示系统，其由干扰信号 $N(s)$ 引起的静态误差为

$$\varepsilon_{sn} = \lim_{s\to 0} sE_N(s) = \lim_{s\to 0} \frac{-sG_2(s)H(s)N(s)}{1+G_1(s)G_2(s)H(s)}$$

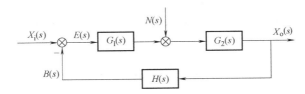

Fig. 5-23　The system with the disturbance signal（干扰信号作用下的系统）

由此可见，干扰信号引起的静态误差不仅与系统的开环传递函数 $G_1(s)G_2(s)H(s)$ 的特性和干扰输入信号 $N(s)$ 的形式有关，而且还与干扰信号的作用点有关，也即与 $G_2(s)H(s)$ 有关。换言之，它与扰动量作用点前的前向通路的积分环节个数和增益有关，扰动量作用点前的积分环节个数愈多，增益愈大，则抗扰动的稳态精度愈高。

例 5-16　设有一随动系统如图 5-24 所示，已知输入信号 $x_i(t)=t$ 及干扰信号 $n(t)=1$，

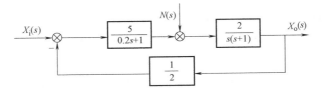

Fig. 5-24　Block diagram of example 5-16（例 5-16 的框图）

change law of steady-state error with time. In fact, the time required for the system to reach steady state after the transition process ends is not certain to be infinite. Generally speaking, when $t \geq t_s$, the system can be considered to be in steady state. Furthermore, for some functions that do not satisfy the final value theorem, such as sine function, the above method is not applicable. Thus there are certain limitations in solving the errors by using the final value theorem. A general application method for solving the steady-state error is introduced as follows. It can not only calculate the final value of the steady-state error (i.e. the static error), but also the change law of the steady-state error with time (i.e. the dynamic error).

For the control system shown in Fig. 5-23, the error transfer function is

$$\Phi_{X_i}(s) = \frac{E(s)}{X_i(s)} = \frac{1}{1+G(s)H(s)} \tag{5-59}$$

where $G(s) = G_1(s) G_2(s)$. According to the final value theorem of the Laplace transform, time $t \to \infty$ corresponds to $s = 0$. The above equation is expanded into a Taylor series in the neighborhood of $s = 0$ as

$$\Phi_{X_i}(s) = \frac{1}{k_0} + \frac{1}{k_1} s + \frac{1}{k_2} s^2 + \cdots \tag{5-60}$$

where k_0, k_1, k_2, $\cdots$ are defined as the dynamic error coefficients. The reciprocal of these coefficients is taken to form the relation that the larger the dynamic error coefficient is, the smaller the dynamic error is, which is the same as the relation between the static error coefficient and the static error. The coefficients can be calculated as

$$\frac{1}{k_j} = \frac{1}{j!} \frac{d^j \Phi_{X_i}(s)}{ds^j} \bigg|_{s=0} \quad (j = 0, 1, 2, \cdots) \tag{5-61}$$

Thus, the Laplace transform of the system static error is obtained by equations (5-61) and equation (5-62) as

$$E(s) = \frac{1}{k_0} X_i(s) + \frac{1}{k_1} s X_i(s) + \frac{1}{k_2} s^2 X_i(s) + \cdots \tag{5-62}$$

The convergence of this series is a certain neighborhood of $s = 0$ (equivalent to a neighborhood of $t \to \infty$). Therefore, when the initial condition is zero, the time-domain expression of dynamic error can be obtained by taking the inverse Laplace transform of the above equation as

$$\varepsilon(t) = \frac{1}{k_0} x_i(t) + \frac{1}{k_1} \frac{dx_i(t)}{dt} + \frac{1}{k_2} \frac{d^2 x_i(t)}{dt^2} + \cdots \tag{5-63}$$

To obtain the static deviation of the control system, the limit value of the equation should be solved when $t \to \infty$, i.e. $\lim_{t \to \infty} \varepsilon(t) = \varepsilon_{si}$.

It can be seen that the dynamic error of the control system is related to the input signal and its derivatives. Therefore, the coefficients corresponding to the input signals and its corresponding derivatives in the dynamic error are respectively defined as: k_0 is called the position dynamic error coefficient, k_1 is called the speed dynamic error coefficient, and k_2 is called the acceleration dynamic error coefficient.

求系统的静态误差。

解：1) 假设干扰信号为零，计算在给定输入信号作用下的静态误差。该系统是 I 型系统，开环增益 $K=5$，其偏差 $\varepsilon(s)=1/5$。根据 $H(s)=1/2$，有

$$\varepsilon_{si} = \frac{1}{5} \Big/ \frac{1}{2} = \frac{2}{5}$$

2) 设输入信号为零，计算在干扰信号作用下的静态误差。仅在干扰信号 $N(s)$ 作用下，系统的输出为

$$X_{oN}(s) = \frac{\dfrac{2}{s(s+1)}}{1+\dfrac{5}{s(s+1)(0.2s+1)}} \cdot \frac{1}{s} = \frac{2(0.2s+1)}{s^2(s+1)(0.2s+1)+5s}$$

因此，系统的静态误差为

$$\begin{aligned}
e_{sn} &= \lim_{s\to 0} sE_N(s) = -\lim_{s\to 0} sX_{oN}(s) \\
&= -\lim_{s\to 0} s\frac{2(0.2s+1)}{s^2(s+1)(0.2s+1)+5s} = -\lim_{s\to 0}\frac{2(0.2s+1)}{s(s+1)(0.2s+1)+5} \\
&= -\frac{2}{5}
\end{aligned}$$

3) 计算系统的总误差，有

$$e_s = e_{si} + \varepsilon_{sn} = 0$$

5.2.3 系统动态误差分析与计算

上述有关系统静态误差的计算公式，是利用终值定理导出的。其优点是求解方便快捷，不用解微分方程也不用求导数。但是，由于这种求解方法的结果不是非零的有限值，就是无穷大或者等于零，因而不能反映出稳态误差随时间的变化规律。事实上，系统在过渡过程结束后即进入稳态，其所需时间并不一定无穷大，一般来说，当 $t \geq t_s$，即可认为系统进入了稳态。再者对于一些不满足终值定理的函数，如正弦函数等，上述方法就不适用，所以用终值定理求解误差存在一定的局限性。下面介绍一种有关稳态误差求解的通用方法，它不仅可以求出稳态误差的终值（即静态误差），还可以求出稳态误差随时间变化的规律（即动态误差）。

对于图 5-23 所示的控制系统，其误差传递函数为

$$\Phi_{X_i}(s) = \frac{E(s)}{X_i(s)} = \frac{1}{1+G(s)H(s)} \tag{5-59}$$

式中，$G(s)=G_1(s)G_2(s)$。依据拉氏变换的终值定理，时间 $t\to\infty$ 对应于 $s=0$。将上式在 $s=0$ 的邻域展开成泰勒级数，得

$$\Phi_{X_i}(s) = \frac{1}{k_0} + \frac{1}{k_1}s + \frac{1}{k_2}s^2 + \cdots \tag{5-60}$$

式中，k_0，k_1，k_2，$\cdots$ 为动态误差系数，取其倒数是为了形成动态误差系数越大、对应的动态误差越小的关系，从而和静态误差系数与静态误差的关系相同，其计算公式为

Example 5-17 The open-loop transfer functions of two feedback systems are respectively

$$G_1(s) = \frac{10}{s(s+1)}, \quad G_2(s) = \frac{10}{s(2s+1)}$$

Try to calculate static error coefficients and dynamic error coefficients of them.

Solution: 1) Calculating the static error coefficients, we have

$$K_p = \infty, \quad K_v = 10, \quad K_a = 0$$

2) Calculating the dynamic error coefficients, we have

$$\Phi_{X_{i1}}(s) = \frac{1}{1+G_1(s)} = \frac{s^2+s}{s^2+s+10}, \quad \Phi_{X_{i2}}(s) = \frac{1}{1+G_2(s)} = \frac{2s^2+s}{2s^2+s+10}$$

According to the equation (5-61), the dynamic error coefficients of system 1 are respectively

$$\frac{1}{k_0} = \Phi_{X_{i1}}(s)\Big|_{s=0} = 0, \quad k_0 = \infty$$

$$\frac{1}{k_1} = \frac{d\left(\frac{s^2+s}{s^2+s+10}\right)}{ds}\Bigg|_{s=0} = \frac{10(2s+1)}{(s^2+s+10)^2}\Bigg|_{s=0} = \frac{1}{10}, \quad k_1 = 10$$

$$\frac{1}{k_2} = \frac{1}{2!}\frac{d^2\left(\frac{s^2+s}{s^2+s+10}\right)}{ds^2}\Bigg|_{s=0} = \frac{9}{100}, \quad k_2 = 11.1$$

Similarly, the dynamic error coefficients of system 2 are calculated as

$$k_0 = \infty, \quad k_1 = 10, \quad k_2 = 5.26$$

Obviously, two different systems with the same static error coefficients have different dynamic error coefficients. Therefore, when they are subjected to the same input, their static errors are identical, but the dynamic errors are different. That is, for the same type of systems, the static errors may be the same, but the dynamic errors are often different.

Example 5-18 For system 1 and system 2 given in example 5-17, when the control input signals are respectively $x_{i1}(t) = 1+2t$ and $x_{i2}(t) = 1+2t+t^2$, try to calculate the static error and dynamic error.

Solution: 1) Calculation of the static error. When the input signal is $x_{i1}(t)$, the static error of system 1 and system 2 are respectively

$$\varepsilon_{si1} = \frac{1}{1+k_p} + \frac{2}{k_v} = 0 + \frac{2}{10} = 0.2$$

$$\varepsilon_{si2} = \frac{1}{1+k_p} + \frac{2}{k_v} = 0 + \frac{2}{10} = 0.2$$

When the input signal is $x_{i2}(t)$, the static error of system 1 and system 2 are respectively

$$\varepsilon_{si1} = \frac{1}{1+k_p} + \frac{2}{k_v} + \frac{2}{k_a} = 0 + \frac{2}{10} + \infty = \infty$$

$$\varepsilon_{si2} = \frac{1}{1+k_p} + \frac{2}{k_v} + \frac{2}{k_a} = 0 + \frac{2}{10} + \infty = \infty$$

2) Calculation of the dynamic error. For the control input signal $x_{i1}(t)$, we have

$$\frac{1}{k_j} = \frac{1}{j!} \left. \frac{d^j \Phi_{X_i}(s)}{ds^j} \right|_{s=0} \quad (j=0,1,2,\cdots) \tag{5-61}$$

因此，由式（5-59）和式（5-60）可得系统静态误差的拉氏变换为

$$E(s) = \frac{1}{k_0} X_i(s) + \frac{1}{k_1} s X_i(s) + \frac{1}{k_2} s^2 X_i(s) + \cdots \tag{5-62}$$

这个级数的收敛域是 $s=0$ 的某个邻域（相当于 $t \to \infty$ 的某邻域），所以，当初始条件为零时，对上式求拉氏反变换，可得到动态误差的时域表达式为

$$\varepsilon(t) = \frac{1}{k_0} x_i(t) + \frac{1}{k_1} \frac{d x_i(t)}{dt} + \frac{1}{k_2} \frac{d^2 x_i(t)}{dt^2} + \cdots \tag{5-63}$$

为了得到控制系统的静态偏差，需要求解此方程在 $t \to \infty$ 的极限值，即 $\lim_{t \to \infty} \varepsilon(t) = \varepsilon_{si}$。

由此可见，控制系统的动态误差与输入信号及其各阶导数有关。因此，将动态误差中的输入信号及其各阶导数所对应的系数分别定义为：k_0 称为位置动态误差系数，k_1 称为速度动态误差系数，k_2 称为加速度动态误差系数。

例 5-17 有两个反馈系统，其开环传递函数分别为

$$G_1(s) = \frac{10}{s(s+1)}, \quad G_2(s) = \frac{10}{s(2s+1)}$$

试计算其静态误差系数和动态误差系数。

解： 1）求静态误差系数，可得

$$K_p = \infty, \quad K_v = 10, \quad K_a = 0$$

2）求动态误差系数，可得

$$\Phi_{X_{i1}}(s) = \frac{1}{1+G_1(s)} = \frac{s^2+s}{s^2+s+10}, \quad \Phi_{X_{i2}}(s) = \frac{1}{1+G_2(s)} = \frac{2s^2+s}{2s^2+s+10}$$

根据式（5-61），系统 1 的动态误差系数为

$$\frac{1}{k_0} = \Phi_{X_{i1}}(s) \big|_{s=0} = 0, \quad k_0 = \infty$$

$$\frac{1}{k_1} = \left. \frac{d\left(\frac{s^2+s}{s^2+s+10}\right)}{ds} \right|_{s=0} = \left. \frac{10(2s+1)}{(s^2+s+10)^2} \right|_{s=0} = \frac{1}{10}, \quad k_1 = 10$$

$$\frac{1}{k_2} = \frac{1}{2!} \left. \frac{d^2\left(\frac{s^2+s}{s^2+s+10}\right)}{ds^2} \right|_{s=0} = \frac{9}{100}, \quad k_2 = 11.1$$

同理计算出系统 2 的动态误差系数为

$$k_0 = \infty, \quad k_1 = 10, \quad k_2 = 5.26$$

由此可见，两个具有相同静态误差系数的不同系统却具有不相同的动态误差系数，因而，它们在相同输入的作用下，静态误差是相同的，但动态误差是不一样的。也就是说，同一类型的系统，其静态误差可能是相同的，而其动态误差则往往是不相同的。

例 5-18 对例 5-17 所给定的系统 1 和系统 2，试计算当控制输入信号分别为 $x_{i1}(t) = 1 + 2t$ 和 $x_{i2}(t) = 1 + 2t + t^2$ 时的静态误差和动态误差。

$$x_{i1}(t) = 1+2t, \quad \frac{dx_{i1}(t)}{dt} = 2, \quad \frac{d^2 x_{i1}(t)}{dt^2} = 0$$

The dynamic error coefficients are calculated in example 5-17 as $k_0 = \infty$, $k_1 = 10$, $k_2 = 11.1$ for system 1, and $k_0 = \infty$, $k_1 = 10$, $k_2 = 5.26$ for system 2. Thus, according to equation (5-63), the dynamic errors of system 1 and system 2 are respectively

$$\varepsilon_1(t) = \frac{1}{k_0} x_i(t) + \frac{1}{k_1}\frac{dx_i(t)}{dt} + \frac{1}{k_2}\frac{d^2 x_i(t)}{dt^2} + \cdots = 0 + \frac{2}{10} + 0 = 0.2$$

$$\varepsilon_2(t) = \frac{1}{k_0} x_i(t) + \frac{1}{k_1}\frac{dx_i(t)}{dt} + \frac{1}{k_2}\frac{d^2 x_i(t)}{dt^2} + \cdots = 0 + \frac{2}{10} + 0 = 0.2$$

For the control input signal $x_{i2}(t)$, we have

$$x_{i2}(t) = 1+2t+t^2, \quad \frac{dx_{i2}(t)}{dt} = 2+2t, \quad \frac{d^2 x_{i2}(t)}{dt^2} = 2, \quad \frac{d^3 x_{i2}(t)}{dt^3} = 0$$

Thus, according to equation (5-63), the dynamic errors of system 1 and system 2 are respectively

$$\varepsilon_1(t) = \frac{1}{k_0} x_i(t) + \frac{1}{k_1}\frac{dx_i(t)}{dt} + \frac{1}{k_2}\frac{d^2 x_i(t)}{dt^2} + \cdots$$

$$= 0 + \frac{1}{10}(2+2t) + 2 \times \frac{1}{11.1} = 0.2t + 0.38$$

$$\varepsilon_2(t) = \frac{1}{k_0} x_i(t) + \frac{1}{k_1}\frac{dx_i(t)}{dt} + \frac{1}{k_2}\frac{d^2 x_i(t)}{dt^2} + \cdots$$

$$= 0 + \frac{1}{10}(2+2t) + \frac{1}{5.26} \times 2 = 0.2t + 0.58$$

Obviously, for the control input signals $x_{i1}(t)$ and $x_{i2}(t)$, the limit value when $t \to \infty$ of dynamic error is consistent with the static error of each system. It can be seen that the changing law of steady-state error with time can be obtained by using dynamic error coefficients, instead of solving the differential equation when $t \to \infty$.

5.2.4 The relationship between open-loop characteristics and steady-state error

For a given input signal, the steady-state error of a control system is related to the type of system and the open-loop amplification factor. When the open-loop frequency characteristic curve of the system is given, the position error coefficient K_p, speed error coefficient K_v and acceleration error coefficient K_a in the steady state can be determined in terms of the slope of its low frequency band. According to the previous analysis, the low frequency band of the logarithmic amplitude-frequency characteristic is characterized by the expression $K/(j\omega)^v$, where v is usually 0, 1, or 2 for an actual control system. Therefore, the correspondence between the type of the system and the slope of the low frequency asymptotic curve in the logarithmic amplitude-frequency characteristic curve is analyzed, and the method for determining values K_p, K_v, K_a are given as follows.

Assume that the open-loop frequency characteristic of a control system is

解： 1) 求静态误差。当输入信号为 $x_{i1}(t)$ 时，系统 1 和系统 2 的静态误差分别为

$$\varepsilon_{si1} = \frac{1}{1+k_p} + \frac{2}{k_v} = 0 + \frac{2}{10} = 0.2$$

$$\varepsilon_{si2} = \frac{1}{1+k_p} + \frac{2}{k_v} = 0 + \frac{2}{10} = 0.2$$

当输入信号为 $x_{i2}(t)$ 时，系统 1 和系统 2 的静态误差分别为

$$\varepsilon_{si1} = \frac{1}{1+k_p} + \frac{2}{k_v} + \frac{2}{k_a} = 0 + \frac{2}{10} + \infty = \infty$$

$$\varepsilon_{si2} = \frac{1}{1+k_p} + \frac{2}{k_v} + \frac{2}{k_a} = 0 + \frac{2}{10} + \infty = \infty$$

2) 求动态误差。对于控制输入信号 $x_{i1}(t)$，由于

$$x_{i1}(t) = 1+2t, \quad \frac{dx_{i1}(t)}{dt} = 2, \quad \frac{d^2 x_{i1}(t)}{dt^2} = 0$$

根据例 5-17 中的动态误差系数计算结果，即系统 1 中 $k_0 = \infty$，$k_1 = 10$，$k_2 = 11.1$，系统 2 中 $k_0 = \infty$，$k_1 = 10$，$k_2 = 5.26$，按照式（5-63）计算，系统 1 和系统 2 的动态误差分别为

$$\varepsilon_1(t) = \frac{1}{k_0} x_i(t) + \frac{1}{k_1} \frac{dx_i(t)}{dt} + \frac{1}{k_2} \frac{d^2 x_i(t)}{dt^2} + \cdots = 0 + \frac{2}{10} + 0 = 0.2$$

$$\varepsilon_2(t) = \frac{1}{k_0} x_i(t) + \frac{1}{k_1} \frac{dx_i(t)}{dt} + \frac{1}{k_2} \frac{d^2 x_i(t)}{dt^2} + \cdots = 0 + \frac{2}{10} + 0 = 0.2$$

对于控制输入信号 $x_{i2}(t)$，由于

$$x_{i2}(t) = 1+2t+t^2, \quad \frac{dx_{i2}(t)}{dt} = 2+2t, \quad \frac{d^2 x_{i2}(t)}{dt^2} = 2, \quad \frac{d^3 x_{i2}(t)}{dt^3} = 0$$

因此按照式（5-63）计算，系统 1 和系统 2 的动态误差分别为

$$\varepsilon_1(t) = \frac{1}{k_0} x_i(t) + \frac{1}{k_1} \frac{dx_i(t)}{dt} + \frac{1}{k_2} \frac{d^2 x_i(t)}{dt^2} + \cdots$$

$$= 0 + \frac{1}{10}(2+2t) + 2 \times \frac{1}{11.1} = 0.2t + 0.38$$

$$\varepsilon_2(t) = \frac{1}{k_0} x_i(t) + \frac{1}{k_1} \frac{dx_i(t)}{dt} + \frac{1}{k_2} \frac{d^2 x_i(t)}{dt^2} + \cdots$$

$$= 0 + \frac{1}{10}(2+2t) + \frac{1}{5.26} \times 2 = 0.2t + 0.58$$

显然，对于控制输入信号 $x_{i1}(t)$ 和 $x_{i2}(t)$，各系统的动态误差在 $t \to \infty$ 时的极限值与各系统的静态误差一致。可见，利用动态误差系数，无需求解微分方程，就可得稳态误差随时间变化的规律。

5.2.4 开环特性与稳态误差的关系

对于给定的输入信号，控制系统的稳态误差与系统的类型、开环放大倍数有关。在给定系统开环频率特性曲线的情况下，可根据其低频段的斜率确定系统处于稳态时的位置误差系

$$G_K(j\omega) = \frac{K}{(j\omega)^v\left(1+\dfrac{j\omega}{1/T}\right)} \tag{5-64}$$

The corresponding logarithmic amplitude-frequency characteristic is

$$L(\omega) = 20\lg K - 20v\lg\omega - 20\lg\sqrt{1+\left(\frac{\omega}{1/T}\right)^2} \tag{5-65}$$

From equation (5-65), the low frequency asymptotic curves of different types of systems can be plotted, as is shown in Fig. 5-25.

1. Type-0 system

For the type-0 system, let $v=0$, $K=K_p$ in equations (5-64) and equation (5-65), the asymptote of the logarithmic amplitude-frequency characteristic curve of the system is drawn in Fig. 5-26. It can be seen from the figure that the low frequency band of the logarithmic amplitude-frequency characteristic curve of the type-0 system has the following characteristics.

1) The asymptotic slope of the low frequency band is 0dB/dec, and the height is $20\lg K_p$.

2) If the height of the low frequency band of the amplitude-frequency characteristic curve is known, the value of the position error coefficient K_p can be obtained according to the equation $L(\omega) = 20\lg K_p$, and then the steady-state error of the system can be calculated.

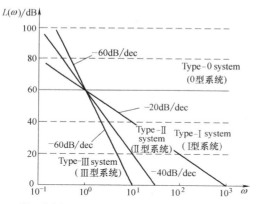

Fig. 5-25 Low frequency asymptotic curves of different types of systems
（不同类型系统的低频渐近线）

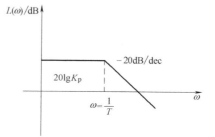

Fig. 5-26 Logarithmic amplitude-frequency characteristic curve of type-0 system
（0型系统的对数幅频特性曲线）

2. Type-I system

Let $v=1$, $K=K_v$ in equations (5-64) and equation (5-65), the asymptote of the logarithmic amplitude-frequency characteristic curve of the system is drawn in Fig. 5-27. The logarithmic amplitude-frequency characteristics of the system have the following characteristics.

1) The slope of the low frequency asymptote is -20 dB/dec.

2) The ordinate value of the low-frequency asymptote (or its extension line) is $20\lg K_v$ at $\omega=1$, thus the steady-state speed error coefficient K_v can be obtained.

3) The open-loop gain, i.e. the speed error coefficient in a steady state, is equal to the fre-

数 K_p、速度误差系数 K_v 和加速度误差系数 K_a。根据前文分析可知，对数幅频特性的低频段是由式 $K/(j\omega)^v$ 来表征的，对于实际的控制系统，v 通常为 0、1 或 2。下面分析系统的类型与对数幅频特性曲线低频渐近线斜率的对应关系，并给出 K_p、K_v 和 K_a 值的确定方法。

设控制系统的开环频率特性为

$$G_K(j\omega) = \frac{K}{(j\omega)^v \left(1 + \dfrac{j\omega}{1/T}\right)} \qquad (5\text{-}64)$$

相应的对数幅频特性为

$$L(\omega) = 20\lg K - 20v\lg\omega - 20\lg\sqrt{1 + \left(\frac{\omega}{1/T}\right)^2} \qquad (5\text{-}65)$$

由式（5-65）可以画出不同类型系统的低频渐近线如图 5-25 所示。

1. 0 型系统

对于 0 型系统，令式（5-64）和式（5-65）中 $v = 0$，$K = K_p$，可绘出系统的对数幅频特性曲线的渐近线，如图 5-26 所示。由图可见，0 型系统的对数幅频特性的低频段具有如下特点。

1）低频段的渐近线斜率为 0dB/dec，高度为 $20\lg K_p$。

2）如果已知幅频特性曲线低频段的高度，即可根据 $L(\omega) = 20\lg K_p$ 求出位置误差系数 K_p 的值，进而计算系统的稳态误差。

2. Ⅰ 型系统

令式（5-64）及式（5-65）中 $v = 1$，$K = K_v$，可绘出系统的对数幅频特性曲线渐近线，如图 5-27 所示。系统对数幅频特性有如下特点。

1）低频渐近线的斜率为 -20dB/dec。

2）低频渐近线（或其延长线）在 $\omega = 1$ 处纵坐标值为 $20\lg K_v$，由此可求出稳态速度误差系数 K_v。

3）开环增益即稳态速度误差系数 K_v，等于低频渐近线（或其延长线）与 0dB/dec 线相交点的频率值。

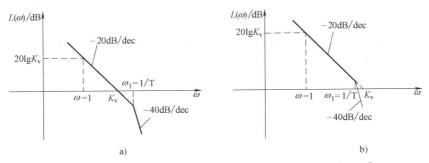

Fig. 5-27 Logarithmic amplitude-frequency characteristic curve of type-Ⅰ system
（Ⅰ型系统的对数幅频特性曲线）

a）$\omega_1 > K_v$ b）$\omega_1 < K_v$

quency at which the low frequency asymptote (or its extension) intersects the 0dB/dec line.

3. Type-II system

Let $v=2$, $K=K_a$ in equations (5-64) and equation (5-65), the asymptote of the logarithmic amplitude-frequency characteristic curve of the system is drawn in Fig. 5-28. It can be seen from the figure that the logarithmic amplitude-frequency characteristic of the type-II system has the following characteristics.

1) The slope of the low frequency asymptote is -40dB/dec.

2) The ordinate value of the low-frequency asymptote (or its extension line) is $20\lg K_a$ at $\omega=1$, thus the acceleration error coefficient K_a in a steady state can be obtained.

3) The open-loop gain of the system, i.e. the acceleration error coefficient, is equal to the square of the frequency at which the low frequency asymptote (or its extension) intersects the 0dB/dec line.

From above analysis, it can be seen that the steeper the slope of the low-frequency curve of the open-loop logarithmic amplitude-frequency characteristic of the system is, the smaller the steady-state error is, and the better the steady-state accuracy of the system is.

5.3 Dynamic Performance Analysis of Control Systems

For a control system that has met the requirements of stability, it requires not only better steady-state performance, but also better dynamic performance for the system with higher requirements. The most direct indicators to analyze the dynamic performance of the control system are the time-domain performance indicators introduced in Chapter 3. However, some important characteristic values involved in the frequency domain analysis method, such as phase margin and amplitude margin in the open-loop frequency characteristic and resonant peak, bandwidth and resonant frequency in the closed-loop frequency characteristic are directly or indirectly related to the transient response and the static error of the control system. For a second-order system, there is a definite correspondence between the characteristic values and the time domain performance indicators; for a higher-order system, as long as there is a pair of dominant closed-loop poles, they also have an approximate correspondence.

5.3.1 The relationship between dynamic performance and open-loop frequency characteristics

For a second-order system, the open-loop frequency characteristic is

$$G_K(j\omega) = \frac{\omega_n^2}{j\omega(j\omega+2\xi\omega_n)}$$

When $\omega=\omega_c$, we have $|G_K(j\omega_c)|=1$, i.e.

$$\frac{\omega_n^2}{\sqrt{\omega_c^4+4\xi^2\omega_c^2\omega_n^2}} = 1$$

3. II型系统

令式（5-64）及式（5-65）中 $v=2$，$K=K_a$，可绘出系统的对数幅频特性曲线的渐近线如图 5-28 所示。由图可见，II型系统的对数幅频特性具有如下特点。

1）低频渐近线的斜率为 -40dB/dec。
2）低频渐近线（或其延长线）在 $\omega=1$ 处坐标值为 $20\lg K_a$，可由此求出稳态加速度误差系数 K_a。
3）系统的开环增益即加速度误差系数 K_a，等于低频渐近线（或其延长线）与 0dB/dec 线相交点的频率值的平方。

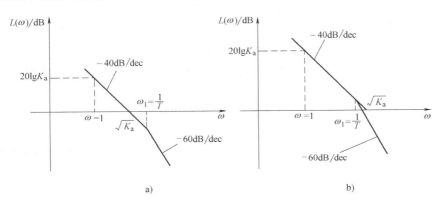

Fig. 5-28 Logarithmic amplitude-frequency characteristic curve of type-II system
（II型系统的对数幅频特性曲线）
a) $\omega_1 > K_a$ b) $\omega_1 < K_a$

由以上分析可以看出，系统的开环对数幅频特性曲线低频段的斜率愈陡，系统的稳态误差愈小，系统的稳态精度愈好。

5.3 控制系统的动态性能分析

对于满足稳定性要求的控制系统，若要达到更高要求，则应具有较好的动态性能。控制系统动态性能最直接的指标是第 3 章介绍的时域性能指标。频域分析法中介绍的一些重要特征值，如开环频率特性中的相位裕度、幅值裕度，闭环频率特性中的谐振峰值、频带宽度和谐振频率等，与控制系统的瞬态响应和静态误差存在着间接或直接的关系。对于二阶系统而言，这些特征值与时域性能指标间有着明确的对应关系；在高阶系统中，只要存在一对闭环主导极点，则它们也有着近似的对应关系。

5.3.1 动态性能与开环频率特性的关系

对于二阶系统，其开环频率特性为

$$G_K(j\omega) = \frac{\omega_n^2}{j\omega(j\omega + 2\xi\omega_n)}$$

当 $\omega = \omega_c$ 时，$|G_K(j\omega_c)| = 1$，即

Solving the above equation, the open-loop cutoff frequency ω_c of the second-order system can be obtained as

$$\omega_c = \omega_n \sqrt{\sqrt{1+4\xi^4} - 2\xi^2} \tag{5-66}$$

Then, the phase margin γ of the system is obtained as

$$\gamma = 180° + \varphi(\omega_c) = 180° - 90° - \arctan\frac{\omega_c}{2\xi\omega_n}$$

$$= \arctan\frac{2\xi}{\sqrt{\sqrt{1+4\xi^4} - 2\xi^2}} \tag{5-67}$$

The equation (5-67) shows that there is a one-to-one correspondence between the phase margin γ and the damping ratio ξ in the second-order system. The phase margin γ increases as the damping ratio ξ increases. Fig. 5-29 shows the relationship between γ and ξ, and when $0<\xi<0.6$, we have $\gamma \approx 100 \times \xi$.

1. The relationship between the maximum overshoot and the open-loop frequency characteristic

According to the time-domain analysis of the control system, the maximum overshoot M_p of the second-order system is

$$M_p = e^{-\xi\pi/\sqrt{1-\xi^2}} \times 100\% \tag{5-68}$$

It can be seen that the maximum overshoot M_p decreases as the damping ratio ξ increases, as is shown in Fig. 5-30.

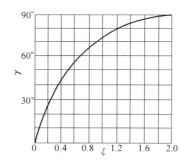

Fig. 5-29 Relationship between phase margin γ and damping ratio ξ
(相位裕度 γ 与阻尼比 ξ 间的关系)

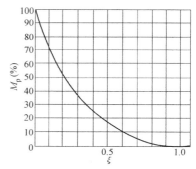

Fig. 5-30 Relationship between maximum overshoot M_p and damping ratio ξ
(最大超调量 M_p 与阻尼比 ξ 间的关系)

According to above analysis, the phase margin of the system is inversely proportional to the maximum overshoot as the damping ratio increases in the underdamping state, as is shown in Fig. 5-31, which indicates that the phase margin reflects the stability of the system. In other words, the larger the phase margin of the system open-loop frequency characteristic is, the smaller the maximum overshoot is, and the better the stability of the system is. In order to make the system have good dynamic characteristics when designing a control system, it is generally desirable to determine the damping ratio after selecting the phase margin, and then determine the maximum overshoot and

$$\frac{\omega_n^2}{\sqrt{\omega_c^4+4\xi^2\omega_c^2\omega_n^2}}=1$$

解上式可得二阶系统开环截止频率 ω_c 为

$$\omega_c=\omega_n\sqrt{\sqrt{1+4\xi^4}-2\xi^2} \tag{5-66}$$

然后求得系统的相位裕度 γ 为

$$\gamma=180°+\varphi(\omega_c)=180°-90°-\arctan\frac{\omega_c}{2\xi\omega_n}$$

$$=\arctan\frac{2\xi}{\sqrt{\sqrt{1+4\xi^4}-2\xi^2}} \tag{5-67}$$

式（5-67）表明，二阶系统的相位裕度 γ 与阻尼比 ξ 之间存在一一对应的关系，相位裕度 γ 随着阻尼比 ξ 的增大而增大，图 5-29 为 γ 与 ξ 的关系曲线，当 $0<\xi<0.6$ 时，$\gamma\approx100\times\xi$。

1. 最大超调量与开环频率特性的关系

由控制系统的时域分析可知，二阶系统最大超调量 M_p 为

$$M_p=\mathrm{e}^{-\xi\pi/\sqrt{1-\xi^2}}\times100\% \tag{5-68}$$

可见，最大超调量 M_p 随着阻尼比 ξ 的增大而减小，如图 5-30 所示。

由以上分析可见，二阶系统在欠阻尼状态下，随着阻尼比的增加，系统的相位裕度与最大超调量成反比，如图 5-31 所示，这说明相位裕度可以反映系统的稳定性。换言之，系统开环频率特性的相位裕度愈大，最大超调量愈小，系统的稳定性愈好。在设计控制系统时，为使系统具有良好的动态特性，一般选定相位裕度后再确定阻尼比，最后确定最大超调量和调整时间。

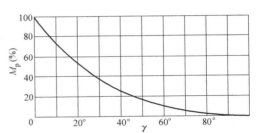

Fig. 5-31 Relationship between maximum overshoot M_p and phase margin γ

（最大超调量 M_p 与相位裕度 γ 间的关系）

2. 调整时间与开环频率特性的关系

由第 3 章可得，二阶系统的相位裕度 γ 与调整时间 t_s 之间的定量关系为

$$\begin{cases} t_s|_{\Delta=\pm5\%}\approx\dfrac{3}{\xi\omega_n} & (0<\xi<0.9) \\ t_s|_{\Delta=\pm2\%}\approx\dfrac{4}{\xi\omega_n} & (0<\xi<0.9) \end{cases} \tag{5-69}$$

将式（5-69）与式（5-66）联合得

$$\begin{cases} t_s\omega_c|_{\Delta=\pm5\%}=\dfrac{3}{\xi}\sqrt{\sqrt{1+4\xi^4}-2\xi^2} \\ t_s\omega_c|_{\Delta=\pm2\%}=\dfrac{4}{\xi}\sqrt{\sqrt{1+4\xi^4}-2\xi^2} \end{cases} \tag{5-70}$$

由此可见，开环截止频率 ω_c 反映了系统的快速性。即截止频率越大，则系统的快速性

settling time according to the damping ratio.

2. The relationship between the settling time and the open-loop frequency characteristic

According to Chapter 3, the quantitative relationship between the phase margin γ of the second-order system and the settling time t_s is

$$\begin{cases} t_s \mid_{\Delta=\pm5\%} \approx \dfrac{3}{\xi\omega_n} & (0<\xi<0.9) \\ t_s \mid_{\Delta=\pm2\%} \approx \dfrac{4}{\xi\omega_n} & (0<\xi<0.9) \end{cases} \quad (5\text{-}69)$$

Combining the equation (5-69) with the equation (5-66), we have

$$\begin{cases} t_s\omega_c \mid_{\Delta=\pm5\%} = \dfrac{3}{\xi}\sqrt{\sqrt{1+4\xi^4}-2\xi^2} \\ t_s\omega_c \mid_{\Delta=\pm2\%} = \dfrac{4}{\xi}\sqrt{\sqrt{1+4\xi^4}-2\xi^2} \end{cases} \quad (5\text{-}70)$$

It can be seen that the open-loop cutoff frequency ω_c reflects the rapidity of the system. That is, the greater the cutoff frequency is, the better the system's rapidity is.

Then, according to equations (5-67) and equation (5-70), we have

$$\begin{cases} t_s\omega_c \mid_{\Delta=\pm5\%} = \dfrac{6}{\tan\gamma} \\ t_s\omega_c \mid_{\Delta=\pm2\%} = \dfrac{8}{\tan\gamma} \end{cases} \quad (5\text{-}71)$$

The functional relation of equation (5-71) is drawn as a curve, as is shown in Fig. 5-32.

If two systems have the same phase margin, their maximum overshoot is approximately the same but their settling time is not likely to be the same. It can be seen from equation (5-71) that the settling time is inversely proportional to the open-loop cutoff frequency, that is, the higher the cutoff frequency is, the shorter the settling time in the time domain is. Therefore, the cutoff frequency is a very special important parameter in frequency characteristics. It not only affects the phase margin of the system, but also affects the settling time of the dynamic process.

5.3.2 The relationship between dynamic performance and closed-loop frequency characteristics

For a second-order system, there is an exact correspondence between the time domain response and the frequency domain response. When $0 \leq \xi \leq 0.707$, the resonance will be generated in the system. According to Chapter 4, the resonant frequency and resonant peak can be obtained as

$$\omega_r = \omega_n\sqrt{1-2\xi^2} \quad (5\text{-}72)$$

$$M_r = \dfrac{1}{2\xi\sqrt{1-\xi^2}} \quad (5\text{-}73)$$

According to equations (5-72) and equation (5-73), we have

越好。

再由式（5-67）和式（5-70）可得

$$\begin{cases} t_s\omega_c \big|_{\Delta=\pm 5\%} = \dfrac{6}{\tan\gamma} \\ t_s\omega_c \big|_{\Delta=\pm 2\%} = \dfrac{8}{\tan\gamma} \end{cases} \quad (5\text{-}71)$$

将式（5-71）的函数关系式绘制成曲线，如图5-32所示。

如果有两个系统，其相位裕度相同，那么它们的最大超调量大致相同，但它们的调整时间并不一定相同。由式（5-71）可知，调整时间与开环截止频率成反比，即截止频率越大，时域的调整时间越短。所以截止频率在频率特性中是一个很特殊的重要参数，它不仅影响系统的相位裕度，还影响动态过程的调整时间。

5.3.2 动态性能与闭环频率特性的关系

对于二阶系统，其时域响应与频域响应之间有着确切的对应关系。当 $0 \leq \xi \leq 0.707$ 时，系统会发生谐振。由第4章可得，其谐振频率和谐振峰值为

$$\omega_r = \omega_n\sqrt{1-2\xi^2} \quad (5\text{-}72)$$

$$M_r = \dfrac{1}{2\xi\sqrt{1-\xi^2}} \quad (5\text{-}73)$$

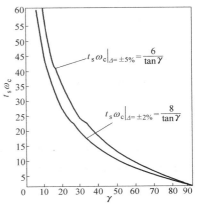

Fig. 5-32 Relationship between $t_s\omega_c$ and γ（$t_s\omega_c$ 与 γ 的关系）

根据式（5-72）和（5-73），解得

$$\xi = \sqrt{\dfrac{1-\sqrt{1-1/M_r^2}}{2}} \quad (5\text{-}74)$$

1. 谐振峰值和最大超调量的关系

为便于比较谐振峰值 M_r 和最大超调量 M_p，分别画出 M_r 和 M_p 与 ξ 的关系曲线，如图5-33所示。由图可见，M_p 和 M_r 均随 ξ 的减小而增大。显然，对于同一个系统，若时域内的 M_p 较大，则频域中的 M_r 必然较大，反之亦然。为了使系统具有良好的相对稳定性，在设计系统时，通常取 ξ 值在0.4~0.7之间，则对应的 M_r 在1~1.4之间。

把式（5-74）代入式（5-68），则得

$$M_p = e^{-\pi\sqrt{\dfrac{M_r - \sqrt{M_r^2-1}}{M_r + \sqrt{M_r^2-1}}}} \quad (5\text{-}75)$$

如果已知 M_r，则可由上式求得对应的 M_p。

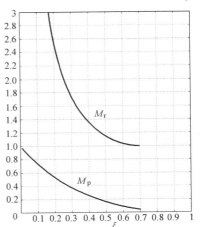

Fig. 5-33 Relationship between the resonant peak M_r and maximum overshoot M_p（谐振峰值 M_r 与最大超调量 M_p 之间的关系）

$$\xi = \sqrt{\frac{1-\sqrt{1-1/M_r^2}}{2}} \qquad (5\text{-}74)$$

1. The relationship between resonant peak and maximum overshoot

In order to compare the resonant peak M_r and the maximum overshoot M_p, the relationship of M_r-ξ and M_p-ξ are plotted respectively in Fig. 5-33. As can be seen from the figure, both M_r and M_p increase as ξ decreases. Obviously, for the same system, if M_p is large in the time domain, M_r must be large in the frequency domain, vice versa. In order to make the system have good relative stability, the value ξ is usually within $0.4 \sim 0.7$ and the corresponding value M_r is within $1 \sim 1.4$ when designing a system.

Substituting equation (5-74) into equation (5-68), we have

$$M_p = e^{-\pi\sqrt{\frac{M_r - \sqrt{M_r^2-1}}{M_r + \sqrt{M_r^2-1}}}} \qquad (5\text{-}75)$$

If M_r is known, the corresponding M_p can be obtained from the above equation.

2. The relationship between resonant peak and settling time and peak time

The peak time t_p of a second-order system is

$$t_p = \frac{\pi}{\omega_n \sqrt{1-\xi^2}}$$

Substituting the above equation into equation (5-72), the relationship between the resonant frequency ω_r and the system peak time t_p is obtained as

$$\omega_r t_p = \pi \sqrt{\frac{1-2\xi^2}{1-\xi^2}} \qquad (5\text{-}76)$$

If the second-order system has a range of allowable error adjustments as $\Delta = \pm 2\% \sim \pm 5\%$, the settling time t_s is

$$t_s = \frac{4 \sim 3}{\xi \omega_n}$$

Thus, the relationship between the resonant frequency ω_r of the system and its settling time t_s is

$$\omega_r t_s = \frac{4 \sim 3}{\xi} \sqrt{1-2\xi^2} \qquad (5\text{-}77)$$

According to above analysis, when the damping ratio ξ is constant, the resonant frequency ω_r is inversely proportional to the peak time t_p and the settling time t_s. The larger the value of ω_r is, the smaller the values of t_p and t_s are, and the faster the system responds in time domain.

Then, substituting equation (5-74) into equations (5-76) and equation (5-77), we have

$$\omega_r t_p = \pi \sqrt{\frac{2\sqrt{M_r^2-1}}{M_r + \sqrt{M_r^2-1}}} \qquad (5\text{-}78)$$

$$\omega_r t_s = (4 \sim 3) \sqrt{\frac{2\sqrt{M_r^2-1}}{M_r - \sqrt{M_r^2-1}}} \qquad (5\text{-}79)$$

2. 谐振峰值和调整时间、峰值时间的关系

二阶系统的峰值时间 t_p 为

$$t_p = \frac{\pi}{\omega_n \sqrt{1-\xi^2}}$$

将上式代入式（5-72）得谐振频率 ω_r 与系统峰值时间 t_p 的关系为

$$\omega_r t_p = \pi \sqrt{\frac{1-2\xi^2}{1-\xi^2}} \tag{5-76}$$

若二阶系统所允许的误差调整范围为 $\Delta = \pm 2\% \sim \pm 5\%$，则调整时间 t_s 为

$$t_s = \frac{4 \sim 3}{\xi \omega_n}$$

由此可得系统的谐振频率 ω_r 与调整时间 t_s 的关系为

$$\omega_r t_s = \frac{4 \sim 3}{\xi} \sqrt{1-2\xi^2} \tag{5-77}$$

由以上分析可见，当阻尼比 ξ 为常数时，谐振频率 ω_r 与峰值时间 t_p 及调整时间 t_s 均成反比。ω_r 值越大，t_p 和 t_s 值均越小，表示系统时间响应越快。

再将式（5-74）代入式（5-76）和式（5-77）得

$$\omega_r t_p = \pi \sqrt{\frac{2\sqrt{M_r^2-1}}{M_r + \sqrt{M_r^2-1}}} \tag{5-78}$$

$$\omega_r t_s = (4 \sim 3) \sqrt{\frac{2\sqrt{M_r^2-1}}{M_r - \sqrt{M_r^2-1}}} \tag{5-79}$$

将式（5-78）和式（5-79）的函数关系用曲线表示，如图 5-34 所示。由图可见，调整时间 t_s 与谐振峰值 M_r 成正比。若已知 M_r 和 ω_r，就能从上述关系中求出 t_p 和 t_s。

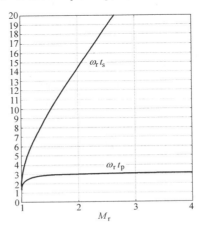

Fig. 5-34 Relationship between the resonant peak M_r,

peak time t_p and settling time t_s

（谐振峰值 M_r 与峰值时间 t_p 及调整时间 t_s 之间的关系）

The functional relationships of equations (5-78) and equation (5-79) are represented by curves, as is shown in Fig. 5-34. As can be seen from the figure, the settling time t_s is proportional to the resonant peak M_r. If M_r and ω_r are known, t_p and t_s can be obtained from the above relations.

3. The relationship between frequency bandwidth and peak time, settling time

According to Chapter 4, the cutoff frequency ω_b of a second-order system is

$$\omega_b = \omega_n \sqrt{1 - 2\xi^2 + \sqrt{2 - 4\xi^2 + 4\xi^4}}$$

Similar to equation (5-76) and equation (5-77), the relationships between the cutoff frequency and the peak time, the settling time of the second-order system can be obtained respectively as

$$\omega_b t_p = \frac{\pi \sqrt{1 - 2\xi^2 + \sqrt{2 - 4\xi^2 + 4\xi^4}}}{\sqrt{1 - \xi^2}} \tag{5-80}$$

$$\omega_b t_s = \frac{4 \sim 3}{\xi} \sqrt{1 - 2\xi^2 + \sqrt{2 - 4\xi^2 + 4\xi^4}} \tag{5-81}$$

It can be seen that when the damping ratio ξ is determined, the cutoff frequency ω_b is inversely proportional to the peak time t_p and the settling time t_s, that is, the greater the value of ω_b is, the wider the bandwidth is, the faster the response speed of the system. However, if the bandwidth is too large, the system's ability to resist high-frequency disturbance will decrease.

Similarly, the relationship between the system resonant peak M_r and the peak time t_p, and the relationship between M_r and the settling time t_s are respectively

$$\omega_b t_p = \pi \sqrt{\frac{2\left(\sqrt{M_r^2 - 1} + \sqrt{2M_r^2 - 1}\right)}{M_r + \sqrt{M_r^2 - 1}}} \tag{5-82}$$

$$\omega_r t_s = (4 \sim 3) \sqrt{\frac{2\left(\sqrt{M_r^2 - 1} + \sqrt{2M_r^2 - 1}\right)}{M_r - \sqrt{M_r^2 - 1}}} \tag{5-83}$$

It can be seen that if M_r and ω_b are known, t_p and t_s can be obtained from the above relationships.

The relationships between the dynamic performance and the frequency characteristics of the second-order system are analyzed above. The relationships between the step response and frequency response are more complex for higher-order systems. If the control performance of a higher-order system is dominated by a pair of conjugate poles, the relationships between its frequency-domain performance indicators and time-domain performance indicators can be approximated treated as those of a second-order system.

3. 频带宽度与峰值时间、调整时间的关系

由第 4 章可知，二阶系统截止频率 ω_b 为

$$\omega_b = \omega_n \sqrt{1 - 2\xi^2 + \sqrt{2 - 4\xi^2 + 4\xi^4}}$$

对照式（5-76）与式（5-77）的推导，可得二阶系统的截止频率与峰值时间和调整时间的关系分别为

$$\omega_b t_p = \frac{\pi \sqrt{1 - 2\xi^2 + \sqrt{2 - 4\xi^2 + 4\xi^4}}}{\sqrt{1 - \xi^2}} \tag{5-80}$$

$$\omega_b t_s = \frac{4 \sim 3}{\xi} \sqrt{1 - 2\xi^2 + \sqrt{2 - 4\xi^2 + 4\xi^4}} \tag{5-81}$$

由此可见，当阻尼比 ξ 确定后，截止频率 ω_b 与峰值时间 t_p 及调整时间 t_s 成反比关系，即 ω_b 越大，带宽越宽，系统的响应速度越快。但带宽过大，系统抗高频干扰的能力就会下降。

同理，系统谐振峰值 M_r 与峰值时间 t_p 及调整时间 t_s 之间的关系分别为

$$\omega_b t_p = \pi \sqrt{\frac{2\left(\sqrt{M_r^2 - 1} + \sqrt{2M_r^2 - 1}\right)}{M_r + \sqrt{M_r^2 - 1}}} \tag{5-82}$$

$$\omega_r t_s = (4 \sim 3) \sqrt{\frac{2\left(\sqrt{M_r^2 - 1} + \sqrt{2M_r^2 - 1}\right)}{M_r - \sqrt{M_r^2 - 1}}} \tag{5-83}$$

由此可见，如果已知 M_r 和 ω_b，就能从上述关系式中求出 t_p 和 t_s。

上述分析均为二阶系统的动态性能与其频率特性之间的关系，高阶系统的阶跃响应与频率响应之间的关系较复杂。如果高阶系统的控制性能主要由一对共轭复数主导极点来支配，则其频域性能指标与时域性能指标之间的关系就可近似按照二阶系统进行处理。

Chapter 6　Comprehensive Compensation of Control Systems

第6章　控制系统的综合校正

The contents of previous chapters mainly focus on the analysis of the system, i. e. to obtain the performance indicators of the system and the relationship between performance indicators and system parameters based on the known system structure and parameters. This chapter mainly introduces the comprehensive compensation of the control system. Specifically, according to the performance indicators of the control system, we will seek the compensation methods that can fully meet the requirements of these performance indicators and determines the appropriate parameter values of the compensation system.

6.1 Introduction

In general, to ensure that a stable system works properly, the system must meet certain performance indicators. As can be seen from the previous chapters, performance indicators are mainly divided into time-domain performance indicators and frequency-domain performance indicators. Generally speaking, from the perspective of use, the time-domain performance indicators are relatively intuitive, and the requirements for the system are often presented in the form of time-domain performance indicators. In the frequency characteristic-based design, time-domain performance indicators are often converted to frequency domain performance indicators.

If the designed system does not fully meet the requirements of performance indicators, we should consider adjusting the parameters for the selected system. If such adjustments still fail to meet the performance requirements, other necessary elements should be added (based on the original system) to enable the system to fully meet the requirements of performance indicators, which is called the compensation of the system. In other words, the so-called compensation refers to the method of adding new elements in the system to improve the performance of the system.

According to the connection patterns of the compensation element $G_c(s)$ in the system, it can be divided into cascade compensation and parallel compensation.

If the compensation element $G_c(s)$ is in series in the forward path of the original transfer function block diagram, the compensation is referred as cascade compensation, as is shown in Fig. 6-1. Usually, in order to reduce power consumption, the cascade compensation element is generally placed at the front end of the forward path.

If the compensation element $G_c(s)$ is paralleled with some elements of the forward path to improve the system performance, the compensation is called parallel compensation. According to the parallel pattern of the compensation element $G_c(s)$, it can be divided into feedback compensation and combined compensation. If the compensation element $G_c(s)$ is located in the feedback channel of the local feedback loop of the original transfer function block diagram, the compensation is called feedback compensation, as is shown in Fig. 6-2. If a feed forward path is added to the feedback control loop, the compensation is called combined compensation, as is shown in Fig. 6-3. The combined compensation can be applied to the open-loop control system alone or as an additional compensation of the feedback control system to form a compound control system.

第6章 控制系统的综合校正

前面几章的内容主要聚焦于系统的分析，即根据已知的系统结构和参数求取系统的性能指标，以及性能指标与系统参数之间的关系。本章主要介绍控制系统的综合校正。具体而言，我们将根据控制系统的性能指标，寻求能够全面满足这些性能指标的校正方法并确定校正系统的参数值。

6.1 概述

通常，要确保稳定系统能正常工作，则要使其满足一定的性能指标。由前面章节可知，性能指标主要分为时域性能指标和频域性能指标。一般从使用的角度来看，时域性能指标比较直观，因此对系统的要求常常以时域性能指标的形式提出。而在基于频率特性的设计中，常常将时域性能指标转换为频域性能指标来考虑。

若设计的系统不能全面地满足性能指标要求，则应考虑对原已选定的系统进行参数调整。如果调整参数后的系统仍然无法达到性能指标的要求，则需在原系统的基础上增加必要的环节，使系统能够全面地满足所要求的性能指标，这就是系统校正。换言之，所谓的校正就是在系统中增加新的环节，以改善系统性能的方法。

根据校正环节 $G_c(s)$ 在系统中的连接模式，可分为串联校正、并联校正。

如果校正环节 $G_c(s)$ 串联在原传递函数框图的前向通道中，则称为串联校正，如图 6-1 所示。通常，为减少功率消耗，串联校正环节一般都置于前向通道的前端。

如果校正环节 $G_c(s)$ 与前向通道某些环节并联，以达到改善系统性能的目的，则称这种校正为并联校正。按校正环节 $G_c(s)$ 的并联方式，又可分为反馈校正和复合校正。若校正环节 $G_c(s)$ 位于原传递函数框图的局部反馈回路的反馈通道中，则称为反馈校正，如图 6-2 所示。若在反馈控制回路中，加入前馈通路，组成一个有机整体，则称为复合校正，如图 6-3 所示。复合校正可以单独作用于开环控制系统，也可以作为反馈控制系统的附加校正而组成复合控制系统。

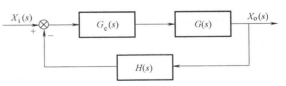

Fig. 6-1　Cascade compensation（串联校正）

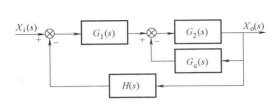

Fig. 6-2　Feedback compensation（反馈校正）

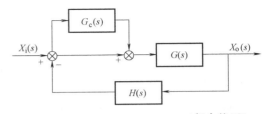

Fig. 6-3　Combined compensation（复合校正）

6.2 控制系统的串联校正

基于频率响应法的串联校正，根据校正环节 $G_c(s)$ 的特性，可分为相位超前校正、相位滞后校正和滞后-超前校正。用频率法对系统进行校正的基本思路是通过所加校正装置，

6.2 Cascade Compensation of Control Systems

According to the characteristics of the compensation element $G_c(s)$, the cascade compensation of the system based on the frequency response method can be divided into phase-lead compensation, phase-lag compensation and lag-lead compensation. The basic idea of compensation based on the frequency method is to change the shape of the open-loop frequency characteristic curve of the system by adding compensators. It is required to have the following characteristics for the open-loop frequency characteristic of the system after compensation.

1) It requires a sufficiently large gain for the low frequency band to meet the requirements of steady state accuracy.

2) It requires a wider frequency band for the medium frequency band to have satisfactory dynamic performance. Here, the slope of amplitude-frequency characteristic curve of medium frequency band is -20dB/dec.

3) It requires a rapid attenuation of the amplitude for the high frequency band to reduce the influence of noise.

6.2.1 Phase-lead compensation

1. Phase-lead compensation network

In general, when the open-loop gain of the control system is increased to the required value to meet its static performance, the system may be unstable, or even if it is stable, its dynamic performance is generally not ideal. In this case, a phase-lead compensator need to be added to the forward path of the system, so that the dynamic performance of the system can meet the design requirements under the premise that the open-loop gain is unchanged.

Fig. 6-4 shows the RC phase-lead compensation network, and its transfer function is

$$G_c(s) = \frac{U_o(s)}{U_i(s)} = \frac{R_2}{R_1+R_2} \frac{R_1 Cs+1}{\frac{R_2}{R_1+R_2} R_1 Cs+1} = \alpha \frac{Ts+1}{\alpha Ts+1} \quad (6\text{-}1)$$

where α is an attenuation coefficient, and $\alpha = \dfrac{R_2}{R_1+R_2} < 1$; $T = R_1 C$.

According to equation (6-1), when the passive phase-lead network is used for cascade compensation, the open-loop gain of the whole system is reduced by $1/\alpha$ times, so it is necessary to increase the amplifier gain for compensation, as is shown in Fig. 6-5. Hence, the transfer function of the compensation network is

$$\frac{1}{\alpha} G_c(s) = \frac{Ts+1}{\alpha Ts+1} \quad (6\text{-}2)$$

Since the attenuation coefficient of the phase-lead network $\alpha < 1$, its negative real zero $-\dfrac{1}{T}$ is

改变系统开环频率特性曲线的形状。校正后系统的开环频率特性应具有如下特点。

1) 低频段具有足够大的增益,以满足稳态精度的要求。
2) 中频段具有较宽的频带,以获得满意的动态性能。此处,中频段的幅频特性曲线的斜率为 $-20\mathrm{dB/dec}$。
3) 高频段幅值迅速衰减,以减少噪声的影响。

6.2.1 相位超前校正

1. 相位超前校正网络

一般而言,当控制系统的开环增益增大到满足其静态性能所要求的数值时,系统有可能不稳定,或者即使能稳定,其动态性能一般也不会理想。在这种情况下,需在系统的前向通路中增加超前校正装置,使系统的动态性能在开环增益不变的前提下满足设计要求。

图 6-4 所示为 RC 相位超前校正网络,其传递函数为

$$G_c(s) = \frac{U_o(s)}{U_i(s)} = \frac{R_2}{R_1+R_2} \cdot \frac{R_1 Cs+1}{\frac{R_2}{R_1+R_2}R_1 Cs+1} = \alpha \frac{Ts+1}{\alpha Ts+1} \tag{6-1}$$

式中,α 为衰减系数,且 $\alpha = \dfrac{R_2}{R_1+R_2} < 1$;$T = R_1 C$。

由式 (6-1) 可知,采用无源相位超前网络进行串联校正时,整个系统的开环增益要缩小 $1/\alpha$ 倍,因此需要提高放大器增益加以补偿,如图 6-5 所示。因此,校正网络的传递函数为

$$\frac{1}{\alpha} G_c(s) = \frac{Ts+1}{\alpha Ts+1} \tag{6-2}$$

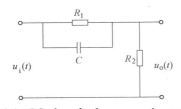

Fig. 6-4　RC phase-lead compensation network
（RC 超前校正网络图）

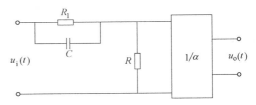

Fig. 6-5　Passive phase-lead compensation network with an additional amplifier
（带有附加放大器的无源相位超前校正网络）

由于相位超前网络的衰减系数 $\alpha<1$,故在复平面上其负实零点 $-\dfrac{1}{T}$ 总是位于负实极点 $-\dfrac{1}{\alpha T}$ 右侧,两者之间的距离由常数 α 决定。可知改变 α 和 T (即电路的参数 R_1、R_2 和 C) 的数值,超前网络的零、极点可在复平面的负实轴任意移动。

由式 (6-2) 可得校正网络的频率特性为

$$\begin{cases} L_c(\omega) = 20\lg\left|\dfrac{1}{\alpha}G_c(s)\right| = 20\lg\sqrt{1+(T\omega)^2} - 20\lg\sqrt{1+(\alpha T\omega)^2} \\ \varphi_c(\omega) = \arctan T\omega - \arctan \alpha T\omega \end{cases} \tag{6-3}$$

always located on the right side of the negative real pole $-\dfrac{1}{\alpha T}$ of the s-plane, and the distance between them is determined by the constant α. It can be seen that when the values of α and T (i.e. the circuit parameters R_1, R_2 and C) are changed, the poles of the phase-lead network can move arbitrarily on the negative real axis of the s-plane.

According to equation (6-2), the frequency characteristic of the compensation network is obtained as

$$\begin{cases} L_c(\omega) = 20\lg\left|\dfrac{1}{\alpha}G_c(s)\right| = 20\lg\sqrt{1+(T\omega)^2} - 20\lg\sqrt{1+(\alpha T\omega)^2} \\ \varphi_c(\omega) = \arctan T\omega - \arctan \alpha T\omega \end{cases} \quad (6\text{-}3)$$

The Bode plot of the phase-lead compensation network is shown in Fig. 6-6. It can be seen from the figure that: ① the phase-lead network has a positive slope segment for the asymptotic curve of the amplitude-frequency characteristic of the frequency between $\dfrac{1}{T}$ and $\dfrac{1}{\alpha T}$, indicating that its input signal has an obvious differential effect; ② the phase-frequency characteristic has a positive phase shift in the frequency range of $\dfrac{1}{T} \sim \dfrac{1}{\alpha T}$, and the phase angle of the output signal is ahead of the phase angle of the input signal, indicating that the steady-state output voltage of the network with the sinusoidal signal leads the input voltage regarding the phase. Thus it is called the phase-lead network.

After completing the derivative of the second formula in equation (6-3) and making it zero, the maximum frequency of the phase-lead angle provided by the phase-lead network is

$$\omega_m = \dfrac{1}{\sqrt{\alpha}\,T} = \dfrac{1}{\sqrt{\alpha T}}\dfrac{1}{\sqrt{T}} \quad (6\text{-}4)$$

Taking the logarithm of both sides of the equation, we have

$$\lg\omega_m = \dfrac{1}{2}\left(\lg\dfrac{1}{T} + \lg\dfrac{1}{\alpha T}\right)$$

It can be seen that ω_m is the geometric center of the two corner frequencies $\dfrac{1}{T}$ and $\dfrac{1}{\alpha T}$. Substituting equation (6-4) into the second formula in equation (6-3), the maximum phase-lead angle can be obtained as

$$\varphi_m = \arcsin\dfrac{1-\alpha}{1+\alpha} \quad (6\text{-}5)$$

As can be seen from Fig. 6-6, the phase-lead network is essentially a high-pass filter. Substituting the equation (6-4) into the first formula of equation (6-3), we have

$$L_c(\omega_m) = 20\lg\sqrt{\dfrac{1+(T\omega_m)^2}{1+(\alpha T\omega_m)^2}}$$

$$= -20\lg\sqrt{\alpha} = -10\lg\alpha \quad (6\text{-}6)$$

The phase-lead compensation can produce a sufficiently large lead phase angle, which can im-

相位超前校正网络的伯德图如图 6-6 所示。由图可见：①超前网络对频率在 $\frac{1}{T}$ 至 $\frac{1}{\alpha T}$ 之间的对数幅频特性渐近线具有正斜率段，表明其输入信号有明显的微分作用；②在该频率范围内，相频特性具有正相移，输出信号相角比输入信号相角超前，表明网络在正弦信号作用下的稳态输出电压在相位上超前于输入，故称相位超前网络。

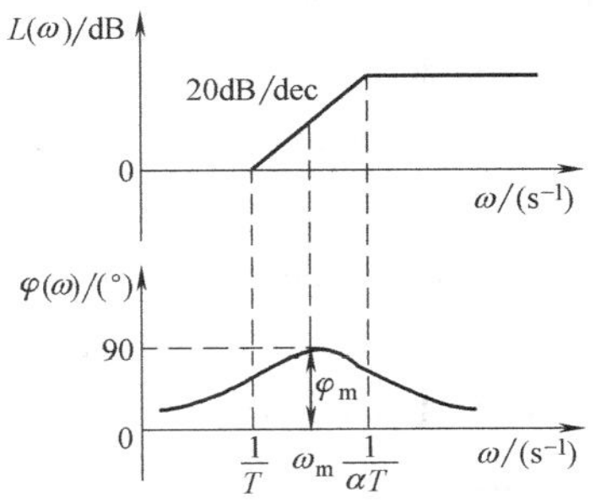

完成对式（6-3）中第二式求导并令其为零，可得超前网络所提供的最大超前角频率为

$$\omega_m = \frac{1}{\sqrt{\alpha} T} = \frac{1}{\sqrt{\alpha T}} \frac{1}{\sqrt{T}} \tag{6-4}$$

Fig. 6-6 Bode plot of the phase-lead compensation network
（超前校正网络的伯德图）

对上式两边进行对数运算有

$$\lg \omega_m = \frac{1}{2}\left(\lg \frac{1}{T} + \lg \frac{1}{\alpha T}\right)$$

可见 ω_m 是两个转折频率 $\frac{1}{T}$ 和 $\frac{1}{\alpha T}$ 的几何中心。把式（6-4）代入式（6-3）中第二式，可求得最大超前角为

$$\varphi_m = \arcsin \frac{1-\alpha}{1+\alpha} \tag{6-5}$$

由图 6-6 可以看出，超前网络实质是一个高通滤波器。将式（6-4）代入式（6-3）第一式得

$$L_c(\omega_m) = 20\lg \sqrt{\frac{1+(T\omega_m)^2}{1+(\alpha T\omega_m)^2}}$$

$$= -20\lg\sqrt{\alpha} = -10\lg\alpha \tag{6-6}$$

相位超前校正能够产生足够大的超前相位角，使瞬态响应得到改善。低频段幅频特性的斜率为零，因此对稳态精度提高较少。

2. 基于伯德图进行相位超前校正

基于频率法对系统进行相位超前校正的基本原理，是利用校正网络的相位超前特性来增大系统的相位裕度，以达到改善系统瞬态响应的目的。为此，在对截止频率没有特别要求时，要求校正网络的最大相位超前角出现在系统的截止频率（或幅值穿越频率）处。利用伯德图进行相位超前校正的步骤如下。

1）根据稳态性能要求，确定开环增益 K。

2）利用已确定的开环增益 K，绘制未校正系统的开环伯德图，确定校正前的相位裕度 γ 和幅值裕度 L_g。

3）计算相位超前校正装置应提供的最大超前相位角 φ_m，设期望的相位裕度值为 $[\gamma]$，有

$$\varphi_m = [\gamma] - \gamma + \varepsilon \tag{6-7}$$

式中，ε 是一个补偿值，用于补偿相位超前校正装置的引入使系统截止频率增大而产生的相角滞后量。如果未校正系统的开环对数幅频特性曲线在截止频率处的斜率为 -40dB/dec，一

prove the transient response. The slope of the amplitude-frequency characteristic curve in the low-frequency is zero, thus the steady-state accuracy is slightly improved.

2. Phase-lead compensation network based on Bode plot

The basic principle of the phase-lead compensation of the system based on the frequency method is to use the phase-lead characteristic of the compensation network for increasing the phase margin of the system, such that the goal of improving the transient response of the system can be achieved. Hence, when there is no special requirement for the cutoff frequency, the maximum phase-lead angle of the compensation network is required to appear at the cutoff frequency (or the amplitude crossing frequency) of the system. The steps for the phase-lead compensation using the Bode plot are as follows.

1) Determine the open-loop gain K according to the performance requirements at a steady state.

2) Draw the open-loop Bode plot of the uncompensated system using the determined open-loop gain K, and determine the phase margin γ and amplitude margin L_g of the uncompensated system.

3) Calculate the corresponding maximum phase-lead angle φ_m that the phase-lead compensator should provide. Let the desired phase margin value be $[\gamma]$, we have

$$\varphi_m = [\gamma] - \gamma + \varepsilon \qquad (6\text{-}7)$$

where ε is used to compensate the phase angle lag caused by the increase of the cutoff frequency of the system due to the introduction of the phase-lead compensator. If the open-loop logarithmic amplitude-frequency characteristic of the uncompensated system has a slope of -40dB/dec at the cutoff frequency, it is generally taken as $\varepsilon = 5° \sim 10°$; if it has a slope of -60dB/dec, it is taken as $\varepsilon = 15° \sim 20°$.

4) Calculate the value of α of the phase-lead network according to φ_m. Here, $\varphi_m = \arcsin\dfrac{1-\alpha}{1+\alpha}$, we have

$$\alpha = \frac{1-\sin\varphi_m}{1+\sin\varphi_m} \qquad (6\text{-}8)$$

5) Determine the open-loop cutoff frequency of the compensated system. The open-loop cutoff frequency ω_c of the compensated system is set to the maximum lead phase angle.

If the open-loop transfer function of the compensated system in parallel is

$$G_K(s) = \frac{1}{\alpha} G_c(s) G(s)$$

where $G(s)$ is the open-loop transfer function of the uncompensated system; $G_c(s)$ is the transfer function of the compensator.

The cutoff frequency ω_c of the compensated system should meet the relation as

$$\left| \frac{1}{\alpha} G_c(j\omega_c) G(j\omega_c) \right| = 1 \text{ or } L_c(\omega_c) + 20\lg |G(j\omega_c)| = 0$$

Let $\omega_c = \omega_m$, we have

$$20\lg |G(j\omega_c)| = -L_c(\omega_c) = -L_c(\omega_m) = 10\lg\alpha \qquad (6\text{-}9)$$

According to the above formula, the amplitude crossing frequency ω_c of the compensated system can

般取 $\varepsilon = 5° \sim 10°$；如果为 -60dB/dec，则取 $\varepsilon = 15° \sim 20°$。

4) 根据 φ_m 计算相位超前网络的 α 值。此处，$\varphi_m = \arcsin \dfrac{1-\alpha}{1+\alpha}$，则有

$$\alpha = \frac{1-\sin\varphi_m}{1+\sin\varphi_m} \tag{6-8}$$

5) 确定校正后系统的开环截止频率。将校正后系统的开环截止频率 ω_c 设置在取得最大超前相位角处。

若并联校正后的系统开环传递函数为

$$G_K(s) = \frac{1}{\alpha} G_c(s) G(s)$$

式中，$G(s)$ 为未校正系统的开环传递函数；$G_c(s)$ 为校正装置的传递函数。

校正后系统的截止频率 ω_c 应满足

$$\left| \frac{1}{\alpha} G_c(j\omega_c) G(j\omega_c) \right| = 1 \quad \text{或} \quad L_c(\omega_c) + 20\lg|G(j\omega_c)| = 0$$

取 $\omega_c = \omega_m$，得

$$20\lg|G(j\omega_c)| = -L_c(\omega_c) = -L_c(\omega_m) = 10\lg\alpha \tag{6-9}$$

可由此求出校正后系统的幅值穿越频率 ω_c。

6) 确定相位超前校正部件的转折频率 $\omega_1 = \dfrac{1}{T}$，$\omega_2 = \dfrac{1}{\alpha T}$。

7) 画出校正后系统的伯德图，并验算相位裕度和幅值裕度是否满足要求。如果不满足，则需增大 ε 值重新进行计算，直到满足要求。

相位超前校正能够使系统相位裕度增大，从而降低了系统的超调量。与此同时，当系统的带宽增加时，系统的响应速度加快，但系统的低频段没有改变，即稳态精度没有改变。

例 6-1 某控制系统如图 6-7a 所示，要求系统在单位斜坡输入下的稳态误差 $\varepsilon_{si} = 0.05$，频域性能指标满足：相位裕度 $\gamma \geq 50°$，幅值裕度 $L_g \geq 10\text{dB}$。试设计系统的校正环节。

解：1) 确定开环增益 K。因为该系统是 I 型系统，所以

$$K = \frac{1}{\varepsilon_{si}} = \frac{1}{0.05} = 20$$

2) 计算未校正系统的相位裕度和幅值裕度。绘制未校正系统的开环伯德图，如图 6-7b 中的曲线 $G(j\omega)$ 所示。由图可知，未校正系统的相位裕度 $\gamma = 17°$，幅值裕度 $L_g = \infty$，系统稳定。

3) 确定所需要增加的超前相位角。因为相位裕度小于 $50°$，故相对稳定性不满足要求。为了在不减小幅值裕度的前提下，将相位裕度从 $17°$ 提高到 $50°$，需要采用相位超前校正，其超前相位角应为 $33°$，但这会使得开环截止频率向右移动。因此，在考虑超前相位角时，要增加 $5°$ 左右，以补偿这一移动。由式 (6-7) 得

$$\varphi_m = 50° - 17° + 5° = 38°$$

4) 确定校正环节中的 α 值。由式 (6-8) 求得 $\alpha = 0.24$，代入式 (6-6) 得

$$L_c(\omega_c) = -10\lg\alpha = 6.2\,(\text{dB})$$

将上式代入式 (6-9) 计算，或者在图 6-7b 上找到曲线 $G(j\omega)$ 的幅值为 -6.2dB 时的频率，

be obtained.

6) Determine the corner frequency of the phase-lead compensation components as $\omega_1 = \dfrac{1}{T}$, $\omega_2 = \dfrac{1}{\alpha T}$.

7) Draw the Bode plot of the compensated system and check whether the phase margin and amplitude margin meet the requirements. If not, increase the value of ε and recalculate until the requirements are met.

The phase-lead compensation can increase the system's phase margin, which reduces the system's overshoot. At the same time, as the bandwidth of the system increases, the response speed of the system is increased, but the low frequency band of the system is unchanged, that is, the steady-state accuracy is not changed.

Example 6-1 The control system is shown in Fig. 6-7a. It is required that the system has a steady-state error of $\varepsilon_{si} = 0.05$ with the unit ramp input, and the frequency domain performance indicators satisfy that phase margin $\gamma \geqslant 50°$, amplitude margin $L_g \geqslant 10\text{dB}$. Please design the compensation elements of the system.

Solution: 1) Determine the open-loop gain K. Since the system is a type-I system, we have

$$K = \dfrac{1}{\varepsilon_{si}} = \dfrac{1}{0.05} = 20$$

2) Calculate the phase margin and amplitude margin of the uncompensated system. Draw an open-loop Bode plot of the uncompensated system, i.e. the curve of $G(j\omega)$ as is shown in Fig. 6-7b. As can be seen from the figure, the phase margin of the uncompensated system is $\gamma = 17°$, and the amplitude margin is $L_g = \infty$, thus the system is stable.

3) Determine the lead phase angle that needs to be increased. Since the phase margin is less than $50°$, the relative stability is not satisfactory. In order to increase the phase margin from $17°$ to $50°$ without reducing the gain margin, it is necessary to use the phase-lead compensation, and the lead phase angle should be $33°$, but this will cause the open-loop cutoff frequency to move to the right. Therefore, when considering the lead phase angle, it is necessary to increase by about $5°$ to compensate for this movement. According to equation (6-7), we have

$$\varphi_m = 50° - 17° + 5° = 38°$$

4) Determine the value of α in the compensation element. According to equation (6-8), $\alpha = 0.24$. Substituting α into equation (6-6), we have

$$L_c(\omega_c) = -10\lg\alpha = 6.2 \ (\text{dB})$$

Substituting the above equation into equation (6-9), or figuring out the frequency at which the amplitude of the curve of $G(j\omega)$ is -6.2dB in Fig. 6-7b, we can have the open-loop cutoff frequency of the compensated system as $\omega_c = 9\text{s}^{-1}$.

5) Determine the corner frequency of the lead compensation element. According to $\omega_1 = \dfrac{1}{T}$, $\omega_2 = \dfrac{1}{\alpha T}$, $\omega_c = \omega_m = \dfrac{1}{\sqrt{\alpha}\,T} = 9\text{s}^{-1}$, we have

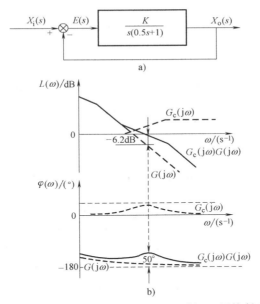

Fig. 6-7　Control system of example 6-1（例 6-1 图控制系统）
a）Block diagram of a system（系统框图）　b）Bode plot of the system（系统的伯德图）

所得即为校正后系统的开环截止频率 $\omega_c = 9 \text{s}^{-1}$。

5）确定超前校正环节的转折频率。根据 $\omega_1 = \dfrac{1}{T}$，$\omega_2 = \dfrac{1}{\alpha T}$，$\omega_c = \omega_m = \dfrac{1}{\sqrt{\alpha}\, T} = 9 \text{s}^{-1}$，可得

$$T = 0.23 \text{s}, \quad \alpha T = 0.055 \text{s}$$

由此得相位超前校正环节的频率特性为

$$G_c(j\omega) = \alpha \frac{1+jT\omega}{1+j\alpha T\omega} = 0.24 \frac{1+j0.23\omega}{1+j0.055\omega}$$

为了补偿超前校正所造成的幅值衰减，原开环增益需放大 K_1 倍。使 $K_1 \alpha = 1$，故

$$K_1 = \frac{1}{0.24} = 4.17$$

又因为 $G_c(s) = \alpha \dfrac{1+Ts}{1+\alpha Ts}$，所以校正后系统的开环传递函数为

$$G_K(s) = K_1 G_c(s) G(s) = \frac{1+0.23s}{1+0.055s} \frac{20}{s(0.5s+1)}$$

通过以上分析可知，相位超前校正有如下特点。

1）相位超前校正主要对系统中频段进行校正，使校正后幅频特性曲线中频段的斜率为 -20dB/dec，且相频特性有足够大的相位裕度，从而达到改善系统动态性能的目的。

2）相位超前校正会使系统瞬态响应的速度变快，校正后系统的截止频率比校正前的要大。这表明校正系统的频带变宽，瞬态响应速度变快；但系统抗高频噪声的能力变差。对此，在设计校正装置时，必须注意这种不足。

3）相位超前校正虽能较为有效地改善动态性能，但对相频特性在截止频率附近急剧下降的系统，单级超前校正网络的改进效果不明显。因为校正后系统的截止频率向高频段移

$$T = 0.23\text{s}, \quad \alpha T = 0.055\text{s}$$

Thus, the frequency characteristic of the phase-lead compensation element is

$$G_c(j\omega) = \alpha\frac{1+jT\omega}{1+j\alpha T\omega} = 0.24\frac{1+j0.23\omega}{1+j0.055\omega}$$

In order to compensate for the amplitude attenuation caused by the phase-lead compensation, the original open-loop gain need to be increased by K_1 times. Let $K_1\alpha = 1$, we have

$$K_1 = \frac{1}{0.24} = 4.17$$

According to $G_c(s) = \alpha\dfrac{1+Ts}{1+\alpha Ts}$, so the open-loop transfer function of the compensated system is

$$G_K(s) = K_1 G_c(s) G(s) = \frac{1+0.23s}{1+0.055s}\frac{20}{s(0.5s+1)}$$

According to above analysis, the phase-lead compensation has the following characteristics.

1) The phase-lead compensation mainly compensates the medium-frequency band in the system, so that the medium-frequency slope of the compensated system amplitude-frequency characteristic curve is -20dB/dec, and there is a large enough phase margin to achieve the purpose of improving the dynamic performance of the system.

2) The phase-lead compensation will make the system transient response speed faster, and the cutoff frequency of the compensated system is larger than that of the uncompensated system. It indicates that the frequency band of the compensated system becomes wider and the transient response speed becomes faster; however, the ability of the system to resist high frequency noise is deteriorated. Hence, the deficiency must be taken into account when designing the compensator.

3) The phase-lead compensation generally improves the dynamic performance effectively. But for an uncompensated system which phase frequency characteristic drops sharply near the cutoff frequency, the compensation effect of a single-level lead compensation network is not significant. The reason is that the cutoff frequency of the compensated system moves to the high-frequency band. At the new cutoff frequency, the phase angle lag of the uncompensated system is too large and it is not enough for ε to compensate for the phase lag caused by the introduction of the lead compensator. Therefore, the single-level lead compensation may not lead to satisfactory results, meanwhile a multi-level cascade compensation should be used.

6.2.2 Phase-lag compensation

1. Phase-lag compensation network

Fig. 6-8 shows the RC phase-lag compensation network, and its transfer function is

$$G_c(s) = \frac{U_o(s)}{U_i(s)} = \frac{R_2 Cs+1}{(R_1+R_2)Cs+1} = \frac{\beta Ts+1}{Ts+1} \tag{6-10}$$

where β is an attenuation coefficient, and $\beta = \dfrac{R_2}{R_1+R_2} < 1$; $T = (R_1+R_2)C$.

The logarithmic frequency characteristics of the compensator are

动，在新的截止频率处，未校正系统的相角滞后量过大，用 ε 不足以补偿因引入超前校正装置而产生的相位滞后，因此用单级超前校正网络可能无法获得满意的效果，此时宜采用多级串联校正。

6.2.2 相位滞后校正

1. 相位滞后校正网络

图 6-8 所示为 RC 滞后校正网络，其传递函数为

$$G_c(s) = \frac{U_o(s)}{U_i(s)} = \frac{R_2 Cs+1}{(R_1+R_2)Cs+1} = \frac{\beta Ts+1}{Ts+1} \tag{6-10}$$

式中，β 为衰减系数，且 $\beta = \dfrac{R_2}{R_1+R_2} < 1$；$T = (R_1+R_2)C$。

校正装置的对数频率特性为

$$\begin{cases} L_c(\omega) = 20\lg|G_c(j\omega)| = 20\lg\sqrt{1+(\beta T\omega)^2} - 20\lg\sqrt{1+(T\omega)^2} \\ \varphi_c(\omega) = \arctan\beta T\omega - \arctan T\omega \end{cases} \tag{6-11}$$

相位滞后校正网络的伯德图如图 6-9 所示。由图可见，其对数幅频特性渐近线具有负斜率段，相频特性具有负相移。这表明在正弦信号作用下，校正网络的稳态输出电压在相位上滞后于输入电压，故称为相位滞后网络。

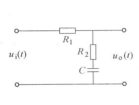

Fig. 6-8 RC phase-lag compensation network
（RC 滞后校正网络）

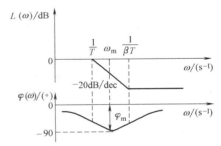

Fig. 6-9 Bode plot of the phase-lag compensation network
（相位滞后校正网络的伯德图）

由式（6-11）得相位滞后网络所提供的最大滞后角及所在频率为

$$\begin{cases} \varphi_m = -\arcsin\dfrac{1-\beta}{1+\beta} \\ \omega_m = \dfrac{1}{\sqrt{\beta}T} \end{cases} \tag{6-12}$$

式中，φ_m 出现在转折频率 $\dfrac{1}{T}$ 和 $\dfrac{1}{\beta T}$ 的几何中心。

由图 6-9 可以看出，相位滞后网络实质上是一个低通滤波器。相位滞后网络在 $\omega < \dfrac{1}{T}$ 时，对信号没有衰减作用；在 $\dfrac{1}{T} < \omega < \dfrac{1}{\beta T}$ 时，对信号有积分作用，呈滞后特性；在 $\omega > \dfrac{1}{\beta T}$ 时，对信号有衰减作用，衰减幅度为 $20\lg\beta$，β 越小，衰减作用越强。在设计过程中，应尽力避免最

$$\begin{cases} L_c(\omega) = 20\lg|G_c(j\omega)| = 20\lg\sqrt{1+(\beta T\omega)^2} - 20\lg\sqrt{1+(T\omega)^2} \\ \varphi_c(\omega) = \arctan\beta T\omega - \arctan T\omega \end{cases} \quad (6\text{-}11)$$

The Bode plot of the phase-lag compensation network is shown in Fig. 6-9. It can be seen from the figure that the logarithmic amplitude-frequency characteristic asymptotic curve has a negative slope and the phase-frequency characteristic has a negative phase shift. It indicates that the phase of the steady-state output voltage of the compensation network lags behind the input voltage with the sinusoidal signal, thus it is called phase-lag network.

The maximum lag angle and corresponding frequency provided by the phase-lag network according to equation (6-11) are

$$\begin{cases} \varphi_m = -\arcsin\dfrac{1-\beta}{1+\beta} \\ \omega_m = \dfrac{1}{\sqrt{\beta}T} \end{cases} \quad (6\text{-}12)$$

where φ_m occurs at the geometric center of the corner frequencies of $\dfrac{1}{T}$ and $\dfrac{1}{\beta T}$.

As can be seen from Fig. 6-9, the phase-lag network is essentially a low-pass filter. When the phase-lag network is at $\omega < \dfrac{1}{T}$, it has no attenuation effect on the signal; when $\dfrac{1}{T} < \omega < \dfrac{1}{\beta T}$, it has integral effect on the signal and shows lag characteristic; when $\omega > \dfrac{1}{\beta T}$, it has attenuation effect and the attenuation amplitude on the signal is $20\lg\beta$. The smaller β is, the stronger the attenuation effect is. In the design process, it is ought to avoid the maximum lag angle occurring near the open-loop cutoff frequency ω_c of the compensated system. Hence, we have

$$L_c(\omega_c) = 20\lg\beta \quad (6\text{-}13)$$

The role of phase-lag compensation is described as: we use the compensator mainly for its negative slope segment of logarithmic amplitude-frequency characteristics to attenuate the high frequency band amplitude of the compensated system, and to shift the open-loop cutoff frequency to the left in order to obtain sufficient phase margin.

2. Phase-lag compensation based on Bode plot

Since the phase-lag compensation network has the characteristic of a low-pass filter, it will make the medium-frequency and high-frequency gain of the system open-loop frequency characteristic be reduced and the cutoff frequency ω_c be decreased that the compensation network is connected in series with the immutable part of a system. Therefore, it is possible for the system to obtain a large enough phase margin meanwhile the low-frequency band of the frequency characteristic is not affected. Thus, it can be seen that the phase-lag compensation can satisfy both the dynamic and static requirements of the system under certain conditions.

The steps for the phase-lag compensation using the Bode plot are as follows.

1) Determine the open-loop gain K according to the steady-state performance requirements.

2) Draw the open-loop Bode plot of the uncompensated system by using the determined open-

大滞后角出现在校正系统开环截止频率 ω_c 附近。可取

$$L_c(\omega_c) = 20\lg\beta \tag{6-13}$$

相位滞后校正的作用可以描述为：主要利用校正装置所产生的对数幅频特性负斜率段，使被校正系统高频段幅值衰减，开环截止频率左移，从而获得充足的相位裕度。

2. 基于伯德图进行相位滞后校正

由于相位滞后校正网络具有低通滤波特性，因而它与系统的不可变部分相串联，会使系统开环幅频特性的中频和高频段增益降低，且截止频率 ω_c 减小，从而有可能使系统获得足够大的相位裕度，且不影响频率特性的低频段。由此可见，相位滞后校正在一定的条件下，也能使系统同时满足动态和静态性能要求。

利用伯德图进行相位滞后校正的步骤如下。

1) 根据稳态性能要求，确定开环增益 K。

2) 利用已确定的开环增益 K，绘制未校正系统的开环伯德图，求出未校正系统的相位裕度和幅值裕度。

3) 确定校正后系统的开环截止频率 ω_c。若系统的相位裕度和幅值裕度不满足要求，应考虑新的开环截止频率。新的开环截止频率 ω_c 应选在相角等于 $-180°$ 加必要的相位裕度所对应的频率上，即

$$\varphi(\omega_c) = -180° + [\gamma] + \varepsilon \tag{6-14}$$

式中，$[\gamma]$ 为校正后系统的期望相位裕度值；ε 用于补偿因引入滞后校正装置而在截止频率 ω_c 处产生的相位滞后量，一般工程上取 $\varepsilon = 5° \sim 15°$。

4) 确定校正装置的 β 值。在未校正系统的幅频特性曲线上，在新的开环截止频率 ω_c 处取对应需要的衰减量，并让衰减量等于 $-20\lg\beta$，根据这一衰减量确定 β 值。

此外，未校正系统的频率特性 $G(j\omega_c)$ 与校正装置的频率特性 $G_c(j\omega_c)$ 之间的关系可描述为

$$|G_c(j\omega_c)G(j\omega_c)| = 1 \text{ 或 } L_c(\omega_c) + L(\omega_c) = 0$$

解得

$$L(\omega_c) = -L_c(\omega_c) = -20\lg\beta \tag{6-15}$$

5) 确定相位滞后校正环节的转折频率。由于在截止频率 ω_c 处，相位滞后校正装置本身会产生一定的相位滞后，设计时应尽可能减小滞后角。为此可使校正装置的转折频率 $\omega_1 = \dfrac{1}{T}$ 和 $\omega_2 = \dfrac{1}{\beta T}$ 相对 ω_c 越小越好。但考虑到工程上的可行性，一般取 $\omega_2 = \dfrac{1}{\beta T} = \left(\dfrac{1}{5} \sim \dfrac{1}{10}\right)\omega_c$，然后根据 $\omega_1 = \dfrac{1}{T}$ 确定另一个转折频率。

6) 计算相位滞后校正环节的传递函数。若全部指标都满足要求，把 T 和 β 值代入 $G_c(s)$ 表达式中，求出相位滞后校正环节的传递函数。

一方面，由于相位滞后校正的衰减作用，开环截止频率左移到较低的频率上，而且是在斜率为 -20dB/dec 的特性区段之内，因此可以满足相位裕度的要求。另一方面，也正是它的衰减作用，使系统的带宽减小，导致系统动态响应时间变长。

例 6-2 设单位反馈系统的开环传递函数为

$$G(s) = \dfrac{K}{s(0.2s+1)(0.5s+1)}$$

loop gain K, and then obtain the phase margin and amplitude margin of the uncompensated system.

3) Determine the open-loop cutoff frequency ω_c of the compensated system. If the system's phase margin and amplitude margin do not meet the requirements, a new open-loop cutoff frequency should be considered. The new open-loop cutoff frequency ω_c should be chosen at a frequency corresponding to the phase angle that is equal to $-180°$ plus the necessary phase margin, i. e.

$$\varphi(\omega_c) = -180° + [\gamma] + \varepsilon \qquad (6\text{-}14)$$

where $[\gamma]$ is the expected phase margin value of the compensated system; ε is used to compensate the phase-lag value which is generated at the cutoff frequency ω_c due to the introduction of the phase-lag compensator, and $\varepsilon = 5° \sim 15°$ is generally taken in engineering.

4) Determine the value of β of the compensator. Take the required attenuation at the new open-loop cutoff frequency ω_c in the amplitude-frequency characteristic curve of the uncompensated system, and let the attenuation be equal to $-20\lg\beta$. And then the value of β can be determined according to the attenuation.

In addition, the relationship between the frequency characteristic $G(j\omega_c)$ of the uncompensated system and the frequency characteristic $G_c(j\omega_c)$ of the compensator can be described as

$$|G_c(j\omega_c)G(j\omega_c)| = 1 \text{ or } L_c(\omega_c) + L(\omega_c) = 0$$

Thus, we have

$$L(\omega_c) = -L_c(\omega_c) = -20\lg\beta \qquad (6\text{-}15)$$

5) Determine the corner frequency of the phase-lag compensation element. Since the phase-lag compensator itself generates a certain phase-lag value at the cutoff frequency ω_c, the lag angle should be reduced as much as possible in the design process. To this end, compared with ω_c, the corner frequencies as $\omega_1 = \dfrac{1}{T}$ and $\omega_2 = \dfrac{1}{\beta T}$ of the compensator should be as small as possible. But considering the feasibility of engineering practice, it is generally taken as $\omega_2 = \dfrac{1}{\beta T} = \left(\dfrac{1}{5} \sim \dfrac{1}{10}\right)\omega_c$ and then another corner frequency can be determine according to $\omega_1 = \dfrac{1}{T}$.

6) Figure out the transfer function of the phase-lag compensation element. If all the indicators meet the requirements, we substitute the values T and β into the expression of $G_c(s)$ to figure out the transfer function of the phase-lag compensation element.

On one hand, due to the attenuation effect of the phase-lag compensation, the open-loop cutoff frequency is shifted to a lower frequency and is within the characteristic section with a slope of -20dB/dec, so it can satisfy the phase margin requirement. On the other hand, also due to its attenuation effect, the bandwidth of the system is reduced, resulting in an increase in the dynamic response time of the system.

Example 6-2 The open-loop transfer function of a unit feedback system is

$$G(s) = \dfrac{K}{s(0.2s+1)(0.5s+1)}$$

The performance indicators are required as: steady-state error coefficient $K_v = 20\text{s}^{-1}$, phase margin

要求其性能指标满足：稳态误差系数 $K_v = 20s^{-1}$，相位裕度 $\gamma \geq 35°$，幅值裕度 $L_g \geq 10dB$。试求串联相位滞后校正环节的传递函数。

解：1）根据稳态指标要求求出 K 值。依据 $G(s) = \dfrac{K}{s(0.2s+1)(0.5s+1)}$，有

$$K_v = \lim_{s \to 0} sG(s) = K = 20$$

因此，未校正系统的开环频率特性为

$$G(j\omega) = \dfrac{20}{j\omega(1+j0.2\omega)(1+j0.5\omega)}$$

2）未校正系统的伯德图如图 6-10 所示，求出相角裕度为 $-30.6°$，幅值裕度为 $-10dB$。这表明满足稳态性能指标后系统不稳定，因此需要对系统进行校正。

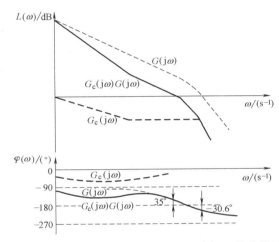

Fig. 6-10 The Bode plot of example 6-2（例 6-2 的伯德图）

3）根据相位裕度 $\gamma \geq 35°$ 的要求，取 $[\gamma] = 35°$，为补偿相位滞后校正环节的相角滞后，取 $\varepsilon = 12°$，所需相位裕度计算为 $35° + 12° = 47°$。由式（6-14）得 $\varphi(\omega_c) = -180° + [\gamma] + \varepsilon = -180° + 47° = -133°$。然后选择使相角为 $-133°$ 的频率为校正后系统的开环截止频率，即

$$\varphi_c(\omega_c) = -90° - \arctan 0.2\omega_c - \arctan 0.5\omega_c = -133°$$

由图 6-10 可求得当 $\omega_c = 1.16$ 时，$\varphi(1.16) = -133°$，即可选择 $\omega_c = 1.16$。

4）原系统在 $\omega_c = 1.16$ 处的对数幅频特性为

$$L(\omega_c) = 20\lg|G(j\omega_c)| = 20\lg \dfrac{20}{1.16 \times \sqrt{1+(0.2 \times 1.16)^2}\sqrt{1+(0.5 \times 1.16)^2}}$$

$$= 23.244 dB$$

校正后系统伯德图在 $\omega = \omega_c$ 处应为 $0dB$。所以有

$$L_c(\omega_c) = -L(\omega_c) = -23.244 dB = 20\lg\beta$$

由此可求出校正环节参数（衰减系数）为 $\beta = 0.058$。

5）根据 $\dfrac{1}{\beta T} = \left(\dfrac{1}{5} \sim \dfrac{1}{10}\right)\omega_c$ 求得 T。为使相位滞后校正环节的时间常数 T 不过分大，取 $\dfrac{1}{\beta T} = \dfrac{1}{5}\omega_c$，求出 $T = 74.32$。这样，相位滞后校正环节的传递函数为

$\gamma \geqslant 35°$, magnitude margin $L_g \geqslant 10\text{dB}$. Try to figure out the transfer function of the cascade phase-lag compensation element.

Solution: 1) Calculate the value of K in terms of the requirements of the steady-state indicators. According to $G(s) = \dfrac{K}{s(0.2s+1)(0.5s+1)}$, we have

$$K_v = \lim_{s \to 0} sG(s) = K = 20$$

So, the frequency characteristic of the uncompensated system is

$$G(j\omega) = \dfrac{20}{j\omega(1+j0.2\omega)(1+j0.5\omega)}$$

2) The Bode plot of the uncompensated system is shown in Fig. 6-10. The phase margin is $-30.6°$ and the amplitude margin is -10dB. This indicates that the system is unstable after satisfying the steady-state performance indicators, so the system needs to be compensated.

3) According to the requirement of the phase margin as $\gamma \geqslant 35°$, let $[\gamma] = 35°$. To compensate for the lag angle of the phase-lag compensation element, let $\varepsilon = 12°$. So the required phase margin is calculated as $35° + 12° = 47°$. According to equation (6-14), we have $\varphi(\omega_c) = -180° + [\gamma] + \varepsilon = -180° + 47° = -133°$. Then select the frequency with the phase angle of $-133°$ as the open-loop cutoff frequency of the compensated system, i.e.

$$\varphi_c(\omega_c) = -90° - \arctan 0.2\omega_c - \arctan 0.5\omega_c = -133°$$

According to Fig. 6-10, we have $\omega_c = 1.16$, then we have $\varphi(1.16) = -133°$. So it can be chosen as $\omega_c = 1.16$.

4) The logarithmic amplitude-frequency characteristic of the original system at $\omega_c = 1.16$ is

$$L(\omega_c) = 20\lg|G(j\omega_c)| = 20\lg \dfrac{20}{1.16 \times \sqrt{1+(0.2 \times 1.16)^2}\sqrt{1+(0.5 \times 1.16)^2}}$$
$$= 23.244\text{dB}$$

The Bode plot of the compensated system at $\omega = \omega_c$ should be 0dB. Then we have

$$L_c(\omega_c) = -L(\omega_c) = -23.244\text{dB} = 20\lg\beta$$

Thus, the compensation element parameter (attenuation coefficient) can be obtained as $\beta = 0.058$.

5) According to $\dfrac{1}{\beta T} = \left(\dfrac{1}{5} \sim \dfrac{1}{10}\right)\omega_c$, T can be obtained. In order to make the time constant T of the phase-lag compensation element not too large, let $\dfrac{1}{\beta T} = \dfrac{1}{5}\omega_c$, and then find $T = 74.32$. Thus, the transfer function of the phase-lag compensation element is

$$G_c(s) = \dfrac{\beta Ts + 1}{Ts + 1} = \dfrac{4.3s + 1}{74.32s + 1}$$

Hence, the open-loop transfer function of the compensated system is

$$G_c(s)G(s) = \dfrac{20(4.3s+1)}{s(0.2s+1)(0.5s+1)(74.32s+1)}$$

6) The Bode plot of the compensated system is shown in Fig. 6-10. According to Fig. 6-10, the phase margin and amplitude margin of the compensated system can be obtained as $\gamma = 35°$ and $L_g =$

$$G_c(s) = \frac{\beta Ts+1}{Ts+1} = \frac{4.3s+1}{74.32s+1}$$

因此，校正后系统的开环传递函数

$$G_c(s)G(s) = \frac{20(4.3s+1)}{s(0.2s+1)(0.5s+1)(74.32s+1)}$$

6）绘出校正后系统的伯德图，如图 6-10 所示，由图 6-10 可求出校正后系统相位裕度 $\gamma = 35°$，幅值裕度 $L_g = 12\text{dB}$，且 $K_v = K = 20$。说明校正后系统的稳态、动态性能均满足指标的要求。

与相位超前校正相比，相位滞后校正有如下特点。

1）由于相位滞后校正装置的低通特性，校正后系统的截止频率 ω_c 减小，频带变窄。

2）由于相位滞后校正装置在滤波性质上与积分环节具有相似性，因此它的引入会增加系统阻尼比，减小超调量，并使瞬态响应速度变慢。

3）不同于相位超前校正，相位滞后校正装置的转折频率选取不是十分苛刻，因此其设计相对简单。

4）由于相位滞后校正装置是利用低通特性使系统的截止频率前移的，因此该校正方法可能使系统的相位裕度超过 90°。

6.2.3 相位滞后-超前校正

相位超前校正可以增加频宽，提高快速性，以及改善相对稳定性，但由于有增益损失而不利于稳态精度。相位滞后校正可以提高稳定性及稳态精度，但降低快速性。

相位滞后-超前校正方法兼有相位滞后校正和相位超前校正的优点，即校正后系统响应速度快，超调量小，能够抑制高频噪声，可全面改善系统的控制性能。当未校正系统不稳定，且对校正后系统的动态和静态性能（如响应速度、相位裕度和稳态误差）均有较高要求时，仅采用相位超前校正或相位滞后校正，均难以达到预期效果，此时宜采用相位滞后-超前校正。

1. 相位滞后-超前校正网络

图 6-11 所示为 RC 相位滞后-超前校正网络，其传递函数为

$$\begin{aligned}G_c(s) &= \frac{U_o(s)}{U_i(s)} = \frac{(R_1C_1s+1)(R_2C_2s+1)}{(R_1C_1s+1)(R_2C_2s+1)+R_1C_2s} \\ &= \frac{(T_1s+1)(T_2s+1)}{(\beta T_1s+1)\left(\dfrac{T_2}{\beta}s+1\right)}\end{aligned} \quad (6\text{-}16)$$

式中，$T_1 = R_1C_1$；$T_2 = R_2C_2$（$T_1 > T_2$）；$R_1C_1 + R_2C_2 + R_1C_2 = \dfrac{T_2}{\beta} + \beta T_1$，$\beta > 1$。

相位滞后-超前校正网络的伯德图如图 6-12 所示。由图可见，转折频率分别为 $\dfrac{1}{\beta T_1}$、$\dfrac{1}{T_1}$、$\dfrac{1}{T_2}$ 和 $\dfrac{\beta}{T_2}$。显然相位滞后校正在先，相位超前校正在后，且高频段和低频段均无衰减，故称相位滞后-超前网络。

12dB, and then $K_v = K = 20$. It means that the steady-state and dynamic performance of the compensated system meet the requirement of the indicators.

Compared with the phase-lead compensation, the phase-lag compensation has the following characteristics.

1) Due to the low-pass characteristic of the phase-lag compensator, the cutoff frequency ω_c of the compensated system is reduced, and the frequency band is narrowed.

2) Since the phase-lag compensator is similar to the integral element about the filtering property, its introduction will increase the damping ratio of the system, reduce the overshoot value, and make the transient response speed slower.

3) Unlike the phase-lead compensation, the selection of the corner frequency of the phase-lag compensator is not very critical, so it is relatively simple to design.

4) Since the phase-lag compensator makes the cutoff frequency of the system move forward out of the low-pass characteristic, the compensation method may cause the phase margin of the system to exceed 90°.

6.2.3 Phase lag-lead compensation

Phase-lead compensation can increase the bandwidth, improve the rapidity, and enhance the relative stability, but it is not good for steady-state accuracy due to the gain loss. Phase-lag compensation can improve the stability and steady-state accuracy, but it reduces the response speed.

The phase lag-lead compensation method has the advantages of phase-lag compensation and phase-lead compensation, i.e. the compensated system has fast response speed, small overshoot, the ability to suppress high frequency noise, and can comprehensively improve the control performance of the system. When the non-compensated system is unstable and the dynamic and static performance (like response speed, phase margin, and steady-state error) of the compensated system are required to be high, it is obvious that it is difficult to use only the phase-lead compensation or phase-lag compensation to achieve the desired compensation effect, so it is appropriate to use the phase lag-lead compensation.

1. Phase lag-lead compensation network

Fig. 6-11 shows the RC phase lag-lead compensation network, and its transfer function is

$$G_c(s) = \frac{U_o(s)}{U_i(s)} = \frac{(R_1 C_1 s+1)(R_2 C_2 s+1)}{(R_1 C_1 s+1)(R_2 C_2 s+1)+R_1 C_2 s} = \frac{(T_1 s+1)(T_2 s+1)}{(\beta T_1 s+1)\left(\dfrac{T_2}{\beta}s+1\right)} \quad (6\text{-}16)$$

where $T_1 = R_1 C_1$; $T_2 = R_2 C_2$, $T_1 > T_2$; $R_1 C_1 + R_2 C_2 + R_1 C_2 = \dfrac{T_2}{\beta} + \beta T_1$, $\beta > 1$.

The Bode plot of the phase lag-lead compensation network is shown in Fig. 6-12. According to the figure, the corner frequencies are $\dfrac{1}{\beta T_1}$, $\dfrac{1}{T_1}$, $\dfrac{1}{T_2}$, $\dfrac{\beta}{T_2}$, respectively. Obviously, the phase-lag

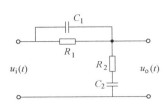

Fig. 6-11　RC phase lag-lead
compensation network

（RC 相位滞后-超前校正网络）

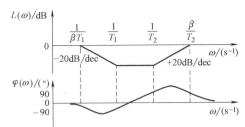

Fig. 6-12　Bode plot of the phase lag-lead
compensation network

（相位滞后-超前校正网络的伯德图）

相位滞后-超前校正装置的对数频率特性为

$$\begin{cases} L(\omega) = \dfrac{\sqrt{1+(\omega T_1)^2}\sqrt{1+(\omega T_2)^2}}{\sqrt{1+(\omega\beta T_1)^2}\sqrt{1+\left(\dfrac{\omega T_2}{\beta}\right)^2}} \\ \varphi_c(\omega) = \arctan T_1\omega + \arctan T_2\omega - \arctan\beta T_1\omega - \arctan\dfrac{T_2}{\beta}\omega \end{cases} \tag{6-17}$$

令

$$\varphi_c(\omega_1) = \arctan T_1\omega_1 + \arctan T_2\omega_1 - \arctan\beta T_1\omega_1 - \arctan\dfrac{T_2}{\beta}\omega_1 = 0$$

可求相角为零的频率

$$\omega_1 = \dfrac{1}{\sqrt{T_1 T_2}} \tag{6-18}$$

可见，当 $\omega<\omega_1$ 时，校正网络具有相位滞后特性；当 $\omega>\omega_1$ 时，校正网络具有相位超前特性。

2. 相位滞后-超前校正方法

相位滞后-超前校正实质上综合了相位滞后和相位超前校正各自的特点，利用校正装置的相位超前部分来增大系统的相位裕度，以改善其动态性能，利用校正装置的相位滞后部分来改善系统的静态性能。采用伯德图进行相位滞后-超前校正的步骤如下。

1）根据稳态性能要求，确定开环增益 K。

2）将 K 值作为开环增益，绘制原系统的对数幅频特性图，并求出原系统的截止频率 ω_c、相角裕度 γ 及幅值裕度 L_g。

3）以未校正系统斜率从 $-20\mathrm{dB/dec}$ 变为 $-40\mathrm{dB/dec}$ 的转折频率作为校正网络相位超前部分的转折频率 $\omega_b(\omega_b=1/T_2)$。这种选择不是唯一的，但它可以降低校正后系统的阶次，并能在中频段产生 $-20\mathrm{dB/dec}$ 斜率频段。

4）根据对响应速度的要求，计算出校正后系统的截止频率 ω_c'，以校正后系统的对数幅频特性 $L_c(\omega_c') + L(\omega_c') = 0\mathrm{dB}$ 为条件，求出衰减因子 $1/\beta$，即

$$-20\lg\beta + 20\lg T_2\omega_c' + L(\omega_c') = 0 \tag{6-19}$$

5）根据校正后系统的相位裕度要求，估算校正网络相位滞后部分的转折频率 $\omega_a(\omega_a=1/T_1)$。

6）验算性能指标。

compensation effect appears first, while the phase-lead compensation effect appears later, and there is no attenuation effect neither in the high-frequency band nor in the low-frequency band, so it is called the phase-lag-lead compensation network.

The logarithmic frequency characteristic of the phase lag-lead compensator is

$$\begin{cases} L(\omega) = \dfrac{\sqrt{1+(\omega T_1)^2}\sqrt{1+(\omega T_2)^2}}{\sqrt{1+(\omega \beta T_1)^2}\sqrt{1+\left(\dfrac{\omega T_2}{\beta}\right)^2}} \\ \varphi_c(\omega) = \arctan T_1\omega + \arctan T_2\omega - \arctan \beta T_1\omega - \arctan \dfrac{T_2}{\beta}\omega \end{cases} \quad (6\text{-}17)$$

Let

$$\varphi_c(\omega_1) = \arctan T_1\omega_1 + \arctan T_2\omega_1 - \arctan \beta T_1\omega_1 - \arctan \dfrac{T_2}{\beta}\omega_1 = 0$$

The frequency with zero phase angle can be obtained as

$$\omega_1 = \dfrac{1}{\sqrt{T_1 T_2}} \quad (6\text{-}18)$$

It can be seen that the compensation network has a phase-lag characteristic when $\omega<\omega_1$ and a phase-lead characteristic when $\omega>\omega_1$.

2. Phase lag-lead compensation method

The phase lag-lead compensation essentially combines the characteristics of phase-lag and phase-lead compensation, which uses the phase-lead part of the compensator to improve its dynamic performance by increasing the phase margin of the system, and uses the phase-lag part of the compensator to improve static performance. The steps for the phase-lag-lead compensation using the Bode plot are as follows.

1) Determine the open loop gain K according to the steady state performance requirements.

2) Draw the logarithmic amplitude-frequency characteristic of the original system using the obtained value of K of the compensated system as the open-loop gain, and calculate the cutoff frequency ω_c, phase margin γ and amplitude margin L_g of the original system.

3) Select the corner frequency of the non-compensated system at which the slope of the amplitude-frequency characteristic asymptote changes from -20dB/dec to -40dB/dec as the corner frequency ω_b ($\omega_b = 1/T_2$) for compensating the phase-lead part of the network. The selection is not unique, but it can reduce the order of the compensated system and generate a slope band with -20dB/dec in the medium-frequency.

4) Calculate the cutoff frequency ω_c of the compensated system according to the requirements of the response speed, and determine the attenuation factor $1/\beta$ based on the condition that the compensated logarithmic amplitude-frequency characteristic $L_c(\omega_c) + L(\omega_c) = 0\text{dB}$ of the system, i.e.

$$-20\lg\beta + 20\lg T_2\omega_c + L(\omega_c) = 0 \quad (6\text{-}19)$$

5) Estimate the corner frequency ω_a ($\omega_a = 1/T_1$) of the phase-lag part of the compensated network according to the requirements of the phase margin.

例 6-3 设某单位反馈系统，其开环传递函数为

$$G(s) = \frac{K}{s(s+1)(0.125s+1)}$$

要求系统性能指标满足：稳态误差系数 $K_v = 20\text{s}^{-1}$，相角裕度 $\gamma = 50°$，截止频率 $\omega_c \geq 2\text{s}^{-1}$，试设计相位滞后-超前校正装置。

解：根据对 K_v 的要求，可得

$$K_v = \lim_{s \to 0} sG(s) = K = 20$$

以 $K = 20$ 绘制原系统的开环对数渐近幅频特性曲线，如图 6-13 中 $G(\mathrm{j}\omega)$ 曲线所示。由于截止频率为 4.47s^{-1}，相角裕度为 $-16.6°$，说明原系统不稳定。选择 $\omega_b = 1/T_2 = 1$ 作为校正网络相位超前部分的转折频率。根据校正后系统相位裕度及截止频率的要求，确定校正后系统的截止频率 $\omega_c = 2.2\text{s}^{-1}$，原系统在频率 2.2s^{-1} 处的幅值为 12.32dB，系统与校正网络串联后，在频率为 2.2s^{-1} 处幅值为 0dB，则有

$$-20\lg\beta + 20\lg 2.2 + 12.32 = 0$$

得

$$\beta = 9.1, \quad T_2/\beta = 0.11$$

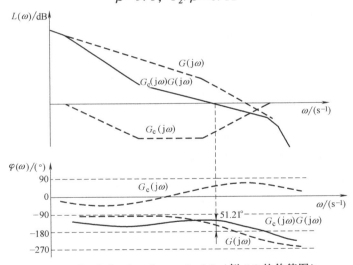

Fig. 6-13 Bode plot of example 6-3（例 6-3 的伯德图）

校正网络的另一个转折频率为 $\quad \beta\omega_b = 9.1 \times 1 = 9.1\text{s}^{-1}$

因此，相位滞后-超前校正网络的传递函数为

$$G_c(s) = \frac{(T_1 s + 1)(T_2 s + 1)}{(\beta T_1 s + 1)\left(\dfrac{T_2}{\beta} s + 1\right)} = \frac{\left(\dfrac{1}{\omega_a} s + 1\right)(s+1)}{\left(\dfrac{9.1}{\omega_a} s + 1\right)(0.11 s + 1)}$$

校正后系统的开环传递函数

$$G_c(s)G(s) = \frac{20\left(\dfrac{1}{\omega_a} s + 1\right)}{s(0.125 s + 1)\left(\dfrac{9.1}{\omega_a} s + 1\right)(0.11 s + 1)}$$

6) Check performance indicators.

Example 6-3 The open-loop transfer function of a unit feedback system is

$$G(s) = \frac{K}{s(s+1)(0.125s+1)}$$

The performance indicators are required as: steady-state error coefficient $K_v = 20\text{s}^{-1}$, phase margin $\gamma = 50°$, cutoff frequency $\omega_c \geqslant 2\text{s}^{-1}$. Try to design a phase lag-lead compensator.

Solution: According to the requirement for K_v, the value of K can be calculated as

$$K_v = \lim_{s \to 0} sG(s) = K = 20$$

The open-loop logarithmic asymptotic amplitude-frequency characteristic curve of the original system is obtained with $K = 20$, as is shown by the curve of $G(j\omega)$ in Fig. 6-13. Since the cutoff frequency of the original system is 4.47s^{-1}, and the phase margin is $-16.6°$, it indicates that the original system is unstable. Select $\omega_b = 1/T_2 = 1$ as the corner frequency of the phase-lead part of compensation network. According to the requirements of the phase margin and cutoff frequency for the compensated system, it is determined that the cutoff frequency of the compensated system $\omega_c = 2.2\text{s}^{-1}$, and the amplitude of the original system at the frequency of 2.2s^{-1} is 12.32dB. After the system is connected with compensation network in series, the amplitude is 0dB at the frequency of 2.2s^{-1}, and we have

$$-20\lg\beta + 20\lg 2.2 + 12.32 = 0$$

Then we have
$$\beta = 9.1, \quad T_2/\beta = 0.11$$

Another corner frequency of the compensation network can be calculated as

$$\beta\omega_b = 9.1 \times 1 = 9.1\text{s}^{-1}$$

Thus, the transfer function of the phase lag-lead compensation network can be written as

$$G_c(s) = \frac{(T_1 s + 1)(T_2 s + 1)}{(\beta T_1 s + 1)\left(\dfrac{T_2}{\beta}s + 1\right)} = \frac{\left(\dfrac{1}{\omega_a}s + 1\right)(s + 1)}{\left(\dfrac{9.1}{\omega_a}s + 1\right)(0.11s + 1)}$$

The open-loop transfer function of the compensated system is

$$G_c(s)G(s) = \frac{20\left(\dfrac{1}{\omega_a}s + 1\right)}{s(0.125s + 1)\left(\dfrac{9.1}{\omega_a}s + 1\right)(0.11s + 1)}$$

According to the requirements of the performance indicators, taking the phase margin of the compensated system $\gamma = 50°$, we have

$$\gamma = 180° + \arctan\frac{2.2}{\omega_a} - 90° - \arctan 0.125 \times 2.2 - \arctan\frac{9.1 \times 2.2}{\omega_a} - \arctan 0.11 \times 2.2$$

$$= 61.01° + \arctan\frac{2.2}{\omega_a} - \arctan\frac{20.02}{\omega_a} = 50°$$

So we have
$$\omega_a = 0.48\text{s}^{-1}$$

根据性能指标的要求，取校正后系统的相位裕度 $\gamma = 50°$，即

$$\gamma = 180° + \arctan\frac{2.2}{\omega_a} - 90° - \arctan 0.125 \times 2.2 - \arctan\frac{9.1 \times 2.2}{\omega_a} - \arctan 0.11 \times 2.2$$

$$= 61.01° + \arctan\frac{2.2}{\omega_a} - \arctan\frac{20.02}{\omega_a} = 50°$$

则

$$\omega_a = 0.48 \text{s}^{-1}$$

得到校正网络的传递函数为

$$G_c(s) = \frac{(2.08s+1)(s+1)}{(18.96s+1)(0.11s+1)}$$

校正后系统的开环传递函数为

$$G_c(s)G(s) = \frac{20(2.08s+1)}{s(0.125s+1)(18.96s+1)(0.11s+1)}$$

校正后系统的对数渐近幅频特性如图 6-13 所示。经校验，校正后系统稳态误差系数为 20s^{-1}，相位裕度为 $51.21°$，截止频率为 2.2s^{-1}，达到了稳态、动态指标要求。

6.2.4　PID 校正

前面所讲的相位超前校正、相位滞后校正及相位滞后-超前校正都采用电阻和电容组成的无源校正环节。无源校正环节结构简单，但是本身没有放大器，而且输入阻抗低，输出阻抗高。事实上，当系统要求较高时，常采用有源校正环节。有源校正环节采用比例（P）、积分（I）和微分（D）等基本控制规律，或者采用比例-微分（PD）、比例-积分（PI）、比例-积分-微分（PID）等复合控制规律，来实现对系统的校正。

1. 比例-微分（PD）校正

图 6-14 所示为有源比例-微分校正环节，其传递函数为

$$G_c(s) = \frac{U_o(s)}{U_i(s)} = \frac{R_2}{R_1}(R_1C_1s+1) = K_p(T_ds+1)$$

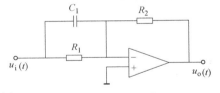

图 6-14　Active PD compensation component
（有源比例-微分校正环节）

式中，$T_d = R_1C_1$；$K_p = \frac{R_2}{R_1}$。其作用对应于式（6-1）的相位超前校正。

比例-微分（PD）校正有如下特点。

1）比例-微分环节使相位超前，可以抵消惯性等环节产生的相位滞后等不良后果，显著改善系统稳定性。

2）比例-微分环节可以提高幅值穿越频率 ω_c，改善系统的快速性，减小系统的调整时间。

3）比例-微分调节器使系统的高频增益增大，因此容易引入高频干扰。

比例-微分校正中的微分控制规律能反映输入信号的变化趋势，产生有效的早期修正信号，以增加系统的阻尼程度，因而可以改善系统的稳定性。在进行串联校正时，可增加一个 $-1/T_d$ 的开环零点到系统中，使系统的相位裕度提高。因此有助于系统动态性能的改善。

2. 比例-积分（PI）校正

图 6-15 所示为有源比例-积分校正环节，其传递函数为

The transfer function of the compensation network is calculated as

$$G_c(s) = \frac{(2.08s+1)(s+1)}{(18.96s+1)(0.11s+1)}$$

Therefore, the open-loop transfer function of the compensated system is

$$G_c(s)G(s) = \frac{20(2.08s+1)}{s(0.125s+1)(18.96s+1)(0.11s+1)}$$

The logarithmic asymptotic amplitude-frequency characteristic curve of the compensated system is shown in Fig. 6-13. After verification, the compensated system has a steady-state error coefficient of $20s^{-1}$, a phase margin of $51.21°$ and a cutoff frequency of $2.2s^{-1}$, which meets the steady-state and dynamic indicator requirements of the system.

6.2.4 PID compensation

The phase-lead compensation, phase-lag compensation and phase lag-lead compensation described above are passive compensation components which are composed of resistors and capacitors. The passive compensation components have simple structures, but have no amplifier itself, and have low input impedance and high output impedance. In fact, when the requirements of the system are high, the active compensation component is often used. The active compensation component uses basic control laws such as proportion (P), integral (I), and derivative (D), or a combination of basic control laws, such as proportion-derivative (PD), proportion-integral (PI), and proportion-integral-derivative (PID) to achieve compensation effect of the system.

1. Proportion-derivative (PD) compensation

Fig. 6-14 shows an active PD compensation component, and its transfer function is

$$G_c(s) = \frac{U_o(s)}{U_i(s)} = \frac{R_2}{R_1}(R_1 C_1 s+1) = K_p(T_d s+1)$$

where $T_d = R_1 C_1$; $K_p = \frac{R_2}{R_1}$. Its effect corresponds to the phase-lead compensation of equation (6-1).

PD compensation has the following characteristics.

1) The proportion-derivative component has phase-lead effect, which can offset the negative phase-lag effects caused by inertia and other components, and the stability of the system is obviously improved.

2) The proportion-derivative component can increase the amplitude crossing frequency to improve the rapidity of the system and reduce the settling time of the system.

3) The proportion-derivative component increases the high-frequency gain of the system, thus it is easy to introduce high-frequency interference.

The derivative control law in the PD compensation can reflect the trend of the input signal and generate an effective early compensation signal to increase the damping degree of the system, so it can improve the stability of the system. In the cascade compensation, an open-loop zero of $-1/T_d$ can be added to the system, which increases its phase margin. To this end, it is helpful to the improvement of the dynamic performance of the system.

$$G_c(s) = \frac{U_o(s)}{U_i(s)} = \frac{R_2}{R_1}\left(1+\frac{1}{R_2 C_2 s}\right) = K_p\left(1+\frac{1}{T_i s}\right)$$

式中，$T_i = R_2 C_2$；$K_p = R_2/R_1$。其作用对应于式（6-10）的相位滞后校正。

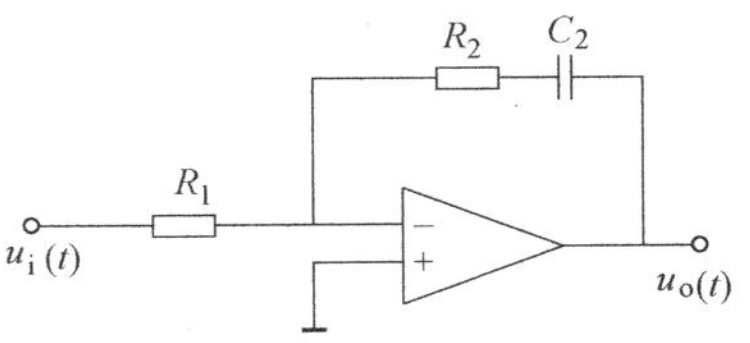

比例-积分（PI）校正有如下特点。

图 6-15 Active PI compensation component
（有源比例-积分校正环节）

1) 系统采用比例-积分校正，可增加开环极点，提高型别，减小稳态误差，使系统的稳态性能得到明显的改善，但使系统的稳定性变差。

2) 右半平面的开环零点可提高系统的阻尼程度，但加剧比例-积分极点对系统产生的不利影响。只要积分时间常数 T_i 足够大，比例-积分校正器对系统的不利影响可大为减小。

3) PI 校正器主要用来改善控制系统的稳态性能。

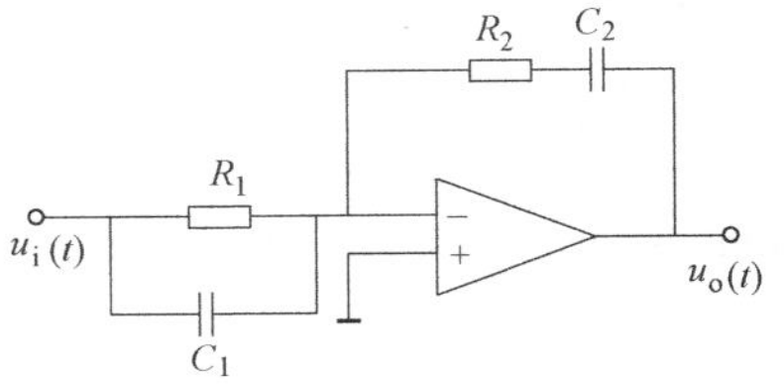

3. 比例-积分-微分（PID）校正环节

图 6-16 所示为有源比例-积分-微分校正环节，其传递函数为

图 6-16 Active PID compensation component
（有源比例-积分-微分校正环节）

$$G_c(s) = \frac{U_o(s)}{U_i(s)} = K_p\left(1+\frac{1}{T_i s}+T_d s\right)$$

式中，$T_i = R_1 C_1 + R_2 C_2$；$T_d = \dfrac{R_1 C_1 R_2 C_2}{R_1 C_1 + R_2 C_2}$；$K_p = \dfrac{R_1 C_1 + R_2 C_2}{R_1 C_2}$。其作用对应于式（6-16）的相位滞后-超前校正。

在比例-积分-微分（PID）校正装置中，若增加一个极点，则可提高型别，改善系统的稳态性能；若增加两个负实零点，则对动态性能的改善会比比例-积分校正更具优越性。此处，积分作用发生在低频段，提高系统稳态性能；微分作用发生在高频段，改善系统动态性能。

比例-积分-微分校正兼顾系统稳态性能和动态性能的改善，因此在要求高系统性能时（或系统中存在积分环节时），多采用比例-积分-微分校正。

例 6-4 某单位反馈系统的开环传递函数为 $G(s) = \dfrac{K}{s(0.15s+1)(0.88\times10^{-3}s+1)(5\times10^{-3}s+1)}$。试设计校正装置，使系统的静态速度误差系数 $K_v \geq 40$，相位裕度 $\gamma \geq 50°$，截止频率 $\omega_c \geq 50\text{s}^{-1}$。

解： 1) 根据静态速度误差系数确定开环增益。该系统为 I 型系统，故 $K = K_v = 40$。

2) 作出未校正系统的伯德图，如图 6-17 所示。可见截止频率 $\omega_{c1} = 15.6\text{s}^{-1}$，相位裕度 $\gamma_1(\omega_{c1}) = 17.9°$，相位穿越频率 $\omega_{g1} = 33.7\text{s}^{-1}$，幅值裕度为 13dB。

3) 确定校正装置。选用串联 PD 校正，其传递函数为
$$G_c(s) = K_p(T_d s + 1)$$
为使原系统结构简单，对未校正部分的高频段小惯性环节做等效处理，即

$$\frac{1}{0.88\times10^{-3}s+1}\cdot\frac{1}{5\times10^{-3}s+1} \approx \frac{1}{(0.88\times10^{-3}+5\times10^{-3})s+1} = \frac{1}{5.88\times10^{-3}s+1}$$

2. Proportion-integral (PI) compensation

Fig. 6-15 shows an active PI compensation component, and its transfer function is as

$$G_c(s) = \frac{U_o(s)}{U_i(s)} = \frac{R_2}{R_1}\left(1 + \frac{1}{R_2 C_2 s}\right) = K_p\left(1 + \frac{1}{T_i s}\right)$$

where $T_i = R_2 C_2$; $K_p = R_2/R_1$. Its effect corresponds to the phase-lag compensation of equation (6-10).

PI compensation has the following characteristics.

1) PI compensation is used in the system to increase the open-loop pole, increase the type level, and reduce the steady-state error, so that it can improve the steady-state performance of the system while the stability of the system will be deteriorated.

2) The open-loop zero on the right half of the s-plane can increase the damping degree of the system, but intensify the adverse effects of the PI pole on the system. As long as the integral time constant is large enough, the adverse effects of the PI compensator on the system can be greatly reduced.

3) The PI compensator is primarily used to improve the steady-state performance of a control system.

3. Proportion-integral-derivative (PID) compensation

Fig. 6-16 shows an active PID compensation component, and its transfer function is

$$G_c(s) = \frac{U_o(s)}{U_i(s)} = K_p\left(1 + \frac{1}{T_i s} + T_d s\right)$$

where $T_i = R_1 C_1 + R_2 C_2$; $T_d = \dfrac{R_1 C_1 R_2 C_2}{R_1 C_1 + R_2 C_2}$; $K_p = \dfrac{R_1 C_1 + R_2 C_2}{R_1 C_2}$. Its effect corresponds to the phase lag-lead compensation of equation (6-16).

In the proportion-integral-derivative (PID) compensator, if one pole is added, the type level can be increased and the steady-state performance of the system can be improved. If two negative real zeros are added, its improvement of system dynamic performance is superior to that of PI. Here, the integral effect occurs in the low-frequency band, which improves the steady-state performance of the system; the derivative effect occurs in the high-frequency band, which improves the dynamic performance.

PID compensation considers both the improvement of the steady-state performance and dynamic performance of the system. Therefore, when the requirement of system performance is high (or there exists an integral component in the system), PID compensation is often used.

Example 6-4 The open-loop transfer function of a unit feedback system is

$$G(s) = \frac{K}{s(0.15s+1)(0.88\times10^{-3}s+1)(5\times10^{-3}s+1)}$$

Try to design a compensation device, making the static speed error coefficient $K_v \geq 10$, cutoff frequency $\omega_c \geq 50 s^{-1}$, and phase margin $\gamma \geq 50°$.

Solution: 1) Determine the open-loop gain according to the static speed error coefficient. The system is a type-I system, so $K = K_v = 40$.

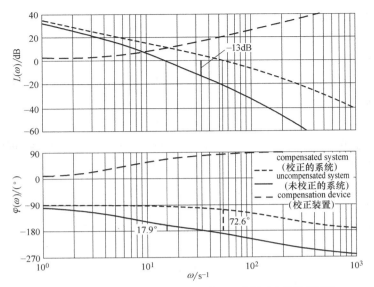

Fig. 6-17 The Bode plot of example 6-4（例 6-4 的伯德图）

则未校正系统的开环传递函数为

$$G(s) = \frac{40}{s(0.15s+1)(5.88\times10^{-3}s+1)}$$

令 $T_d = 0.15$，则校正后的开环传递函数为

$$G(s)G_c(s) = \frac{40}{s(0.15s+1)(5.88\times10^{-3}s+1)} K_p(0.15s+1)$$

$$= \frac{40K_p}{s(5.88\times10^{-3}s+1)}$$

校正后，开环增益与截止频率相等，即 $40K_p = \omega_c$，根据 $\omega_c \geq 50\mathrm{s}^{-1}$，可令 $K_p = 1.4$。则有

$$G(s)G_c(s) = \frac{56}{s(5.88\times10^{-3}s+1)}$$

校正后的截止频率 $\omega_{c2} = 53.4\mathrm{s}^{-1}$，相位裕度为

$$\gamma_2(\omega_{c2}) = 180° - 90° - \arctan 5.88\times10^{-3}\omega_{c2} = 72.6°$$

例 6-5 已知某一控制系统如图 6-18 所示，其中 $G_c(s)$ 为 PID 控制器，它的传递函数为 $G_c(s) = K_p + \dfrac{K_i}{s} + K_d s$。要求校正后系统的闭环极点为 $-10\pm\mathrm{j}10$ 和 -100。确定 PID 控制器的参数 K_p、K_i 和 K_d。

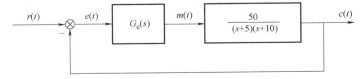

Fig. 6-18 The control system of example 6-5（例 6-5 的控制系统）

2) Plot the Bode plot of the uncompensated system, as is shown in Figure 6-17. It can be seen that the cutoff frequency is $\omega_{c1} = 15.6 \text{s}^{-1}$, phase margin is $\gamma_1(\omega_{c1}) = 17.9°$, phase cross-over frequency is $\omega_{g1} = 33.7 \text{s}^{-1}$, and magnitude margin is 13dB.

3) Determine the compensation device. A serial PD compensation element is considered and its transfer function is

$$G_c(s) = K_p(T_d s + 1)$$

In order to simplify the structure of the original system, the uncompensated part of the small inertia link in the high-frequency band is treated as equivalent, i.e.

$$\frac{1}{0.88 \times 10^{-3} s + 1} \cdot \frac{1}{5 \times 10^{-3} s + 1} \approx \frac{1}{(0.88 \times 10^{-3} + 5 \times 10^{-3})s + 1} = \frac{1}{5.88 \times 10^{-3} s + 1}$$

Thus, the open-loop transfer function of the uncompensated system is

$$G(s) = \frac{40}{s(0.15s + 1)(5.88 \times 10^{-3} s + 1)}$$

Let $T_d = 0.15$, the open-loop transfer function of the compensated system is

$$G(s)G_c(s) = \frac{40}{s(0.15s + 1)(5.88 \times 10^{-3} s + 1)} K_p(0.15s + 1)$$

$$= \frac{40 K_p}{s(5.88 \times 10^{-3} s + 1)}$$

After the compensation, the open-loop gain is equal to the cutoff frequency, i.e. $40K_p = \omega_c$. According to $\omega_c \geq 50 \text{s}^{-1}$, we may let $K_p = 1.4$. Then

$$G(s)G_c(s) = \frac{56}{s(5.88 \times 10^{-3} s + 1)}$$

Therefore, after the compensation, the cutoff frequency is $\omega_{c2} = 53.4 \text{s}^{-1}$, and the phase margin is

$$\gamma_2(\omega_{c2}) = 180° - 90° - \arctan 5.88 \times 10^{-3} \omega_{c2} = 72.6°$$

Example 6-5 A control system is as is shown in Fig. 6-18, where $G_c(s)$ is the PID controller and its transfer function is $G_c(s) = K_p + \frac{K_i}{s} + K_d s$. The closed-loop poles of the compensated system are required as $-10 \pm j10$ and -100. Please determine the parameters as K_p, K_i and K_d of the PID controller.

Solution: Since the required indicators given by the example are the poles of the system, the system can be designed by using the pole configuration principle. The desired closed-loop characteristic polynomial of the system is

$$F^*(s) = (s + 10 - j10)(s + 10 + j10)(s + 100) = s^3 + 120 s^2 + 2200 s + 20000$$

The closed-loop transfer function of the compensated system is

$$\frac{C(s)}{R(s)} = \frac{50(K_d s^2 + K_p s + K_i)}{s(s+5)(s+10) + 50(K_d s^2 + K_p s + K_i)}$$

Its corresponding closed-loop characteristic polynomial is

解：由于本例给出的指标是系统的极点要求，因此可直接用极点配置原理对系统进行设计。系统希望的闭环特征多项式为

$$F^*(s) = (s+10-j10)(s+10+j10)(s+100) = s^3 + 120s^2 + 2200s + 20000$$

校正后系统的闭环传递函数为

$$\frac{C(s)}{R(s)} = \frac{50(K_d s^2 + K_p s + K_i)}{s(s+5)(s+10) + 50(K_d s^2 + K_p s + K_i)}$$

闭环特征多项式为

$$F(s) = s(s+5)(s+10) + 50(K_d s^2 + K_p s + K_i)$$
$$= s^3 + (15 + 50K_d)s^2 + 50(1 + K_p)s + 50K_i$$

令 $F^*(s) = F(s)$，则得

$$\begin{cases} 15 + 50K_d = 120 \\ 50(1+K_p) = 2200 \\ 50K_i = 20000 \end{cases} \Rightarrow \begin{cases} K_d = 2.1 \\ K_p = 43 \\ K_i = 400 \end{cases}$$

由此可见，微分系数远小于比例系数和积分系数，这种情况在实际应用中经常会遇到，尤其是在过程控制系统中。因此，在许多情况下用 PID 校正器就能满足系统性能要求。

6.3 控制系统的并联校正

6.3.1 反馈校正

在反馈校正中，若 $G_c(s) = K$，则称为位置反馈校正；若 $G_c(s) = Ks$，则称为速度反馈校正；若 $G_c(s) = Ks^2$，则称为加速度反馈校正。

1. 位置反馈校正

位置反馈校正的框图如图 6-19 所示。

（1）$G(s)$ 为积分环节　当 $G(s) = \dfrac{K_1}{s}$，则校正后系统的传递函数为

$$\frac{X_o(s)}{X_i(s)} = \frac{\dfrac{K_1}{s}}{1 + \dfrac{KK_1}{s}} = \frac{\dfrac{1}{K}}{\dfrac{s}{KK_1} + 1}$$

用位置反馈包围积分环节，其结果是将原来的积分环节变成惯性环节，降低了原系统的型次，这意味着降低了系统的稳态精度，但有可能提高系统的稳定性。

（2）$G(s)$ 为惯性环节　当 $G(s) = \dfrac{K_1}{Ts+1}$，则校正后系统的传递函数为

$$\frac{X_o(s)}{X_i(s)} = \frac{\dfrac{K_1}{Ts+1}}{1 + \dfrac{KK_1}{Ts+1}} = \frac{\dfrac{K_1}{1+KK_1}}{\dfrac{T}{1+KK_1}s + 1}$$

$$F(s) = s(s+5)(s+10) + 50(K_d s^2 + K_p s + K_i)$$
$$= s^3 + (15+50K_d)s^2 + 50(1+K_p)s + 50K_i$$

Let $F^*(s) = F(s)$, we have

$$\begin{cases} 15+50K_d = 120 \\ 50(1+K_p) = 2200 \\ 50K_i = 20000 \end{cases} \Rightarrow \begin{cases} K_d = 2.1 \\ K_p = 43 \\ K_i = 400 \end{cases}$$

It can be seen that the derivative coefficient is much smaller than the proportion coefficient and the integral coefficient, which is often encountered in practical applications, especially in process control systems. Therefore, PID compensator can be used to meet requirements of the system performance in many applications.

6.3 Parallel Compensation of Control Systems

6.3.1 Feedback compensation

In the feedback compensation, if $G_c(s) = K$, it is called position feedback compensation; if $G_c(s) = Ks$, it is called speed feedback compensation; if $G_c(s) = Ks^2$, it is called acceleration feedback compensation.

1. Position feedback compensation

The block diagram of the position feedback compensation is shown in Fig. 6-19.

(1) $G(s)$ being an integral component   When $G(s) = \dfrac{K_1}{s}$, the transfer function of the compensated system is

$$\frac{X_o(s)}{X_i(s)} = \frac{\dfrac{K_1}{s}}{1+\dfrac{KK_1}{s}} = \frac{\dfrac{1}{K}}{\dfrac{s}{KK_1}+1}$$

Surrounding the integral component with position feedback, the result is to turn the original integral component into an inertia component, which reduces the type level of the original system. It means that the steady-state accuracy of the system is reduced, but it may improve the stability of the system.

(2) $G(s)$ being an inertia component   When $G(s) = \dfrac{K_1}{Ts+1}$, the transfer function of the compensated system is

$$\frac{X_o(s)}{X_i(s)} = \frac{\dfrac{K_1}{Ts+1}}{1+\dfrac{KK_1}{Ts+1}} = \frac{\dfrac{K_1}{1+KK_1}}{\dfrac{T}{1+KK_1}s+1}$$

用位置反馈包围惯性环节，其结果仍是惯性环节，但时间常数下降。换言之，惯性减弱导致过渡过程时间变短，响应速度加快；同时，系统增益下降。此外，反馈系数 K 越大，时间常数越小，系统增益越小。

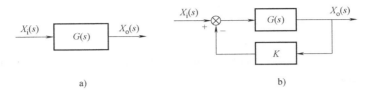

Fig. 6-19　Position feedback compensation（位置反馈校正）

2. 速度反馈校正

速度反馈校正的框图如图 6-20 所示。原系统的传递函数为

$$G(s) = \frac{\omega_n^2}{s^2 + 2\xi\omega_n s + \omega_n^2} \quad (0 < \xi < 1)$$

在加入微分负反馈后，系统的传递函数为

$$\frac{X_o(s)}{X_i(s)} = \frac{\omega_n^2}{s^2 + (2\xi\omega_n + K\omega_n^2)s + \omega_n^2}$$

显然，校正后系统的阻尼比较原系统大为提高，但固有频率不改变。因此，速度负反馈校正可以增加阻尼比，改善系统的相对稳定性。

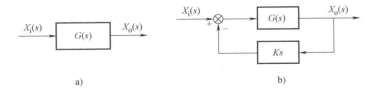

Fig. 6-20　Velocity feedback compensation（速度反馈校正）

6.3.2　复合校正

在复合校正中，由于校正环节 $G_c(s)$ 位于系统回路之外，因此可以先设计系统的回路，保证系统具有较好的动态性能，然后设计校正环节 $G_c(s)$，以提高系统的稳态精度。

图 6-21a 所示为一个一般的单位反馈闭环控制系统，$E(s) \neq 0$。若要使 $E(s) = 0$，应在原系统中加入复合校正环节 $G_c(s)$，如图 6-21b 所示。

校正后系统的误差为

$$E(s) = \left[1 - \frac{G_1(s)G_2(s) + G_c(s)G_2(s)}{1 + G_1(s)G_2(s)} \right] X_i(s) = \frac{1 - G_c(s)G_2(s)}{1 + G_1(s)G_2(s)} X_i(s)$$

根据 $E(s) = 0$，有 $G_c(s) = \dfrac{1}{G_2(s)}$。

由此可知，系统的特征方程没有改变，因此系统虽然加了复合校正，但稳定性并不受影响。采用复合校正，既能使系统满足动态性能的要求，又能保证系统的稳态精度。

Surrounding the inertia component with position feedback, the result is still an inertia component, but the time constant decreases. In other words, the inertia weakens, leading to the shorter transition time and faster response speed; meanwhile, the gain of the system also decreases. In addition, the larger the feedback coefficient K is, the smaller the time constant is, and the smaller the system gain is.

2. speed feedback compensation

The block diagram of the speed feedback compensation is shown in Fig. 6-20. The transfer function of the original system is

$$G(s) = \frac{\omega_n^2}{s^2 + 2\xi\omega_n s + \omega_n^2} \quad (0 < \xi < 1)$$

After adding the speed negative feedback, the transfer function of the system is

$$\frac{X_o(s)}{X_i(s)} = \frac{\omega_n^2}{s^2 + (2\xi\omega_n + K\omega_n^2)s + \omega_n^2}$$

Obviously, the damping ratio of the compensated system is greatly improved compared to that of the original system, but the natural frequency is unchanged. Therefore, the speed negative feedback compensation can increase the damping ratio and improve the relative stability of the system.

6.3.2 Compound compensation

In the compound compensation, since the compensation component $G_c(s)$ is located outside the loop of the system, the loop of the system can be designed first to ensure better dynamic performance, and then to design the compensation component $G_c(s)$ to improve the steady-state accuracy of the system.

Fig. 6-21a shows a common unit feedback closed-loop control system, where $E(s) \neq 0$. If let $E(s) = 0$, the feedback compensation component $G_c(s)$ should be added to the original system, as is shown in Fig. 6-21b.

The error of the compensated system is

$$E(s) = \left[1 - \frac{G_1(s)G_2(s) + G_c(s)G_2(s)}{1 + G_1(s)G_2(s)}\right] X_i(s) = \frac{1 - G_c(s)G_2(s)}{1 + G_1(s)G_2(s)} X_i(s)$$

According to $E(s) = 0$, we have $G_c(s) = \dfrac{1}{G_2(s)}$.

According to above analysis, the characteristic equation of the system does not change. So although the compound compensation is added to the system, the stability is not affected. By using the compound compensation, the system can meet the requirements of dynamic performance and ensure the steady-state accuracy.

In summary, cascade compensation and feedback compensation are widely used in control systems, and in many cases, they are simultaneously applied to obtain better results.

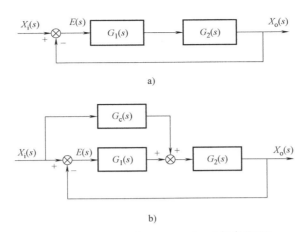

Fig. 6-21　Compound compensation（复合校正）

总之，在控制系统中，串联校正、反馈校正均会得到广泛应用，并且在很多情况下同时应用，以获得更好的效果。

Chapter 7 Analysis and Compensation of Control Systems based on MATLAB

第7章 基于MATLAB的控制系统分析与校正

MATLAB is the most widely used tool software in the control field, almost all control theory and application branches have corresponding MATLAB toolbox. This chapter will combine the basic content of control engineering theory learned in the previous section, and uses control systems toolbox and simulation environment (Simulink) to learn the application of MATLAB.

7.1 Introduction

MATLAB is a high-performance language for engineering calculation, its programming code is very close to the mathematical derivation format, and thus the MATLAB programming is extremely convenient. Its typical applications include mathematical computation, algorithm development, modeling and simulation, data analysis and visualization, engineering mapping, application development (including graphical interfaces), and so on. The user can enter a command in the MATLAB workspace, can also use MATLAB language to write application programs. MATLAB software can translate the command or each statement in the program, and then process it in MATLAB environment, and finally return the results of operations. MATLAB command window and programming window are shown in Fig. 7-1.

Different from other programming languages, the basic variable unit of MATLAB language is complex matrix, and its matrix processing function and graphics processing function are the most significant features. We will introduce some simple and practical MATLAB commands and their operations next.

7.1.1 Basic operations and commands

1. Statements and variables of MATLAB

MATLAB is a descriptive language. It interprets and executes the input expression, just as it executes the statement directly in BASIC. The common format of MATLAB statements is "variable = expression ;", or "expression;". The expression can consist of operators, special symbols, functions, variable names, and so on. The result of the expression is a matrix that is assigned to the left variable and displayed on the screen. If the variable name and the sign " = " are omitted, MATLAB automatically generates a variable named "ans" to represent the result.

One of the features of MATLAB is that the dimensions need not be determined prior to application. In MATLAB, variables are automatically generated as soon as they are adopted (the dimensions of the variables can be changed later if necessary). These variables will remain in memory until the command "exit" or "quit" is entered. To get a list of variables in the workspace, we can enter the command "who" on the keyboard, and the variables currently stored in the workspace will be displayed on the screen. The command "clear" can clear all non-permanent variables from the workspace.

2. Program line beginning with "%"

In MATLAB, a program line that starts with "%" means the comments and explanations, and it is not executed. That is, in the program lines of MATLAB, everything after "%" can be neglec-

第7章 基于MATLAB的控制系统分析与校正

MATLAB 是控制领域目前使用最广的工具软件，几乎所有的控制理论与应用分支中都有与之相匹配的 MATLAB 工具箱。本章将结合前面所学控制工程理论的基本内容，采用控制系统工具箱（Control Systems Toolbox）和仿真环境（Simulink）来学习 MATLAB 的应用。

7.1 概述

MATLAB 是一种用于工程计算的高性能语言，其编程代码很接近数学推导格式，所以编程非常方便。其典型应用包括数学计算、算法开发、建模和仿真、数据分析和可视化、工程绘图、应用开发（包括图形界面）等。用户可以在 MATLAB 的工作空间中键入命令，也可以应用 MATLAB 语言编写应用程序，接着 MATLAB 软件对命令或程序中的各条语句进行翻译，然后在 MATLAB 环境中进行处理，最后返回运算结果。MATLAB 命令窗口与编程窗口如图 7-1 所示。

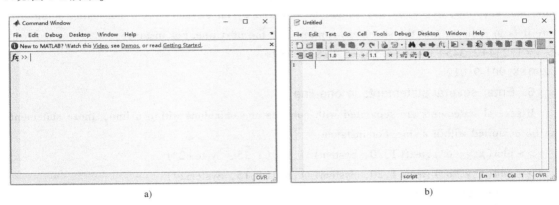

Fig 7-1 MATLAB command window and programming window （MATLAB 命令窗口与编程窗口）
a) Command window （命令窗口）　　b) Programming window （编程窗口）

与其他程序设计语言不同，MATLAB 语言的基本变量单元是复数矩阵，矩阵处理功能和图形处理功能是其最显著的特色。下面将介绍一些简单实用的 MATLAB 命令及其操作。

7.1.1 基本操作及命令

1. MATLAB 的语句与变量

MATLAB 是一种描述性语言。它对输入的表达式边解释边执行，就像 BASIC 语言中直接执行语句一样。MATLAB 语句的常用格式为"变量=表达式;"，或者简化为"表达式;"。表达式可以由操作符、特殊符号、函数、变量名等组成。表达式的结果为一矩阵，它赋给左边的变量，同时显示在屏幕上。如果省略变量名和"="号，则 MATLAB 自动产生一个名为"ans"的变量来表示结果。

MATLAB 的一个特点是在应用之前，维数不必是确定的。在 MATLAB 中，变量一旦被采用，便会自动产生（如果必要，变量的维数还可以改变）。在命令"exit"或"quit"被输入之前，这些变量将保留在存储器中。为了得到工作空间内的变量清单，可以通过键盘输入命令"who"，则当前存放在工作空间内的所用变量便会显示在屏幕上。命令"clear"能

307

ted. If the comment or explanation requires more than one program line, each line should start with "%".

3. Program line ending with ";"

If the last symbol of the statement is a semicolon, the print is cancelled, while the command is still executed and the result is no longer displayed. In addition, when entering a matrix, a semicolon is used to indicate the end of a line.

4. The colon operator

The colon operator is used to establish a vector by assigning values to the subscripts of matrix elements and giving the range of integers. For example, "j: k" represents $[j, j+1, \cdots, k]$, "A(:,j)" represents the jth column of the matrix A, "A(i,:)" represents the ith row of the matrix A.

5. Enter a statement longer than one line

A statement usually ends with an enter or an input key. If the statement entered is too long to be more than one line, the enter key should be followed by an ellipsis "..." consisting of three or more dots to indicate that the statement continues to the next line. For instance,

\>\> x = 1.234+2.345+3.456+4.567+5.678+6.789+...
7.890+8.901-9.012

6. Enter several statements in one line

If several statements are separated with commas or semicolons within a line, these statements can be compiled within a line. For instance,

\>\> plot(x,y,'o'),text(1,20,'System1'),text(1,15,'System2')

\>\> plot(x,y,'o');text(1,20,'System1');text(1,15,'System2')

7. Select the output format

All calculations in MATLAB are done in double precision, but the result displayed can be a fixed-point output with four decimal places. For example, for vector $x = [1/3 \ 0.00002]$, its input and output in MATLAB are as

\>\> x = [1/3 0.00002]

x =

0.3333   0.0000

If at least one element in the matrix is not a strict integer, there are four possible output formats. The output displayed can be controlled using the following commands.

1) The scaled fixed point format with 5 digits: format   short.
2) The scaled fixed point format with 15 digits: format   long.
3) The floating point format with 5 digits: format   short   e.
4) The floating point format with 15 digits: format   long   e.

Once a format is called, the chosen format remains until the format is changed. In control system analysis, formats 1) and 2) are frequently used. Once MATLAB is enabled, MATLAB displays the numerical results in format 1), even if no format command is entered. If all the elements of a matrix or vector are strictly integers, the results of formats 1) and 2) are the same.

从工作空间中清除所有非永久性变量。

2. 以"%"开始的程序行

在MATLAB中,以"%"开始的程序行表示注释,该程序行是不执行的。也就是说,在MATLAB程序行中,出现在"%"之后的一切内容都是可以忽略的。如果注释或说明需要一行以上的程序行,则每一行均需以"%"开始。

3. 以";"结束的程序行

如果语句的最后一个符号是分号";",则打印被取消,但是命令仍会被执行,而结果不再显示。此外,在输入矩阵时,分号用来指示一行的结束。

4. 冒号运算符

冒号运算符用于通过赋予矩阵下标及规定数的范围来建立向量。例如,"j:k"表示 $[j, j+1, \cdots, k]$,"A(:,j)"表示矩阵 A 的第 j 列,"A(i,:)"表示矩阵 A 的第 i 行。

5. 输入超过一行的长语句

一个语句通常以回车或输入键终结。如果输入的语句太长,超出了一行,则回车键后面应跟随由3个或3个以上圆点组成的省略号"...",以表明语句将延续到下一行。例如,

>> x = 1.234+2.345+3.456+4.567+5.678+6.789+...
7.890+8.901-9.012

6. 在一行内输入数个语句

如果在一行内可以把数个语句用逗号或分号隔开,则可以把数个语句放在一行内编译。例如,

>> plot(x,y,'o'),text(1,20,'System1'),text(1,15,'System2')

>> plot(x,y,'o');text(1,20,'System1');text(1,15,'System2')

7. 选择输出格式

MATLAB中的所有计算都是以双精度方式完成的,但是显示结果可以是具有4个小数位的定点输出。例如,对于向量 $x = [1/3 \ 0.00002]$,MATLAB中的输入和输出为

>> x = [1/3 0.00002]

x =

    0.3333   0.0000

如果在矩阵中至少有一个元素不是严格的整数,则有4种可能的输出格式。显示的输出结果可以利用下列命令加以控制。

1)5字长定点数:format short。

2)15字长定点数:format long。

3)5字长浮点数:format short e。

4)15字长浮点数:format long e。

一旦调用了某种格式,则这种被选用的格式将保持,直到对格式进行了更变为止。在控制系统分析中,格式1)和2)是经常采用的格式。一旦启用了MATLAB,即使没有输入格式命令,MATLAB也将以格式1)显示数值结果。如果矩阵或向量的所有元素都是严格的整数,则格式1)和2)的结果是相同的。

8. 退出 MATLAB 时如何保存变量

键入"exit"或"quit"后,MATLAB中的所有变量将会消失。如果在命令"exit"之前

8. How to save variables when exiting MATLAB

After "exit" or "quit" is entered, all variables in MATLAB will disappear. If the command "save" is entered before the command "exit", all variables are saved in the disk file "matlab. mat". When entering MATLAB again, the command "load" will make the workspace restore to its previous state.

7.1.2 MATLAB functions

The typical call format of MATLAB functions can be described as

[Return list of variables] = func_ name (list of input variables)

where variables on the left hand of the expression are return variables; variables on the right hand of the expression are input variables.

MATLAB allows multiple variables to be returned simultaneously during function being called. A function can be called in more than one format. For example, the function bode ( ) can be called in the following format:

[mag, phase] = bode (num, den, w)

where mag returns magnitudes; phase returns phase values; the function bode ( ) is used to find or draw the Bode plot of the system; the system is represented by the numerator "num" and denominator "den" of the transfer function; w is the specified response frequencies.

In addition, the control system toolbox of MATLAB also allows the model G to describe a system. For example

[mag, phase] = bode (G, w)

where mag returns magnitudes; phase returns phase values; the function Bode ( ) is used to find or draw the Bode plot of the system; G is a dynamic system model; w is the specified response frequencies.

MATLAB functions can be different in the format of the return variables. If the above statement does not return variables, the Bode plot of the system will be automatically drawn. For example, with only "bode (G, w)" input, MATLAB will automatically draw the Bode plot of the system.

MATLAB has a number of predefined functions for users to call when solving different types of control problems, as is listed in Table 7-1. Table 7-1 only lists the simplest expression, and other applications of functions can be queried through the help command in MATLAB.

7.1.3 Plotting response curves

MATLAB has a set of plotting functions to obtain graph output. For example, the command "plot" can generate linear x-y plot, the commands "loglog", "semilogy" or "polar" instead of "plot", can generate logarithmic plot and polar plot respectively.

1. x-y plot

If x and y are vectors of the same length, the command "plot (x, y)" plots the value of y against the value of x. For example, enter the command in the command window of MATLAB as

输入命令"save",则所有的变量被保存在磁盘文件"matlab.mat"中。当再次进入 MATLAB 时,键入命令"load"可使工作空间恢复到之前的状态。

7.1.2 MATLAB 函数

MATLAB 函数的典型调用格式可描述为

[返回变量列表] = func_name(输入变量列表)

其中,表达式左边的变量为返回变量;表达式右边的变量为输入变量。

MATLAB 允许在函数调用时同时返回多个变量,而一个函数又可以由多种格式进行调用。例如,bode( )函数可以由下面的格式调用:

[mag,phase] = bode(num,den,w)

其中,mag 是返回幅值;phase 是返回相角值;bode 是求取或绘制系统伯德图的函数;num 是传递函数的分子部分;den 是传递函数的分母部分;w 表示指定的响应频率值。

此外,MATLAB 的控制系统工具箱中还允许用 G 模型来描述系统。例如,

[mag,phase] = bode(G,w)

其中,mag 是返回幅值;phase 是返回相角值;bode 是求取或绘制系统伯德图的函数;G 是动态系统模型;w 表示指定的响应频率值。

MATLAB 函数在返回变量的格式上可以不同,若上面的语句没有返回变量,则将自动地绘制系统伯德图。例如,只输入"bode(G,w)"时,MATLAB 会自动绘制系统的伯德图。

MATLAB 具有许多预先定义的函数,以便在求解不同类型的控制问题时调用,见表 7-1。表 7-1 只给出了最简单的表达方式,其他函数可通过 MATLAB 中 help 命令查询。

7.1.3 绘制响应曲线

MATLAB 具有丰富的获取图形输出的程序集。例如,命令"plot"可以产生线性 x-y 图形,用命令"loglog"、"semilogy"或"polar"取代"plot",分别可以产生对数坐标图或极坐标图。

1. x-y 图

如果 x 和 y 是同一长度的向量,则命令"plot(x,y)"将画出 y 值对于 x 值的关系图。例如,在 MATLAB 命令窗口输入:

>> x = 1:10,y = [1 3 5 6 7 8 10 15 18 20],plot(x,y)

2. 多条曲线的绘制

为了在一幅图上画出多条曲线,可采用具有多个自变量的"plot(x1,y1,x2,y2,…,xn,yn)"命令,每对 (x_i, y_i) $(i = 1, 2, …, n)$ 都可以被图解表示出来,而在一幅图上形成多条曲线。例如,在 MATLAB 命令窗口输入:

>>x1 = 1:10;y1 = x1; x2 = 1:10;y2 = x2.^2; plot(x1,y1,x2,y2)

此外,在一幅图上画一条以上的曲线时,也可以利用"hold"命令保持当前的图形,并且防止删除和修改比例尺,随后生成的一条曲线将会重叠地画在原曲线上。再次输入"hold",会使当前的图形复原。例如,在 MATLAB 命令窗口输入:

>> x1 = 1:10;y1 = x1; plot(x1,y1);hold; x2 = 1:10;y2 = x2.^2; plot(x2,y2)

Table 7-1 Functions commonly used in control systems（控制系统中常用函数）

Function(函数)		Description(说明)
Command window functions（命令窗口函数）	ans	The answer when an expression is not given（表示表达式未给定时的答案）
	clc	Clear the command window（清空命令窗口）
	clear	Clear variables from the workspace（从工作空间中清空变量）
	exit	Terminate MATLAB programs（终止 MATLAB 程序）
Matrix functions（矩阵函数）	det	Find the determinant of the matrix（求矩阵的行列式）
	diag	Find the diagonal matrix of the matrix（求矩阵的对角矩阵）
	eig	Find the characteristic value of the matrix（求矩阵的特征值）
	length	Calculate the length of the vector（计算向量的长度）
	max	Find the maximum value of the vector（求向量中的最大值）
	mean	Find the mean value of the vector（求向量中的平均值）
	min	Find the minimum value of the vector（求向量中的最小值）
	ones	Generate an array with element 1（生成元素为 1 的数组）
	rand	Generate a uniformly distributed pseudo random number matrix（生成均匀分布的伪随机数矩阵）
	rank	Return the rank of the matrix（返回矩阵的秩）
	reshape	Change the dimension of the matrix（改变矩阵维数）
	size	Calculate the dimensions of the array（计算数组维数大小）
	sort	Arrange the elements of an array in ascending or descending order（把数组元素按升序或降序排列）
Mathematical functions（数学函数）	abs	Calculate the absolute value and the complex modulus value（计算绝对值和复数模值）
	acos	Return the arccosine value in radians（以弧度的形式返回反余弦值）
	acosd	Return the arccosine value in angles（以角度的形式返回反余弦值）
	asin	Return the arcsine value in radians（以弧度的形式返回反正弦值）
	asind	Return the arcsine value in angles（以角度的形式返回反正弦值）
	atan	Return the arctangent value in radians（以弧度的形式返回反正切值）
	atand	Return the arctangent value in angles（以角度的形式返回反正切值）
	cos	Return the cosine value in radians（以弧度的形式返回余弦值）
	cosd	Return the cosine value in angles（以角度的形式返回余弦值）
	exp	Calculate the exponent（计算指数）
	log	Calculate the natural logarithm（计算自然对数）

(续)

Function(函数)		Description(说明)
Mathematical functions（数学函数）	sin	Return the sine value in radians(以弧度的形式返回正弦值)
	sind	Return the sine value in angles(以角度的形式返回正弦值)
	sqrt	Calculate the square root(计算平方根)
	tan	Return the tangent value in radians(以弧度的形式返回正切值)
	tand	Return the tangent value in angles(以角度的形式返回正切值)
Equation functions（方程函数）	bode	Plot the Bode plot of a continuous system(绘制连续系统的伯德图)
	feedback	Form a feedback connection(形成反馈连接)
	impulse	Plot the unit impulse response curve of a continuous system(绘制连续系统的单位脉冲响应曲线)
	nyquist	Plot the Nyquist diagram frequency response curve of a continuous system(绘制连续系统的奈奎斯图频率响应曲线)
	parallel	Form a parallel connection(形成串联连接)
	plot	Generate the 2-D line plot(生成二维线图形)
	polar	Generate the polar coordinate plot(生成极坐标图形)
	polyfit	Generate the polynomial fitting curve(生成多项式拟合曲线)
	polyval	Generate the polynomial equation(生成多项式方程)
	residue	Convert between partial fraction expansion and polynmial coefficients(用于在部分分式展开和多项式系数之间转换)
	series	Form a series connection(形成串联连接)
	ss	Generate the state space model(生成状态空间模型)
	ss2tf	Transform the state space form of the system into the transfer function form(将系统状态空间形式变换为传递函数形式)
	ss2zp	Transform the state space form of the system into the pole-zero gain form(将系统状态空间形式变换为零极点增益形式)
	step	Plot the unit step response curve of a continuous system(绘制连续系统的单位阶跃响应曲线)
	tf	Generate the transfer function model(生成传递函数模型)
	tf2ss	Transform the transfer function form of the system into the state space form(将系统传递函数形式变换为状态空间形式)
	tf2zp	Transform the transfer function form of the system into the pole-zero gain form(将系统传递函数形式变换为零极点增益形式)
	zp2ss	Transform the pole-zero gain form of the system into the state space form(将系统零极点增益形式变换为状态空间形式)
	zp2tf	Transform the pole-zero gain form of the system into the transfer function form(将系统零极点增益形式变换为传递函数形式)
	zpk	Generate the pole-zero gain model(生成零极点增益模型)

\>\> x = 1:10, y = [1 3 5 6 7 8 10 15 18 20], plot(x,y)

2. Plot multiple curves

To plot multiple curves on a graph, the command of multiple independent variables "plot (x1, y1, x2, y2, ..., xn, yn)" can be used, each pair of (x_i, y_i) ($i = 1, 2, \cdots, n$) can be graphically represented, thus forming multiple curves on a graph. The advantage of multiple variables is that it allows vectors of different lengths to be displayed on the same graph. For example, entering the command in the command window of MATLAB

\>\>x1 = 1:10; y1 = x1; x2 = 1:10; y2 = x2.^2; plot(x1,y1,x2,y2)

In addition, the command "hold" can also be used when plotting more than one curve on a graph. To maintain the current graph and prevent deleting and modifying the scale. Therefore, a subsequent curve will be drawn superimposed on the original curve. Entering "hold" again will restore the current graph. For example, enter the command in the command window of MATLAB as

\>\> x1 = 1:10; y1 = x1; plot(x1,y1); hold; x2 = 1:10; y2 = x2.^2; plot(x2,y2)

3. Add the grid line, graph title, x-axis label and y-axis label

Once the graph is displayed on the screen, the corresponding grid line, graph title, x-axis label and y-axis label can be given. In MATLAB, the commands on grid line, graph title, x-axis. label and y-axis label are respectively "grid", "title", "xlabel" and "ylabel". For example, enter the command in the command window of MATLAB as

\>\> xlabel('X 坐标'); ylabel('Y 坐标');

4. Write text on a graphic screen

To write text on points (x, y) of the graphic screen, the command "text (x, y, 'text')" can be adopted. For example, enter the command in the command window of MATLAB as

\>\> text(3,5,'hello')

It means that "hello" can be written horizontally from the point (3, 5). In addition, entering the command as

\>\> x1 = 1:10; y1 = x1; x2 = 1:10; y2 = x2.^2; plot(x1,y1,x2,y2); text(x1,y1,'1'); text(x2,y2,'2')

can make two curves be marked so that they can be easily distinguished from each other.

5. Graphic types

Graphic types that MATLAB can provide include line and point. The general calling form of plot function is "plot (x, y, LineSpec)". "LineSpec" is the plotting style specified by the user. The types of lines and points provided by MATLAB are listed in Table 7-2. For example, the statement "plot (x1, y1, 'x')" will draw a plot using the marker symbol 'x', while the statement "plot (x1, y1, ':', x2, y2, '+')" will draw the first curve with the dotted line and the second curve with the plus sign.

6. Colors

The colors provided by MATLAB are listed in Table 7-3. For example, the statement "plot (x, y, 'r')" means using red line to plot a curve, while "plot (x, y, '+g')" means using green line with the plus sign to plot a curve.

3. 添加网格线、图形标题、x 轴标记和 y 轴标记

一旦在屏幕上显示出图形,就可以给出对应的网格线、图形标题、x 轴和 y 轴标记。MATLAB 中关于网格线、标题、x 轴标记和 y 轴标记的命令分别为 "grid" "title" "xlabel" "ylabel"。例如,在 MATLAB 命令窗口输入:

>> xlabel('X 坐标');ylabel('Y 坐标');

4. 在图形屏幕上书写文本

为了在图形屏幕的点 (x,y) 上书写文本,可采用命令 "text(x,y,'text')"。例如,输入命令:

>> text(3,5,'hello')

这表示 "hello" 可以从点 (3,5) 开始水平书写出。另外,输入命令:

>> x1=1:10;y1=x1;x2=1:10;y2=x2.^2;plot(x1,y1,x2,y2);text(x1,y1,'1');text(x2,y2,'2')

这会使两条曲线被标记为 "1" 和 "2",也就使它们很容易地区分开来。

5. 图形类型

MATLAB 能够提供的图形类型包括线和点两类,plot 函数的一般调用形式为 "plot(x,y,LineSpec)","LineSpec" 是用户指定的绘图样式,主要类型见表 7-2。例如,语句 "plot(x1,y1,'x')" 将利用标记符号 "x" 画出一个叉号图,而语句 "plot(x1,y1,':',x2,y2,'+')" 将用虚线画出第一条曲线,用加号画出第二条曲线。

Table 7-2　Types of lines and points(线和点的类型)

Types of lines(线的类型)		Types of points(点的类型)	
—	Solid line(实线)	.	Dot(圆点)
— —	Dashed line(短画线)	+	Plus sign(加号)
……	Dotted line(虚线)	*	Asterisk(星号)
—.—	Dash-dotted line(点画线)	○	Circle(圆圈)
		×	Cross(叉号)

6. 颜色

MATLAB 提供的颜色见表 7-3。例如,语句 "plot(x,y,'r')" 表示采用红线绘制曲线图,而 "plot(x,y,'+g')" 表示采用绿色 "+" 号绘制曲线图。

Table 7-3　Types of colors(颜色类型)

Types of colors(颜色类型)	Description of colors(颜色说明)
r	Red(红色)
g	Green(绿色)
b	Blue(蓝色)
w	White(白色)
i	Colorless(无色)

7. 手工坐标轴定标

如果需要在指定的范围内绘制曲线,则可以采用 "axis(v)" 命令把坐标轴定标在规定

7. Calibrate coordinate axis manually

If a curve needs to be plotted within a specified range, the command "axis (v)" can be adopted to establish coordinate axis calibration. For example, entering the command in the command window of MATLAB that

>>v = [x-min x-max y-min y-max] ; axis(v)

where v is a quaternion vector; x-min, x-max, y-min, y-max are the minimum value and maximum value of x, minimum value and maximum value of y. For logarithmic plots, the elements of v should be the common logarithms of the minimum and maximum values.

Performing "axis (v)" will save the current coordinate axis calibration to the following figure, and typing "axis" will restore the automatic calibration of coordinate axis again.

7.2 Description of Mathematical Models based on MATLAB

When using MATLAB to describe a control system, the common mathematical models have three forms: transfer function, pole-zero gain and state space. Each model can be classified as a continuous or discrete model, and each has its own characteristics. It is sometimes necessary to require transformation among the models. This book mainly introduces how to describe the mathematical model of continuous control system with MATLAB.

7.2.1 Description of mathematical models of continuous systems

1. Transfer function model

When the transfer function of a control system is

$$G(s) = \frac{Y(s)}{X(s)} = \frac{b_0 s^m + b_1 s^{m-1} + \cdots + b_{m-1} s + b_m}{a_0 s^n + a_1 s^{n-1} + \cdots + a_{n-1} s + a_n}$$

the form of the numerator-denominator coefficients can be used to express the model in MATLAB, i.e.

num = $[b_0, b_1, \cdots, b_m]$;   % The vector composed of the numerator coefficients
den = $[a_0, a_1, \cdots, a_n]$;   % The vector composed of the denominator coefficients
G (s) = tf[num, den]   % Output the transfer function

Example 7-1   Describe the system by using MATLAB, whose transfer function is

$$G(s) = \frac{2s+1}{4s^3 + 3s^2 + 2s}$$

Solution: The MATLAB programming code of the example is
num = [2,1];
den = [4,3,2,0];
G = tf(num,den)

The result after the execution is obtained as
G =

    2 s+1

的范围内。例如，输入命令：

>>v=[x-min x-max y-min y-max];axis(v)

其中，v 是一个四元向量；x-min、x-max、y-min、y-max 分别是 x 的最小、最大值和 y 的最小、最大值。对于对数坐标图，v 的元素应为最小值和最大值的常用对数。

执行"axis(v)"会把当前的坐标轴定标保持到后面的图中，再次键入"axis"恢复自动定标。

7.2 基于 MATLAB 的数学模型描述

在用 MATLAB 描述控制系统时，常用的数学模型有三种形式：传递函数、零极点增益和状态空间。每种模型均有连续和离散之分，它们各有特点，有时需在各种模型之间转换。本书主要讲解使用 MATLAB 描述连续型控制系统的数学模型。

7.2.1 连续系统数学模型的描述

1. 传递函数模型

当控制系统的传递函数为

$$G(s) = \frac{Y(s)}{X(s)} = \frac{b_0 s^m + b_1 s^{m-1} + \cdots + b_{m-1} s + b_m}{a_0 s^n + a_1 s^{n-1} + \cdots + a_{n-1} s + a_n}$$

则在 MATLAB 中，可用分子-分母系数的形式表示该模型，即

num = $[b_0, b_1, \cdots, b_m]$; %分子系数组成的向量
den = $[a_0, a_1, \cdots, a_n]$; %分母系数组成的向量
G(s) = tf[num, den] %输出传递函数

例 7-1 用 MATLAB 描述系统，其传递函数为

$$G(s) = \frac{2s+1}{4s^3 + 3s^2 + 2s}$$

解：本例的 MATLAB 程序代码为

num = [2,1];
den = [4,3,2,0];
G = tf(num,den)

则执行后得到的结果为

G =

        2 s + 1

----

4 s^3 + 3 s^2 + 2 s

Continuous-time transfer function. %运行结果说明该函数为连续时间传递函数

此外，传递函数可采用部分分式展开形式进行描述，即

num = $[b_0, b_1, \cdots, b_m]$
den = $[a_0, a_1, \cdots, a_n]$
[r,p,k] = residue(num,den) %输出传递函数的部分分式展开式

```
4 s^3 + 3 s^2 + 2 s
```
Continuous-time transfer function.   % The result shows that the function is a
% continuous-time transfer function

In addition, the transfer function can be described in the form of partial fraction expansion,
num = $[b_0, b_1, \cdots, b_m]$;
den = $[a_0, a_1, \cdots, a_n]$;
[r,p,k] = residue(num,den)   % Output the partial fraction expansion of the transfer function
where r, p, k represent the residue, pole and direct term respectively. The partial fraction expansion of $Y(s)/X(s)$ is

$$\frac{Y(s)}{X(s)} = \frac{r(1)}{s-p_1} + \frac{r(2)}{s-p_2} + \cdots + \frac{r(n)}{s-p_n} + k(s)$$

Conversely, the command "[num,den] = residue(r,p,k)" can be used to obtain the transfer function for the above the partial fraction expansion.

Example 7-2 Figure out the partial fraction expansion of the following transfer function in MATLAB.

$$\frac{Y(s)}{X(s)} = \frac{3s^3 + 6s^2 + 4s + 7}{s^3 + 6s^2 + 11s + 6}$$

Solution: The MATLAB programming code of the example is
num = [3,6,4,7];
den = [1,6,11,6];
[r,p,k] = residue(num,den)
The result after the execution is obtained as
r =
  -16.0000
    1.0000
    3.0000
p =
   -3.0000
   -2.0000
   -1.0000
k =
    3

Thus, the partial fraction expansion of $Y(s)/X(s)$ is

$$\frac{Y(s)}{X(s)} = \frac{3s^3 + 6s^2 + 4s + 7}{(s+1)(s+2)(s+3)} = \frac{-16}{s+3} + \frac{1}{s+2} + \frac{3}{s+1} + 3$$

Example 7-3 Figure out the transfer function model of the following partial fraction expansion in MATLAB.

$$\frac{Y(s)}{X(s)} = \frac{5}{s+4} + \frac{3}{s+3} + \frac{2}{s+2} + 4$$

其中，r、p 和 k 分别表示留数、极点和直接项。$Y(s)/X(s)$ 的部分分式展开式即为

$$\frac{Y(s)}{X(s)} = \frac{r(1)}{s-p_1} + \frac{r(2)}{s-p_2} + \cdots + \frac{r(n)}{s-p_n} + k(s)$$

反之，对如上分式展开式采用命令"[num,den] = residue(r,p,k)"可求出传递函数。

例 7-2 在 MATALAB 中，求如下传递函数的部分分式展开式。

$$\frac{Y(s)}{X(s)} = \frac{3s^3 + 6s^2 + 4s + 7}{s^3 + 6s^2 + 11s + 6}$$

解：本例的 MATLAB 程序为

num = [3,6,4,7];
den = [1,6,11,6];
[r,p,k] = residue(num,den)

执行后得到的结果为

r =
　　-16.0000
　　　1.0000
　　　3.0000
p =
　　-3.0000
　　-2.0000
　　-1.0000
k =
　　3

因此，$Y(s)/X(s)$ 的部分分式展开式为

$$\frac{Y(s)}{X(s)} = \frac{3s^3 + 6s^2 + 4s + 7}{(s+1)(s+2)(s+3)} = \frac{-16}{s+3} + \frac{1}{s+2} + \frac{3}{s+1} + 3$$

例 7-3 在 MATLAB 中，求如下部分分式展开式的传递函数模型。

$$\frac{Y(s)}{X(s)} = \frac{5}{s+4} + \frac{3}{s+3} + \frac{2}{s+2} + 4$$

解：本例的 MATLAB 程序为

r = [5,3,2];
p = [-4,-3,-2];
k = 4;
[num,den] = residue(r,p,k)
G = tf(num,den)

执行后得到的结果为

num =
　　4　　46　　161　　174
den =
　　1　　9　　26　　24

Solution: The MATLAB programming code of the example is

```
r=[5,3,2];
p=[-4,-3,-2];
k=4;
[num,den]=residue(r,p,k)
G=tf(num,den)
```

The result after the execution is obtained as

```
num =
    4    46    161    174
den =
    1    9    26    24
G =
   4 s^3 + 46 s^2 + 161 s + 174
   ----------------------------
     s^3 + 9 s^2 + 26 s + 24
```

Continuous-time transfer function.

2. Pole-zero gain model

When the transfer function of a control system is

$$G(s) = k \frac{(s-z_0)(s-z_1)\cdots(s-z_m)}{(s-p_0)(s-p_1)\cdots(s-p_n)}$$

function "zpk(z,p,k)" can be used to express the model in MATLAB, where "z" is the vector composed of zero elements; "p" is the vector composed of pole elements, "k" is the vector composed of the gain element. The program is

```
z=[z_0,z_1,...,z_m];     % The vector composed of zero elements
p=[p_0,p_1,...,p_m];     % The vector composed of pole elements
k=[k];                   % The vector composed of the gain element
G=zpk(z,p,k)             % Output the transfer function of the pole-zero gain model
```

Example 7-4 Describe the following transfer function of a system by using MATLAB.

$$G(s) = \frac{2(s-1)}{s(s+2)(s-3)}$$

Solution: The MATLAB programming code of the example is

```
z=1;
p=[0,-2,3];
k=2;
G=zpk(z,p,k)
```

The result after the execution is obtained as

```
G =
      2 (s-1)
    -------------
```

G =

    4 s^3+46 s^2+161 s+174

    ----

    s^3+9 s^2+26 s+24

Continuous-time transfer function。

2. 零极点增益模型

当控制系统的传递函数为

$$G(s)=k\frac{(s-z_0)(s-z_1)\cdots(s-z_m)}{(s-p_0)(s-p_1)\cdots(s-p_n)}$$

则在MATLAB中，可用函数"zpk(z,p,k)"表示该模型，"z"表示由零点元素组成的向量，"p"表示由极点元素组成的向量，"k"表示由增益元素组成的向量。程序为

z=[$z_0,z_1,\cdots,z_m$];%由零点元素组成的向量
p=[$p_0,p_1,\cdots,p_m$];%由极点元素组成的向量
k=[k];%由增益元素组成的向量
G=zpk(z,p,k)%输出零极点增益模型的传递函数

例7-4 用MATLAB描述系统的以下传递函数。

$$G(s)=\frac{2(s-1)}{s(s+2)(s-3)}$$

解：本例的MATLAB程序为

z=1;
p=[0,-2,3];
k=2;
G=zpk(z,p,k)

执行后得到的结果为

G =

    2(s-1)

    ----

    s(s+2)(s-3)

Continuous-time zero/pole/gain model.%连续时间零极点增益模型

3. 状态空间模型

对于一个具有 Nx 个状态、Ny 个输出、Nu 个输入的系统，其可以用一个状态空间模型来表示，即

$$\dot{x}=Ax+Bu,\ y=Cx+Du$$

其中，A 是 $Nx\times Nx$ 的矩阵；B 是 $Nx\times Nu$ 的矩阵；C 是 $Ny\times Nx$ 的矩阵；D 是 $Ny\times Nu$ 的矩阵。在MATLAB中，该控制系统可以用函数"ss(A,B,C,D)"来创建。

例7-5 用MATLAB对系统的如下状态空间矩阵建立状态空间模型。

$$A=\begin{bmatrix}0 & 1\\-5 & -2\end{bmatrix}\quad B=\begin{bmatrix}0\\3\end{bmatrix}\quad C=[0\ \ 1]\quad D=[0]$$

解：本例的MATLAB程序为

s (s+2) (s-3)

Continuous-time zero/pole/gain model. % The result shows that the model is a continuous-time
% pole-zero gain model

3. State space model

For a system with Nx states, Ny outputs, and Nu inputs, the system can be represented by a state-space model as

$$\begin{cases} \dot{x} = Ax + Bu \\ y = Cx + Du \end{cases}$$

where A is an $Nx \times Nx$ matrix; B is an $Nx \times Nu$ matrix; C is an $Ny \times Nx$ matrix; D is an $Ny \times Nu$ matrix. In MATLAB, it can be created by the function "ss(A,B,C,D)".

Example 7-5 Create a state-space model of the following state-space matrixes of a system by using MATLAB.

$$A = \begin{bmatrix} 0 & 1 \\ -5 & -2 \end{bmatrix} \quad B = \begin{bmatrix} 0 \\ 3 \end{bmatrix} \quad C = \begin{bmatrix} 0 & 1 \end{bmatrix} \quad D = \begin{bmatrix} 0 \end{bmatrix}$$

Solution: The MATLAB programming code of the example is

A = [0,1;-5,-2];
B = [0;3];
C = [0,1];
D = 0;
sys = ss(A,B,C,D)

The result after the execution is obtained as

sys =

　a =
　　　x1　x2
　x1　 0　 1
　x2　-5　-2

　b =
　　　u1
　x1　 0
　x2　 3

　c =
　　　x1　x2
　y1　 0　 1

　d =
　　　u1

```
A=[0,1;-5,-2];
B=[0;3];
C=[0,1];
D=0;
sys=ss(A,B,C,D)
```
执行后得到的结果为
```
sys =
  a =
         x1   x2
   x1    0    1
   x2   -5   -2
  b =
         u1
   x1    0
   x2    3
  c =
         x1   x2
   y1    0    1
  d =
         u1
   y1    0
Continuous-time state-space model
```
%运行结果说明该函数为连续时间状态空间模型

4. 复杂传递函数的求取

复杂传递函数求解涉及多项式相乘，在 MATLAB 中，一般可用函数"conv(u,v)"来求取，"u"和"v"均是多项式系数的向量。此外，conv() 函数允许任意多层嵌套，从而实现复杂计算。

例 7-6 用 MATLAB 描述系统的以下传递函数。

$$G(s)=\frac{6(2s^2+s+1)}{(2s^2+4s+1)^2(2s^3+7s^2+6s+4)(2s+3)}$$

解：本例的 MATLAB 程序为
```
num=6*[2,1,1];
den=conv(conv(conv([2,4,1],[2,4,1]),[2,7,6,4]),[2,3])
G=tf(num,den)
```
执行后得到的结果为
```
G =
                        12 s^2 + 6 s + 6
  --------------------------------------------------------------------
  16 s^8 + 144 s^7 + 532 s^6 + 1064 s^5 + 1288 s^4 + 996 s^3 + 481 s^2+ 122 s + 12
Continuous-time transfer function.
```
%运行结果说明该函数为连续时间传递函数

y1   0

Continuous-time state-space model. % The result shows that the model is a continuous-time
% state-space model

4. Solving the complex transfer function

Generally, in MATLAB the function "conv (u, v)" can be used for complex transfer function which involves polynomial multiplication, where "u" and "v" are polynomial coefficient vectors. In addition, the function conv ( ) allows any number of levels of nesting, thus enabling the complex calculation.

Example 7-6   Describe the following transfer function of a system by using MATLAB.

$$G(s) = \frac{6(2s^2+s+1)}{(2s^2+4s+1)^2(2s^3+7s^2+6s+4)(2s+3)}$$

Solution: The MATLAB programming code of the example is

num = 6 * [2,1,1];
den = conv(conv(conv([2,4,1],[2,4,1]),[2,7,6,4]),[2,3]);
G = tf(num,den)

The result after the execution is obtained as

G =

            12 s^2 + 6 s + 6
    ----
    16 s^8 + 144 s^7 + 532 s^6 + 1064 s^5 + 1288 s^4 + 996 s^3 + 481 s^2 + 122 s + 12

Continuous-time transfer function. % The result shows that the function is a continuous-time
% transfer function

7.2.2 Transformations between various mathematical models

In the toolbox of signal processing and control system of the MATLAB toolbar, functions of the model transformation are provided as ss2tf, ss2zp, tf2ss, tf2zp, zp2ss, zp2tf, etc. Their relationship can be represented by the structure shown in Fig. 7-2.

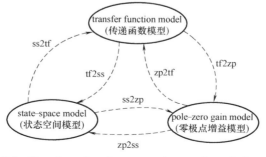

Fig. 7-2   Transformations between various mathematical models
(各种数学模型之间的转换)

7.2.2 各种数学模型之间的转换

MATLAB 工具栏里的信号处理和控制系统工具箱中，提供的模型转换函数有 ss2tf、ss2zp、tf2ss、tf2zp、zp2ss、zp2tf 等，它们的关系可用图 7-2 所示的结构来表示。

1) 将状态空间模型转换成传递函数模型，命令格式为
$$[num, den] = ss2tf(A, B, C, D, iu)$$
其中，iu 为输入信号的序号。转换公式为
$$G(s) = \frac{num(s)}{den(s)} = C(sI-A)^{-1}B + D$$

2) 将状态空间模型转换成零极点增益模型，命令格式为
$$[z, p, k] = ss2zp(A, B, C, D, iu)$$

3) 将传递函数模型转换成状态空间模型，命令格式为
$$[A, B, C, D] = tf2ss(num, den)$$

4) 将传递函数模型转换成零极点增益模型，命令格式为
$$[z, p, k] = tf2zp(num, den)$$

5) 将零极点增益模型转换成状态空间模型，命令格式为
$$[A, B, C, D] = zp2ss(z, p, k)$$

6) 将零极点增益模型转换成传递函数模型，命令格式为
$$[num, den] = zp2tf(z, p, k)$$

以上命令格式的具体使用说明请参照 MATLAB 帮助文档。

7.2.3 基于 MATLAB 的控制系统建模

简单系统的建模可直接采用传递函数、零极点增益和状态空间模型三种基本模型，但实际系统通常是由简单系统通过并联、串联、闭环或反馈等连接而成的复杂系统。设有 sys1 和 sys2 两个系统模型，sys1 = tf(num1, den1)，sys2 = tf(num2, den2)，其中 num1、den1 分别是系统 sys1 传递函数的分子系数和分母系数向量，num2、den2 分别是系统 sys2 传递函数的分子系数和分母系数向量。由它们组成系统时，可构建如图 7-3 所示的连接形式。

1. 串联连接

两个系统按串联方式连接的形式如图 7-3a 所示，在 MATLAB 中可用 series 函数实现，其命令格式为
$$sys = series(sys1, sys2)$$

2. 并联连接

两个系统按并联方式连接的形式如图 7-3b 所示，在 MATLAB 中可用 parallel 函数实现，其命令格式为
$$sys = parallel(sys1, sys2)$$

3. 反馈连接

将两个系统通过正或负反馈连接成闭环系统的形式如图 7-3c 所示，在 MATLAB 中可用 feedback 函数实现，其命令格式为
$$sys = feedback(sys1, sys2, sign)$$

1) Transforming the state-space model to the transfer function model, the command format is
$$[\text{num}, \text{den}] = \text{ss2tf}(A, B, C, D, \text{iu})$$
where iu is the serial number of the input signal. The conversion formula is
$$G(s) = \frac{\text{num}(s)}{\text{den}(s)} = C(sI-A)^{-1}B+D$$

2) Transforming the state-space model to the pole-zero gain model, the command format is
$$[z, p, k] = \text{ss2zp}(A, B, C, D, \text{iu})$$

3) Transforming the transfer function model to the state-space model, the command format is
$$[A, B, C, D] = \text{tf2ss}(\text{num}, \text{den})$$

4) Transforming the transfer function model to the pole-zero gain model, the command format is
$$[z, p, k] = \text{tf2zp}(\text{num}, \text{den})$$

5) Transforming the pole-zero gain model to the state-space model, the command format is
$$[A, B, C, D] = \text{zp2ss}(z, p, k)$$

6) Transforming the pole-zero gain model to the transfer function model, the command format is
$$[\text{num}, \text{den}] = \text{zp2tf}(z, p, k)$$

As for the specific instructions of the above command formats, please refer to the MATLAB help documentation.

7.2.3 Control system modeling based on MATLAB

Three basic models of transfer function, pole-zero gain, and state space model can be directly used in the modeling of simple systems. However, an actual system is usually made up of simple systems connected in parallel, series, closed loop or feedback, thus forming complex system. Two system models sys1 and sys2 are set as sys1 = tf (num1, den1) and sys2 = tf (num2, den2), where num1 and den1 are numerator coefficient vector and denominator coefficient vector of the system transfer function on sys1 respectively, num2 and den2 are numerator coefficient vector and denominator coefficient vector of the system transfer function on sys2 respectively. When they form a system, the connections shown in Fig. 7-3 can be constructed.

1. Series connection

Two systems in series connection is as shown in Fig. 7-3a, and series function can be implemented in MATLAB, whose command format is
$$\text{sys} = \text{series}(\text{sys1}, \text{sys2})$$

2. Parallel connection

Two systems in parallel connection is as shown in Fig. 7-3b, and parallel function can be implemented in MATLAB, whose command format is

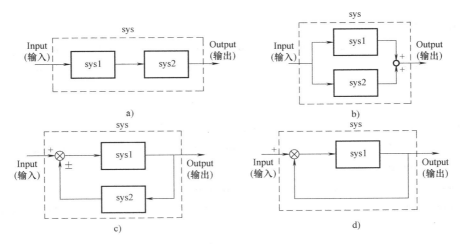

Fig. 7-3 Different connection forms of the control system（控制系统不同连接形式）
a) Parallel connection （并联连接） b) Series connection （串联连接）
c) Feedback connection （反馈连接） d) Unit feedback （单位反馈）

其中，sign 为自定义参数，sign = -1 时为负反馈，而 sign = 1 时为正反馈，缺省值为负反馈。

4. 单位反馈

一个系统按反馈方式连接成闭环系统而形成单位反馈系统的形式如图 7-3d 所示。在 MATLAB 中可以用 cloop 函数实现，其命令格式为

$$sys = cloop(num1, den1, sign)$$

7.2.4 基于 Simulink 的建模方法

在一些实际应用中，如果系统过于复杂，则前面介绍的方法就不适用了。这时功能完善的 Simulink 模块可以用来建立系统的模型。由于篇幅所限，本小节仅给出 Simulink 建模处理的一些简明介绍。

1. 开始准备

若想按 Simulink 格式建立一个系统模型，则应该首先启动 Simulink 模块。可以在 MATLAB 命令窗口的提示符下键入命令"Simulink"来启动 Simulink 模块，这时 Simulink 库浏览器将显示出来，如图 7-4 所示。

2. 打开编辑窗口

单击图 7-4 所示的"新模型"按钮，Simulink 将打开一个空白的模型编辑窗口用来建立新的系统模型，如图 7-5 所示。

3. 画出系统的各个模块

如图 7-6 所示，先双击"Simulink 库"按钮，将库浏览器扩展开来。然后单击"Sources"按钮使"Library Simulink/Sources"选项卡显示出来，最后拖动所需模块。例如，将 Sine Wave 模块从 Simulink 库浏览器中拖至新建模型的编辑窗口中，如图 7-7 所示。将其余的模块用类似的操作从各自的库中拖到模型编辑窗口中，而后可以通过拖动模块的方法将模块移至合适的位置。

$$sys = parallel(sys1, sys2)$$

3. Feedback connection

Two systems can be connected to be a closed-loop system through positive or negative feedback, as is shown in Fig. 7-3c, and feedback function can be implemented in MATLAB, whose command format is

$$sys = feedback(sys1, sys2, sign)$$

where sign is a custom parameter, the feedback is called negative feedback, when sign = -1, and is called positive feedback when sign = 1. The default value leads to negative feedback.

4. Unit feedback

A system can be connected to be a closed-loop system according to the feedback pattern, thus forming a unit feedback system, as is shown in Fig. 7-3d. And cloop function can be implemented in MATLAB, whose command format is

$$sys = cloop(num\ 1, den\ 1, sign)$$

7.2.4 Modeling method based on Simulink

In some practical applications, if the system is too complex, then the methods described above are not applicable. In this circumstance, the fully functional Simulink program can be used to establish the model of the system. Because of the limited space, this section only gives some brief introduction of Simulink modeling.

1. Start to prepare

If a system model is wanted in the Simulink format, the Simulink program should be started first. The Simulink program can be started by typing the command "simulink" at the prompt in the MATLAB command window, and the Simulink Library Browser will be displayed, as is shown in Fig. 7-4.

2. Open the editing window

Click "new model" button as is shown in Fig. 7-4, Simulink will open a blank model editing window to create a new system model, as is shown in Fig. 7-5.

3. Draw the modules of the system

As is shown in Fig. 7-6, double-click "Simulink" button to expand the library browser. Then click the "Sources" button to make "Library Simulink/Sources" tab appear, and finally drag the required module. For example, when the Sine Wave module is dragged from the Simulink Library Browser into the new model editing window, the screen becomes as is shown in Fig. 7-7. Other modules can be copied in a similar way from their respective libraries into the model editing window, and then the modules can be dragged to the appropriate location.

4. Draw connecting lines

When all the modules have been drawn, the necessary connecting lines between the modules can be drawn to form a complete system. After we click the output end of the starting module with the mouse, drag the mouse to the input end of the terminating module and release the mouse, the connecting lines will be drawn automatically with arrows between the two modules.

Fig. 7-4　Simulink Library Browser（Simulink 库浏览器）　　Fig. 7-5　Simulink editing window（Simulink 编辑窗口）

4. 画出连接线

当所有的模块都画出来之后，则可以画出模块间必要的连接线，构成完整的系统。在起始模块的输入端处单击鼠标左键并拖动鼠标至终止模块的输出端处，Simulink 就会自动地在两个模块间画出带箭头的连接线。

5. 给出各个模块参数

假设有如图 7-8 所示系统，各个模块只包含默认的模型参数。若要修改模块的参数，则需用鼠标左键双击模块图标，此时会出现相应的对话框，可根据提示修改模块参数。

7.3　基于 MATLAB 的控制系统性能分析

7.3.1　控制系统的时域分析

MATLAB 提供了求取连续控制系统单位阶跃响应、单位脉冲响应等的仿真函数，利用这些函数可以方便地对控制系统进行时域分析。

1. step 命令

功能：求连续系统的阶跃响应。

若需要返回阶跃响应曲线的图形数据，则可采用的调用格式为

$$[y,x] = step(num,den,t) \quad \text{或} \quad [y,x] = step(num,den)$$

其中，t 为选定的仿真时间向量，若仿真时间范围自动选择，则可以不用加 t。此函数只返回仿真数据而不在屏幕上画出仿真图形，返回值 y 为系统在各个仿真时刻的输出所组成的矩阵，而 x 为自动选择的状态变量的时间响应数据。

若只需要绘制出系统的阶跃响应曲线，则可采用的调用格式为

$$step(num,den,t) \quad \text{或} \quad step(num,den)$$

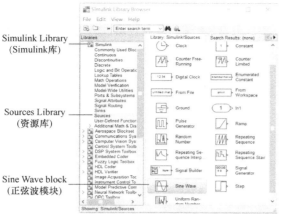

Fig. 7-6 Simulink Sources library（Simulink 资源库）

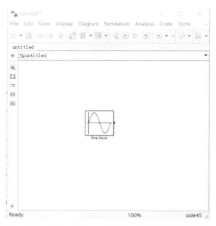

Fig. 7-7 Sine Wave block（正弦波模块）

5. Provide the parameters of each module

Suppose that a control system is established as is shown in Fig. 7-8, and each module contains only default model parameters. To modify module parameters, we need to double-click the module icon, and the corresponding dialog box appears to further indicate support for modifying module parameters.

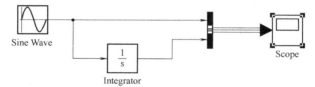

Fig. 7-8 System modeling based on Simulink（基于 Simulink 的系统建模）

7.3　Performance Analysis of Control Systems based on MATLAB

7.3.1　Time-domain analysis of control systems

MATLAB provides emulational functions to solve the unit step response unit ramp response and etc. for the time-continuous control systems, and these functions can be used to facilitate the time-domain analysis of the control system.

1. step command

Function: solve the step response of a continuous system.

If it is necessary to return the graphic data of the step response curve, it can be called in the format as

$$[y,x] = \text{step}(num,den,t) \text{ or } [y,x] = \text{step}(num,den)$$

where t is the selected simulation time vector. If the range of the simulation time is automatically selected, t needs not be added. This function only returns the simulation data instead of drawing the

2. impulse 命令

功能：求连续系统的脉冲响应。

若需要返回脉冲响应曲线的图形数据，则可采用的调用格式为

$$[y,x]=impulse(num,den,t) \quad 或 \quad [y,x]=impulse(num,den)$$

若只需要绘制出系统的脉冲响应曲线，则可采用的调用格式为

$$impulse(num,den,t) \quad 或 \quad impulse(num,den)$$

例 7-7 采用 MATLAB 程序求出如下传递函数系统的单位阶跃响应曲线。

$$\frac{X_o(s)}{X_i(s)}=\frac{s+20}{15s^2+s+1}$$

解：本例的 MATLAB 程序为

\>\> num=[0,1,20];
den=[15,1,1];
step(num,den);
grid

单位阶跃响应曲线如图 7-9 所示。

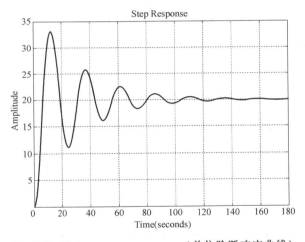

Fig. 7-9 Unit step response curve（单位阶跃响应曲线）

例 7-8 一个系统的传递函数为

$$\frac{X_o(s)}{X_i(s)}=\frac{s+20}{15s^2+s+1}$$

采用 MATLAB 程序求出该系统的单位脉冲响应曲线。

解：本例的 MATLAB 程序为

\>\> num=[0,1,20];
den=[15,1,1];
impulse(num,den);
grid

单位脉冲响应曲线如图 7-10 所示。

simulation graph on the screen. The return value y is the matrix formed by the output of the system at each simulation moment, while x is the time response data of the automatically selected state variable. If it is only wanted to plot the step response curve of the system, it can be called in the format as

$$\text{step (num, den, t) or step (num, den)}$$

2. impulse command

Function: solve the impulse response of the continuous system.

If it is necessary to return the graphic data of the impulse response curve, it can be called in the format as

$$[y,x] = \text{impulse(num,den,t) or } [y,x] = \text{impulse(num,den)}$$

If it is necessary to plot the impulse response curve of the system, only it can be called in the format as

$$\text{impulse(num,den,t) or impulse(num,den)}$$

Example 7-7 Solve the unit step response curve of the system with the following transfer function by MATLAB program.

$$\frac{X_o(s)}{X_i(s)} = \frac{s+20}{15s^2+s+1}$$

Solution: The MATLAB programming code of the example is

```
>> num=[0,1,20];
den=[15,1,1];
step(num ,den);
grid
```

The unit step response curve is shown in Fig. 7-9.

Example 7-8 The transfer function of a system is

$$\frac{X_o(s)}{X_i(s)} = \frac{s+20}{15s^2+s+1}$$

Solve the unit impulse response curve of the system by MATLAB program.

Solution: The MATLAB programming code of the example is

```
>> num=[0,1,20];
den=[15,1,1];
impulse(num ,den);
grid
```

The unit impulse response curve is shown in Fig. 7-10.

In MATLAB, there is no ramp response command, but the ramp response can be obtained by using step response command. First, s is divided by $G(s)$, and then step response command is used. For example, consider the closed-loop system with the transfer function of

$$\frac{X_o(s)}{X_i(s)} = \frac{s+20}{15s^2+s+1}$$

For the unit ramp input, $X_i(s) = 1/s^2$, we have

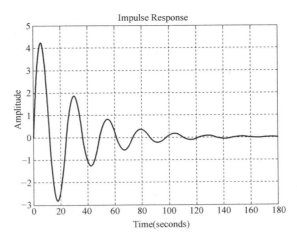

Fig. 7-10　Unit impulse response curve（单位脉冲响应曲线）

在 MATLAB 中没有斜坡响应命令，可以利用阶跃响应命令求斜坡响应，先用 s 除 $G(s)$，再利用阶跃响应命令。例如，考虑一个闭环系统，其传递函数为

$$\frac{X_o(s)}{X_i(s)} = \frac{s+20}{15s^2+s+1}$$

对于单位斜坡输入量，$X_i(s) = 1/s^2$，则

$$\frac{X_o(s)}{X_i(s)} = \frac{s+20}{15s^2+s+1} \cdot \frac{1}{s^2} = \frac{s+20}{(15s^2+s+1)s} \cdot \frac{1}{s} = \frac{s+20}{15s^3+s^2+s} \cdot \frac{1}{s}$$

下列 MATLAB 程序将给出该系统的单位斜坡响应曲线。单位斜坡响应曲线如图 7-11 所示。

```
>> num=[0,0,1,20];
den=[15,1,1,0];
t=0:0.01:100;
step(num,den,t);
grid
```

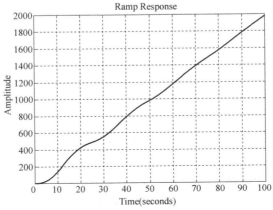

Fig. 7-11　Unit ramp response curve（单位斜坡响应曲线）

$$\frac{X_o(s)}{X_i(s)} = \frac{s+20}{15s^2+s+1} \frac{1}{s^2} = \frac{s+20}{(15s^2+s+1)s} \frac{1}{s} = \frac{s+20}{15s^3+s^2+s} \frac{1}{s}$$

The following MATLAB program will give out the unit ramp response curve of the system. The unit ramp response curve is shown in Fig. 7-11.

```
>> num=[0,1,20];
den=[15,1,1];
t=0:0.01:100;
step(num,den,t);
grid
```

command in MATLAB, In addition, the analytic solution of linear time response can also be easily obtained using the function residue ( ), as is shown in the following examples.

Example 7-9 The transfer function of a system is

$$\frac{C(s)}{R(s)} = \frac{2s^3+8s^2+25s+25}{s^4+10s^3+35s^2+50s+24}$$

Solve the step response of the system by MATLAB program.

Solution: The MATLAB programming code of the example is

```
>> num=[2,8,25,25];
den=[1,10,35,50,24];
[r,p,k]=residue(num,[den,0])
```

After the program is executed, the results are obtained as

```
r =
    -3.1250
     5.3333
    -2.2500
    -1.0000
     1.0417
p =
    -4.0000
    -3.0000
    -2.0000
    -1.0000
     0
k =
    []
```

Thus, the mathematical formula represented by the solution is

$$y(t) = -3.125e^{-4t} + 5.3333e^{-3t} - 2.25e^{-2t} - e^{-t} + 1.0417$$

This is the response to the step input of the system.

In addition, the function residue ( ) is also applicable when complex poles are included in the system.

此外，也可以利用 residue( ) 函数比较方便地求取线性时间响应的解析解，具体如以下例题所示。

例 7-9　一个系统的传递函数为

$$\frac{C(s)}{R(s)}=\frac{2s^3+8s^2+25s+25}{s^4+10s^3+35s^2+50s+24}$$

采用 MATLAB 程序求出该系统的阶跃响应。

解：本例的 MATLAB 程序为

\>\> num=[2,8,25,25];
den=[1,10,35,50,24];
[r,p,k]=residue(num,[den,0])

程序执行后得到的结果为

r =
　　－3.1250
　　　5.3333
　　－2.2500
　　－1.0000
　　　1.0417
p =
　　－4.0000
　　－3.0000
　　－2.0000
　　－1.0000
　　　　　0
k =
　　[ ]

该解所表示的数学公式为

$$y(t)=-3.125\mathrm{e}^{-4t}+5.3333\mathrm{e}^{-3t}-2.25\mathrm{e}^{-2t}-\mathrm{e}^{-t}+1.0417$$

这也就是该系统阶跃输入的响应。

此外，当系统中含有复数极点时，residue( ) 函数照样适用。

例 7-10　对于下列系统传递函数

$$\frac{C(s)}{R(s)}=\frac{2s+4}{2s^4+3s^3+12s^2+19s+18}$$

采用 MATLAB 程序求出该系统的阶跃响应。

解：本题的 MATLAB 程序为

\>\> num=[2,4];
den=[2,3,12,19,18,0];
[r,p,k]=residue(num,den)

程序执行后得到的结果为

r =

Example 7-10 The transfer function of a system is

$$\frac{C(s)}{R(s)} = \frac{2s+4}{2s^4+3s^3+12s^2+19s+18}$$

Solve the step response of the system by MATLAB program.

Solution: The MATLAB programming code of the example is

```
>> num=[2,4];
den=[2,3,12,19,18,0];
[r,p,k]=residue(num,den)
```

After the program is executed, the results are obtained as

r =
  -0.0198 + 0.0446i
  -0.0198 - 0.0446i
  -0.0913 - 0.0183i
  -0.0913 + 0.0183i
   0.2222 + 0.0000i

p =
   0.2388 + 2.2723i
   0.2388 - 2.2723i
  -0.9888 + 0.8639i
  -0.9888 - 0.8639i
   0.0000 + 0.0000i

k =
   [ ]

Thus, the mathematical formula represented by the solution is

$$y(t) = (-0.0198+0.0446j)e^{(0.2388+2.2723j)t} + (-0.0198-0.0446j)e^{(0.2388-2.2723j)t} +$$
$$(-0.0913-0.0183j)e^{(-0.9888+0.8639j)t} + (-0.0913+0.0183j)e^{(-0.9888-0.8639j)t} + 0.2222$$

7.3.2 Frequency-domain analysis of control systems

The frequency-domain analysis with MATLAB mainly uses Bode plot and polar plot, and amplitude margin, phase margin, roots of the characteristic polynomial of the system can be solved as well.

1. bode command

Function: draw the bode plot of a continuous system.

Format: [mag,phase,w]=bode(num,den)

[mag,phase,w]=bode(num,den,w)

2. nyquist command

Function: draw the polar plot of a continuous system

Format: [re,im,w]=nyquist(num,den)

[re,im,w]=nyquist(num,den,w)

```
    −0.0198 + 0.0446i
    −0.0198 − 0.0446i
    −0.0913 − 0.0183i
    −0.0913 + 0.0183i
     0.2222 + 0.0000i
p =
     0.2388 + 2.2723i
     0.2388 − 2.2723i
    -0.9888 + 0.8639i
    -0.9888 − 0.8639i
     0.0000 + 0.0000i
k =
    []
```

该解所表示的数学公式为

$$y(t) = (-0.0198+0.0446j)e^{(0.2388+2.2723j)t} + (-0.0198-0.0446j)e^{(0.2388-2.2723j)t} + \\ (-0.0913-0.0183j)e^{(-0.9888+0.8639j)t} + (-0.0913+0.0183j)e^{(-0.9888-0.8639j)t} + 0.2222$$

7.3.2 控制系统的频域分析

应用 MATLAB 进行频域分析主要应用伯德图和极坐标图，也可求系统的幅值裕度和相位裕度，以及特征多项式的根。

1. bode 命令

功能：绘制连续系统的伯德图。

格式：[mag,phase,w] = bode(num,den)

[mag,phase,w] = bode(num,den,w)

2. nyquist 命令

功能：绘制连续系统极坐标图。

格式：[re,im,w] = nyquist(num,den)

[re,im,w] = nyquist(num,den,w)

其中，输出变量 re、im 分别是系统极坐标图的实部、虚部。

3. margin 命令

功能：求幅值裕度（gm）和相位裕度（pm），以及幅值穿越频率（wcg）和相位穿越频率（wcp）。

格式：[gm,pm,wcg,wcp] = margin(num,den)

[gm,pm,wg,wp] = margin(mag,phase,w)

4. roots 命令

功能：求系统特征多项式的根。

格式：[r] = roots(c)

其中，输入变量 c 为系统特征多项式系数组成的矢量。

where the output variables re, im are real part and imaginary part of the polar plot respectively.

3. margin command

Function: solve the amplitude margin (gm) and phase margin (pm) and the crossing frequency with amplitude (wcg) and crossing frequency with phase (wcp).

Format: [gm,pm,wcg,wcp] = margin(num,den)

[gm,pm,wg,wp] = margin(mag,phase,w)

4. roots command

Function: Solve roots of the characteristic polynomial of the system.

Format: [r] = roots(c)

where the input variable c is the vector composed of the characteristic polynomial coefficients.

Example 7-11 The transfer function of a system is

$$G(s) = \frac{s+20}{15s^2+s+1}$$

Draw the Bode plot of the system by MATLAB program.

Solution: The MATLAB programming code of the example is

```
>> num = [0,1,20];
den = [15,1,1];
bode(num,den);
grid
```

The Bode plot is obtained as is shown in Fig. 7-12a.

If it is wanted that the Bode plot is from 10^{-23} rad/s to 10^{100} rad/s, we can enter

```
>> w = logspace(-23,100);  %Generate a row vector w of 50 logarithmically spaced points
                           %between decades 10^-23 and 10^100
bode(num,den,w);
grid
```

The Bode plot is obtained as is shown in Fig. 7-12b.

Example 7-12 The transfer function of a system is

$$G(s) = \frac{11(s+4)}{s(2s+3)(s^2+3s+5)}$$

Draw the Bode plot of the system by MATLAB program.

Solution: The MATLAB programming code of the example is

```
>> num = [11,44];
den1 = [2,3,0];
den2 = [1,3,5];
den = conv(den1,den2);
w = logspace(-23,100);
bode(num,den,w);
grid
```

The Bode plot is obtained as is shown in Fig. 7-13.

例 7-11 一个系统的传递函数为

$$G(s) = \frac{s+20}{15s^2+s+1}$$

采用 MATLAB 程序求出该系统的伯德图。

解：本例的 MATLAB 程序为

\>\> num = [0,1,20];
den = [15,1,1];
bode(num,den);
grid

所得伯德图如图 7-12a 所示。
如果希望从 10^{-23} rad/s 到 10^{100} rad/s 画伯德图，可输入

\>\> w = logspace(-23,100);%在 10^{-23} 到 10^{100} 间生成由 50 个对数间隔点组成的行向量 w
bode(num,den,w);
grid

所得伯德图如图 7-12b 所示。

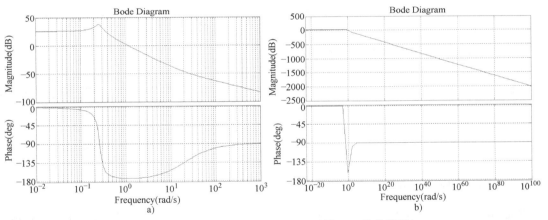

Fig. 7-12 Bode plot of example 7-11 （例 7-11 的伯德图）

例 7-12 一个系统的传递函数为

$$G(s) = \frac{11(s+4)}{s(2s+3)(s^2+3s+5)}$$

采用 MATLAB 程序求出该系统的伯德图。

解：本例的 MATLAB 程序为

\>\> num = [11,44];
den1 = [2,3,0];
den2 = [1,3,5];
den = conv(den1,den2);
w = logspace(-23,100);
bode(num,den,w);
grid

Example 7-13 The transfer function of a system is

$$G(s) = \frac{s+20}{15s^2+s+1}$$

Draw the polar plot of the system by MATLAB program.

Solution: The MATLAB programming code of the example is

```
>> num=[0,1,20];
den=[15,1,1];
nyquist(num,den);
grid
```

The polar plot is obtain as is shown in Fig. 7-14.

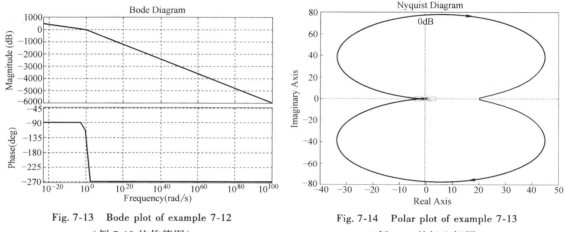

Fig. 7-13   Bode plot of example 7-12
（例 7-12 的伯德图）

Fig. 7-14   Polar plot of example 7-13
（例 7-13 的极坐标图）

7.3.3 Stability analysis of control systems

The simplest way to solve the stability problem of a linear system is to calculate all the poles of the system and observe whether there are any poles whose real part is greater than 0. If there are such poles, the system is unstable, otherwise it is stable; if a stable system has a pole whose real part is equal to 0, such a system is a critical stable system.

In addition, for a given control system, we can evaluate the stability of the system when using MATLAB to do time-domain and frequency-domain analysis, and the phase margin and gain margin of the system can be directly calculated. Moreover, we can determine the stability of the system more directly by obtaining the distribution of characteristic roots in MATLAB. The closed-loop system is stable if all characteristic roots have negative real parts.

Example 7-14 The closed-loop transfer function of a given system is

$$\frac{C(s)}{R(s)} = \frac{2s^3+8s^2+25s+25}{s^4+10s^3+35s^2+50s+24}$$

Evaluate the stability of the system by using MATLAB.

Solution: The MATLAB programming code of the example is

所得伯德图如图 7-13 所示。

例 7-13 一个系统的传递函数为

$$G(s) = \frac{s+20}{15s^2+s+1}$$

采用 MATLAB 程序求出该系统的极坐标图。

解：本例的 MATLAB 程序为

```
>> num = [0,1,20];
den = [15,1,1];
nyquist(num,den);
grid
```

所得极坐标图如图 7-14 所示。

7.3.3 控制系统的稳定性分析

求解线性系统稳定性问题最简单的方法是求出该系统的所有极点，并观察是否含有实部大于 0 的极点。如果有这样的极点，则系统不稳定，否则稳定；若稳定系统中存在实部等于 0 的极点，则这样的系统为临界稳定系统。

另外，对于一个给定的控制系统，可以在利用 MATLAB 进行时域、频域分析时评判系统的稳定性，并可直接求出系统的相位裕度和幅值裕量。此外，在 MATLAB 中，还可以通过求出特征根的分布更直接地判断系统的稳定性。如果闭环系统所有的特征根都有负实部，则系统稳定。

例 7-14 一个给定系统的闭环传递函数为

$$\frac{C(s)}{R(s)} = \frac{2s^3+8s^2+25s+25}{s^4+10s^3+35s^2+50s+24}$$

采用 MATLAB 判断该系统的稳定性。

解：本例的 MATLAB 的程序为

```
>>   num = [2,8,25,25];
den = [1,10,35,50,24];
G = tf(num,den);
roots(G.den{1})       %"{ }"表示维数
```

执行后得到的结果为

```
ans =
   -4.0000
   -3.0000
   -2.0000
   -1.0000
```

此外，还可以使用 zpk( ) 函数来解决该问题，MATLAB 编程代码为

```
>> num = [2,8,25,25];
den = [1,10,35,50,24];
G = tf(num,den);
```

```
>>  num = [2,8,25,25];
den = [1,10,35,50,24];
G = tf(num,den);
roots(G.den{1})     %"{}"represents dimensions
```
The results after execution are obtained as
```
ans =
    -4.0000
    -3.0000
    -2.0000
    -1.0000
```

In addition, the zpk ( ) function can be also used to solve the problem, the MATLAB programming code is
```
>>  num = [2,8,25,25];
den = [1,10,35,50,24];
G = tf(num,den);
G1 = zpk(G);
G1.p{1}
```
The results after execution are obtained as
```
ans =
    -4.0000
    -3.0000
    -2.0000
    -1.0000
```
Because poles of the system all have negative real roots, it is stable.

Example 7-15   The closed-loop transfer function of a given system is
$$\frac{C(s)}{R(s)} = \frac{2s^3 + 8s^2 + 25s + 25}{2s^8 + 3s^7 + 4s^6 + 5s^5 + 6s^4 + 7s^3 + 8s^2 + 9s + 1}$$

Evaluate the stability of the system by using MATLAB.

Solution: The MATLAB programming code of the example is
```
>>  G = tf([2,8,25,25],[2,3,4,5,6,7,8,9,1]);
roots(G.den{1})
```
The results after execution are obtained as
```
ans =
     0.8151 + 0.8930i
     0.8151 - 0.8930i
    -0.0219 + 1.2286i
    -0.0219 - 1.2286i
    -0.8703 + 0.8633i
    -0.8703 - 0.8633i
    -1.2225 + 0.0000i
    -0.1233 + 0.0000i
```

G1 = zpk(G);

G1.p{1}

执行后得到的结果为

ans =

  -4.0000

  -3.0000

  -2.0000

  -1.0000

由于该系统的极点全部为负实根，因此是稳定。

例 7-15 一个给定系统的闭环传递函数为

$$\frac{C(s)}{R(s)} = \frac{2s^3 + 8s^2 + 25s + 25}{2s^8 + 3s^7 + 4s^6 + 5s^5 + 6s^4 + 7s^3 + 8s^2 + 9s + 1}$$

采用 MATLAB 判断该系统的稳定性。

解：本例的 MATLAB 的程序为

```
>> G = tf([2,8,25,25],[2,3,4,5,6,7,8,9,1]);
roots(G.den{1})
```

执行后得到的结果

ans =

  0.8151 + 0.8930i

  0.8151 − 0.8930i

  −0.0219 + 1.2286i

  −0.0219 − 1.2286i

  −0.8703 + 0.8633i

  −0.8703 − 0.8633i

  −1.2225 + 0.0000i

  −0.1233 + 0.0000i

由于该系统的极点中有 2 个具有正实部，因此不稳定。

例 7-16 一个给定系统的闭环传递函数为

$$\frac{C(s)}{R(s)} = \frac{4s^4 + 3s^3 + 2s^2 + 5s + 3}{5s^5 + 6s^4 + 2s^3 + 3s^2 + 3s + 2}$$

采用 MATLAB 判断该系统的稳定性。

解：本例的 MATLAB 的程序为

```
>> num = [4,3,2,5,3];
den = [5,6,2,3,3,2];
[z,p] = tf2zp(num,den)
```

执行后得到的结果为

z =

  0.4121 + 0.9844i

  0.4121 − 0.9844i

Since two of the poles of the system have positive real parts, it is unstable.

Example 7-16 The closed-loop transfer function of a given system is

$$\frac{C(s)}{R(s)}=\frac{4s^4+3s^3+2s^2+5s+3}{5s^5+6s^4+2s^3+3s^2+3s+2}$$

Evaluate the stability of the system by using MATLAB.

Solution: The MATLAB programming code of the example is

```
>> num=[4,3,2,5,3];
den=[5,6,2,3,3,2];
[z,p]=tf2zp(num,den)
```

The results after execution are obtained as

z =
    0.4121 + 0.9844i
    0.4121 - 0.9844i
   -0.7871 + 0.1976i
   -0.7871 - 0.1976i
p =
    0.4511 + 0.7038i
    0.4511 - 0.7038i
   -1.1421 + 0.0000i
   -0.4800 + 0.5204i
   -0.4800 - 0.5204i

Further, the specific pole-zero distribution is obtained, the pole-zero distribution diagram is drawn, the instability of the system is clearly indicated and the specific right limit that causes the instability of the system is indicated as well. The MATLAB programming code is

```
>> num=[4,3,2,5,3];
den=[5,6,2,3,3,2];
pzmap(num,den)              %Draw the pole-zero distribution diagram of the system
>>ii=find(real(p)>0)        %Returns the positions of elements with non-negative real part of
                            %the vector
ii =
    1
    2
>>nl=length(ii)             % Count the number of elements in the array
nl =
    2
>> if(nl>0)
disp(['system is unstable,with'  int2str(nl)  'unstable poles']);   %Output the content in
                                                                    %the MATLAB command window directly
else
```

$$-0.7871 + 0.1976i$$
$$-0.7871 - 0.1976i$$
p =
$$0.4511 + 0.7038i$$
$$0.4511 - 0.7038i$$
$$-1.1421 + 0.0000i$$
$$-0.4800 + 0.5204i$$
$$-0.4800 - 0.5204i$$

进一步求出具体的零极点分布情况并画出零极点分布图,明确指出系统不稳定,并指出引起系统不稳定的具体右限。MATLAB 程序为

```
>> num=[4,3,2,5,3];
den=[5,6,2,3,3,2];
pzmap(num,den)    %绘制系统的零极点分布图
>>ii=find(real(p)>0)   %返回向量中实部非负元素的位置
  ii =
    1
    2
>> nl=length(ii)   %计算数组中的元素个数
  nl =
    2
>> if(nl>0)
     disp([′system is unstable,with ′   int2str(nl)   ′unstable poles′]);
                              %直接将内容输出在 MATLAB 命令窗口中
   else
     disp(′System is stable′);
   end
System is unstable,with 2 unstable poles
>>disp(′The unstable poles are:′),disp(p(ii))
The unstable poles are:
  0.4121 + 0.9844i
  0.4121 - 0.9844i
```

所得零极点分布图如图 7-15 所示。

7.4 基于 MATLAB 的控制系统校正

在 MATLAB 中,主要利用伯德图对控制系统进行校正设计。其基本思想是通过比较校正前、后系统的频率特性,尝试选择恰当的校正结构,确定校正参数,最后对校正后的系统进行校验,并反复设计直至满足设计要求。下面举例说明设计控制系统的过程。

例 7-17 给定的对象环节为

disp('System is stable');
end

System is unstable, with 2 unstable poles

\>\>disp('The unstable poles are:'),disp(p(ii))

The unstable poles are:

0.4121 + 0.9844i

0.4121 − 0.9844i

The pole-zero distribution diagram is obtain as is shown in Fig. 7-15.

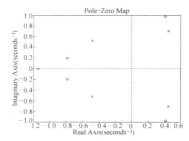

Fig. 7-15　Pole-zero distribution diagram
（零极点分布图）

7.4　Compensation of Control Systems based on MATLAB

In MATLAB, Bode plot is mainly used to design compensators for control systems. The basic idea is that by comparing the frequency characteristics before and after the compensation, try to select the appropriate compensation structure, determine the compensation parameters, and finally check the system after the compensation, and design repeatedly until it meets the design requirements. The following example illustrates the process of designing a control system.

Example 7-17　The given object element is

$$G_0(s) = \frac{K}{s(0.04s+1)}$$

Design a compensator which satisfies the requirements that the static speed error coefficient of the compensated system is $K_v \geq 100$, the crossing frequency is greater than 60rad/s, and the phase margin is $\gamma \geq 45°$.

Solution: 1) First, according to the requirement of static speed error coefficient, the open-loop gain K of the system is determined as $K = 100$.

2) Write down the transfer function of the system and calculate the gain margin and phase margin. The MATLAB programming code is

```
>> G0 = tf(100,conv([1,0],[0.04,1]));
[gm,pm,wcg,wcp] = margin(G0);
[gm,pm,wcg,wcp]
ans =
        Inf    28.0243    Inf    46.9701;
>> w = logspace(-1,3);    %Generate a row vector w of 50 logarithmically spaced points between
                         %decades 10^-1 and 10^3
[m,p] = bode(G0,w);
subplot(2 1 1), semilogx(w,20*log10(m(:)))    % Plot the logarithmic coordinate graph,
%where the logarithmic scale is used for w and the linear scale is used for 20*log10(m(:))
grid
subplot(2 1 2), semilogx(w,p(:))
grid
```

$$G_0(s) = \frac{K}{s(0.04s+1)}$$

设计一个补偿器,使校正后系统的静态速度误差系数 $K_v \geq 100$,穿越频率大于 60rad/s,相位裕度 $\gamma \geq 45°$。

解: 1) 首先根据对静态速度误差系数的要求,确定系统的开环增益 $K=100$。

2) 写出系统传递函数,并计算幅值裕度和相位裕度。MATLAB 程序为

```
>> G0 = tf(100,conv([1,0],[0.04,1]));
[gm,pm,wcg,wcp] = margin(G0);
[gm,pm,wcg,wcp]
ans =
        Inf    28.0243        Inf    46.9701
>> w = logspace(-1,3);     %在 10⁻¹ 到 10³ 间生成由 50 个对数间隔点组成的行向量 w
[m,p] = bode(G0,w);
subplot(2 1 1),semilogx(w,20*log10(m(:)))    %绘制对数坐标图形,w 采用对数刻度,
                                              %20*log10(m(:))采用线性刻度
grid
subplot(2 1 2),semilogx(w,p(:))
grid
```

可以看到,未校正环节的幅值裕度为无穷大,相位裕度为 25°,穿越频率为 45rad/s,不满足要求,其伯德图如图 7-16a 所示。

3) 根据对系统动态性能的要求,可试探性地引入一个超前补偿器来增加相位裕度,为此可假设校正装置的传递函数为

$$G_c(s) = \frac{0.02s+1}{0.01s+1}$$

则可通过下列的 MATLAB 语句得到校正后系统的幅值裕度和相位裕度。

```
>>figure(2)
Gc = tf([0.02,1],[0.01,1]);
bode(Gc,w)
grid
G_o = Gc*G0;
[gm,pm,wcg,wcp] = margin(G_o);
[gm,pm,wcg,wcp]
ans =
Inf    43.5910    Inf    54.2668
```

从而可得到补偿器的伯德图如图 7-16b 所示。可以看出,在频率为 60rad/s 处系统的幅值裕度和相位裕度均增加了。在这样的控制器下,校正后系统的相位裕度增加到 46°,而穿越频率增加到为 60rad/s。

4) 绘制校正后系统的伯德图,如图 7-16a 中的实线所示,用如下的 MATLAB 语句绘制校正前、后系统的阶跃响应曲线,结果如图 7-16c 所示。

It can be seen that the amplitude margin of the uncompensated element is infinite, the phase margin is 25°, and the crossing frequency is 45rad/s, which do not meet the requirements, as is shown in Fig. 7-16a.

3) According to the requirement of dynamic performance of the system, a phase-lead compensator can be tentatively introduced to increase the phase margin, so the transfer function of the compensation device can be assumed to be

$$G_c(s) = \frac{0.02s+1}{0.01s+1}$$

Then the amplitude margin and phase margin of the compensation system can be obtained through the following MATLAB program.

```
>>figure(2)
Gc = tf([0.02,1],[0.01,1]);
bode(Gc,w)
grid
G_o = Gc * G0;
[gm,pm,wcg,wcp] = margin(G_o);
[gm,pm,wcg,wcp]
ans =
        Inf    43.5910    Inf    54.2668
```

Thus, the Bode plot of the compensator can be obtained as is shown in Fig. 7-16b. It can be seen that the amplitude margin and phase margin of the system increase at the frequency of 60rad/s. With such a controller, the phase margin of the compensated system increases to 46° and the crossing frequency increases to 60rad/s.

4) Draw the Bode plot of the compensated system, as is shown with the solid line in Fig. 7-16a, and draw the step response curve of the system before and after the compensation with the following MATLAB program, which is shown in Fig. 7-16c.

```
>> [m,p] = bode(G0,w);
[m1,p1] = bode(G_o,w);
subplot(2 1 1), semilogx(w,20 * log10([m(:),m1(:)]))
grid
subplot(2 1 2), semilogx(w,[p(:),p1(:)])
grid
G_c1 = feedback(G0,1);    % Represent feedback connction, connecting the system with
                          % the model of G0 into a 1-stage negative feedback loop
G_c2 = feedback(G_o,1);
[y,t] = step(G_c1);
y = [y,step(G_c2,t)];
figure(3), plot(t,y)
```

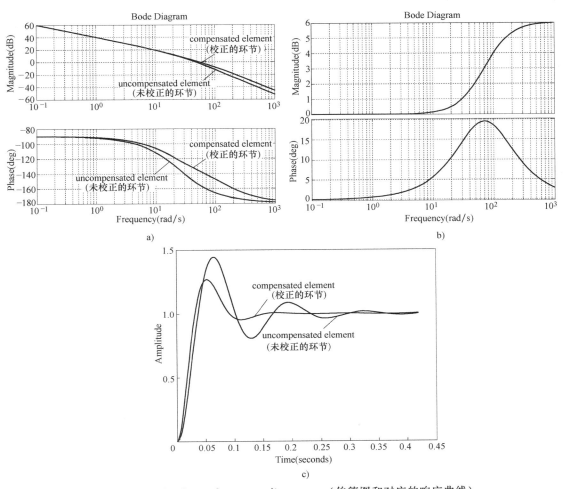

Fig. 7-16 Bode plots and corresponding curves(伯德图和对应的响应曲线)

```
>> [m,p] = bode(G0,w);
[m1,p1] = bode(G_o,w);
subplot(2 1 1),semilogx(w,20*log10([m(:),m1(:)]))
grid
subplot(2 1 2),semilogx(w,[p(:),p1(:)])
grid
G_c1 = feedback(G0,1);    %表示反馈连接,将系统模型G0反馈连接成1级负反馈循环
G_c2 = feedback(G_o,1);
[y,t] = step(G_c1);
y = [y,step(G_c2,t)];
figure(3),plot(t,y)
```

Appendix A   Exercises

课后练习

Chapter 1

1.1 Please describe the characteristics of an open-loop control system and a closed-loop control system respectively, and analyze their advantages and disadvantages.

1.2 Please describe the working principle of closed-loop control system briefly.

1.3 What are the main elements of a closed-loop control system, and what are the functions of each element in the system?

1.4 Fig. A-1 shows the schematic diagram of the automatic control system for the warehouse gate. Try to analyze its control principle when the automatic door is on or off, and draw a schematic block diagram.

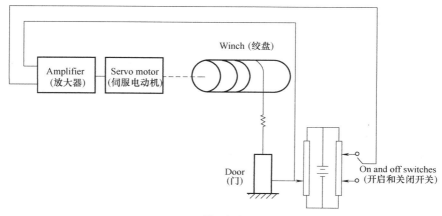

Fig. A-1

1.5 For two liquid level control systems shown in Fig. A-2, try to draw block diagrams of their composition, explain their control process, and determine whether they are open-loop control system or closed-loop control system.

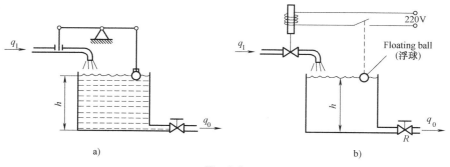

Fig. A-2

Chapter 2

2.1 As is shown in Fig. A-3, it is a mechanical system. Try to figure out the differential equations of the system and its transfer function.

2.2 A RLC network is shown in Fig. A-4. Try to figure out the transfer function of the network.

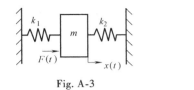
Fig. A-3

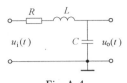
Fig. A-4

2.3 An active network is shown in Fig. A-5. Try to figure out the transfer function of the network.

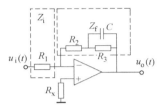

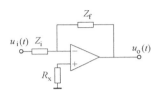

Fig. A-5

2.4 Given that k_1 and k_2 are the spring stiffness, c is the damping coefficient of the damper, $x_i(t)$ is the input variable, and $x_o(t)$ is the output variable, try to figure out the transfer function and unit step response of the spring damping system as is shown in Fig. A-6.

2.5 Fig. A-7 shows the series and parallel circuits of inductor L, resistor R and capacitor C. $u_i(t)$ is the input voltage and $u_o(t)$ is the output voltage. figure out the transfer function of the system.

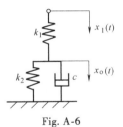
Fig. A-6

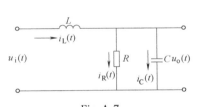
Fig. A-7

2.6 The system is shown in Fig. A-8, where k is the spring stiffness, c is the damping coefficient of the damper, $x_i(t)$ is the input signal of the system, $x_o(t)$ is the output signal of the system. Figure out the transfer function of the system.

2.7 The passive RCL network is shown in Fig. A-9, where $u_i(t)$ is the input voltage, $u_o(t)$ is the output voltage, $i(t)$ is the current, R is the resistance, C is the capacitance, L is the inductance. Please figure out the transfer function.

2.8 The mechanical systems are shown in Fig. A-10. Try to figure out the differential equations of the systems and their transfer functions.

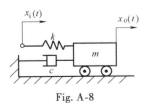
Fig. A-8

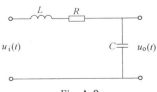
Fig. A-9

2.9 Given that k_1 and k_2 are the spring stiffness, c is the damping coefficient of the damper, $x_i(t)$ is the input value, and $x_o(t)$ is the output value. Please figure out the transfer function of the spring damping system shown in Fig. A-11.

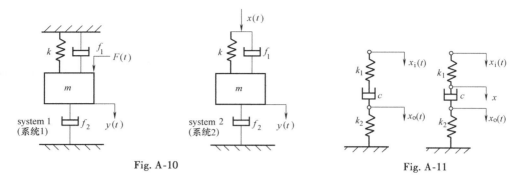

Fig. A-10   Fig. A-11

2.10 The circuit systems are shown in Fig. A-12. Try to figure out the transfer functions of the systems.

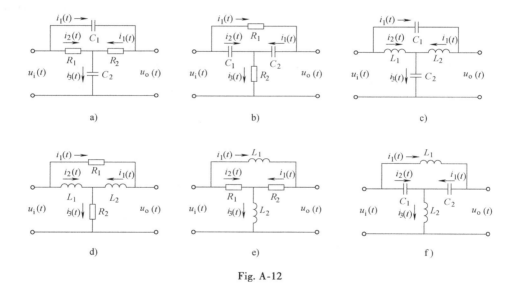

Fig. A-12

2.11 The proportional differential compensation circuit is shown in Fig. A-13. Please figure out the transfer function of the compensation circuit.

2.12 An electrical network is shown in Fig. A-14, where the input is $u_i(t)$, the output is $u_o(t)$. Try to figure out the differential equation.

2.13 As is shown in Fig. A-15, the working circuit consists of inductor L, resistor R and capacitor C. $u_i(t)$ is the input voltage, and $u_o(t)$ is the output voltage. Figure out the transfer function of the circuit.

2.14 The system block diagram is shown in Fig. A-16. Do calculation as follows.

(1) Calculate the closed-loop transfer functions of the output considering $X_o(s)$, $Y(s)$, $B(s)$ and $E(s)$ respectively, when $X_i(s)$ is input variable and $N(s)$ is equal to 0.

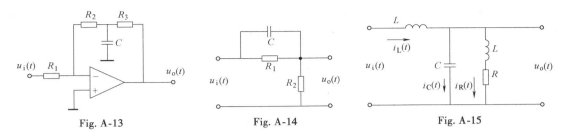

Fig. A-13   Fig. A-14   Fig. A-15

(2) the closed-loop transfer functions of the output considering $X_o(s)$, $Y(s)$, $B(s)$ and $E(s)$ respectively, when $N(s)$ is input variable and $X_i(s)$ is equal to 0.

2.15 As is shown in Fig. A-17, the electronic circuit is composed of resistors R_0, R_1, R_2, capacitor C and amplifier. $u_i(t)$ is the input voltage, and $u_o(t)$ is the output voltage. Figure out the transfer function of the system.

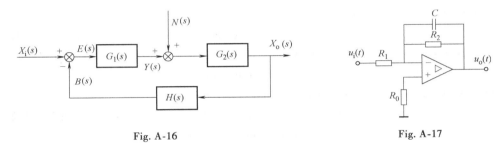

Fig. A-16   Fig. A-17

2.16 Please figure out the transfer function of each block diagram shown in Fig. A-18 by using the equivalent transformation method.

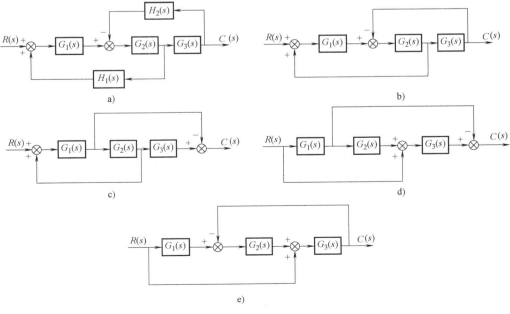

Fig. A-18

2.17 The block diagram of two-stage RC filter network is shown in Fig. A-19. Please simplify the block diagram by using the equivalent transformation method.

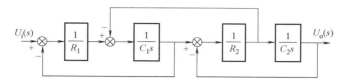

Fig. A-19

Chapter 3

3.1 The unit step response curve of a typical second-order system is shown in Fig. A-20.

(1) Calculate the natural frequency and damping ratio of the system.

(2) Determine the closed-loop transfer function of the system.

3.2 The unit step response curve of a typical second-order system is shown in Fig. A-21. When the control error is in the range of $\Delta = \pm 0.02$, the adjustment time $t_s \leq 0.5$s.

(1) Calculate the natural frequency and damping ratio of the system.

(2) Determine the closed-loop transfer function of the system.

3.3 The unit step response curve of a typical second-order system is shown in Fig. A-22.

(1) Calculate the natural frequency and damping ratio of the system.

(2) Determine the closed-loop transfer function of the system.

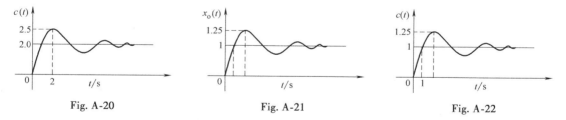

Fig. A-20   Fig. A-21   Fig. A-22

3.4 The block diagrams of two first-order systems are shown in Fig. A-23.

(1) Figure out the transfer functions and time constants of the system.

(2) Figure out the unit step response of the system.

(3) Calculate the adjustment time under the unit step response.

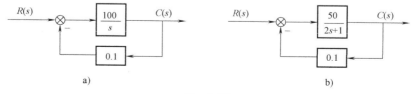

Fig. A-23

3.5 The differential equation of a first-order system is $2\dot{x}_o(t) + x_o(t) = 4\dot{x}_i(t) + x_i(t)$. Figure out the transfer function of the system and its unit step response.

3.6 A resistance-capacitance circuit is shown in Fig. A-24, where $u_i(t)$ is the input voltage, $u_o(t)$ is the output voltage, $i(t)$ is current, R is resistance, and C is capacitance. Please figure out the transfer function and unit step response of the system

3.7 The RC differential circuit is shown in Fig. A-25, where $u_i(t)$ is the input voltage, $u_o(t)$ is the output voltage, $i(t)$ is current, R is resistance, and C is capacitance. Please figure out the transfer function and unit slope response of the system.

3.8 The schematic diagram of hydraulic damper is shown in Fig. A-26, where the spring is rigidly connected with the piston, and the inertial force of the moving part is ignored. $x_i(t)$ is set as the input displacement, $x_o(t)$ is the output displacement, k is spring stiffness, c is the viscous damping coefficient. Try to figure out the transfer function between the output and the input and the unit slope response of the system.

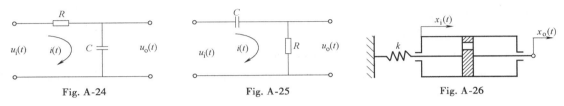

Fig. A-24　　　　　Fig. A-25　　　　　Fig. A-26

3.9 A spring damping system is shown in Fig. A-27, where $x_i(t)$ is set as input displacement, $x_o(t)$ is the output displacement, k is spring stiffness and c is viscous damping coefficient. Try to figure out the transfer function between output and input and unit slope response of the system.

3.10 The structure block diagram of a system is shown in Fig. A-28, and the transfer function is $G(s)=\dfrac{10}{0.2s+1}$. Now adjusting the time t_s to 0.1 times of the original settling time meanwhile ensuring the total amplification times unchanged, try to determine the value of K_h and K_0.

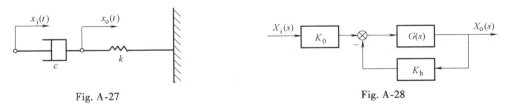

Fig. A-27　　　　　　　　　Fig. A-28

3.11 The unit slope response of a system is $x_{vo}=2t-e^{-2t}+e^{-t}$. Figure out the transfer function and the unit step response function of the system.

3.12 Under zero initial conditions, let the response function of a system with unit impulse signal $\delta(t)$ be $g(t)=K\left[\dfrac{T_2}{T_1}\delta(t)-\dfrac{T_2-T_1}{T_1^2}e^{\frac{t}{T_1}}\right]$.

(1) Figure out the transfer function of the system.

(2) Figure out the unit step response of the system under zero initial conditions.

3.13 The unit step response curve of a system is shown in Fig. A-29. Try to determine the values of K_1, K_2 and a.

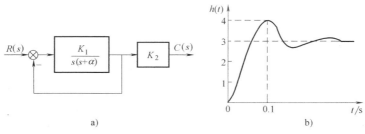

Fig. A-29

3.14 The block diagram of a robot control system is shown in Fig. A-30. Try to determine the values of the parameters K_1 and K_2, when the peak time of the system step response $t_p = 4.0s$, and the maximum overshoot $\sigma\% = 5\%$.

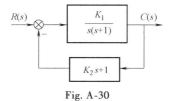

Fig. A-30

3.15 The block diagram of a system is shown in Fig. A-31.

(1) Figure out the closed-loop transfer function of the system.

(2) When $K_f = 0$, $K_a = 10$, try to determine the damping ratio ξ and natural frequency ω_n of the system.

3.16 The block diagram of a system is shown in Fig. A-32, and it is required to meet dynamic performance indicators as the maximum overshoot is 20% and the peak time is 1s. Try to determine the system parameters K and A.

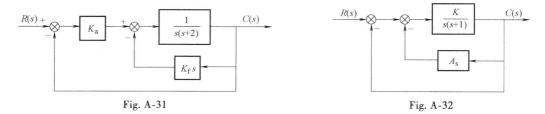

Fig. A-31   Fig. A-32

3.17 If the transfer function of a system is $G(s) = \dfrac{a}{\tau s + a + 1}$.

(1) Calculate the steady-state value of the unit step response function of the system.

(2) Calculate the adjustment time t_s for the transition process.

Chapter 4

4.1 Figure out the frequency characteristics of the following systems.

(1) $G(s) = \dfrac{K}{s} e^{-Ts}$.

(2) $G(s) = \dfrac{Ts}{Ts+1}$.

(3) $G(s) = \dfrac{K}{Ts+1}$.

4.2 The open-loop transfer function of a unit feedback system is $G_K(s) = \dfrac{K}{s(0.5s+1)}$. Try to figure out the phase-frequency crossing frequency and cut-off frequency of the system.

4.3 The open-loop "logarithmic" amplitude-frequency characteristic diagram of a minimum phase system is shown in Fig. A-33.

(1) Figure out the open-loop transfer function of the system.

(2) Calculate $A(10) = |G(j10)|$.

4.4 The open-loop logarithmic amplitude-frequency characteristic diagram of a minimum phase system is shown in Fig. A-34.

(1) Figure out the open-loop transfer function of the system.

(2) Calculate the value of ω_1 when $A(\omega_1) = |G(j\omega_1)| = 9$.

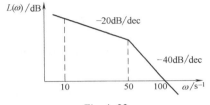

Fig. A-33

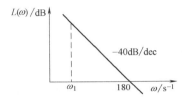

Fig. A-34

4.5 The transfer function of a system is $G(s) = \dfrac{K}{s}$. Try to figure out the frequency response of the input sinusoidal signal as $x_i(t) = \sin \omega t$.

4.6 The transfer function of a system is $G(s) = K$. Try to figure out the frequency response of the input sinusoidal signal as $x_i(t) = \sin \omega t$.

4.7 The transfer function of a system is $G(s) = \dfrac{K}{s+1}$. Try to figure out the steady-state frequency response of the input sinusoidal signal as $x_i(t) = \sin \omega t$.

4.8 The transfer function of the a is $G(s) = \dfrac{Ts}{Ts+1}$. Try to figure out the frequency response of the input sinusoidal signal as $x_i(t) = \sin \omega t$.

4.9 The asymptote of the open-loop logarithmic amplitude-frequency characteristics of a minimum phase system is shown in Fig. A-35. Figure out the open-loop transfer function and amplitude-frequency characteristics of the system.

4.10 The asymptote of the open-loop logarithmic amplitude-frequency characteristics of a minimum phase system is shown in Fig. A-36. Try to figure out the open-loop transfer function of the system.

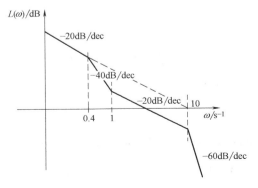

Fig. A-35

4.11 The asymptote of the open-loop logarithmic amplitude-frequency characteristics of a minimum phase system is shown in Fig. A-37. Please figure out the transfer function of the system.

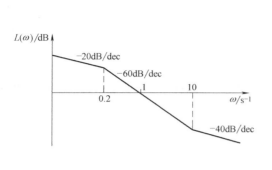
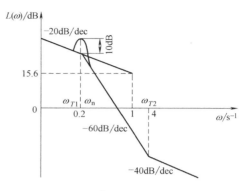

Fig. A-36

Fig. A-37

4.12 The asymptotic logarithmic amplitude-frequency characteristic curve of a minimum phase system is shown in Fig. A-38. Try to figure out the transfer function of the system.

4.13 The open-loop logarithmic amplitude-frequency characteristic asymptote of a unit feedback system is shown in Fig. A-39. Try to figure out the open-loop transfer function of the system.

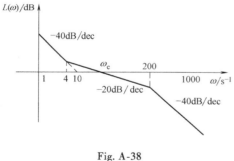
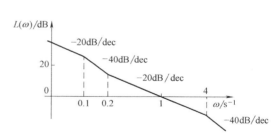

Fig. A-38

Fig. A-39

4.14 The open-loop transfer function of a system is $G(s) = \dfrac{Ks^2}{(0.2s+1)(0.02s+1)}$. Calculate the gain K of the system when $\omega_c = 5\text{s}^{-1}$.

4.15 The open-loop transfer function of a system is $G(s) = \dfrac{Ke^{0.1s}}{s(s+1)(0.1s+1)}$. Calculate the gain K of the system when $\omega_c = 5\text{s}^{-1}$.

4.16 The open-loop transfer function of a unit feedback system is $G(s) = \dfrac{1}{s(1+0.2s)(1+0.05s)}$. Calculate the phase-frequency crossing frequency and cut-off frequency of the system.

4.17 The open-loop transfer function of a unit feedback system is $G_K(s) = \dfrac{K}{s(0.2s+1)(0.5s+1)}$. Calculate the phase-frequency and amplitude-frequency crossing frequency of the system.

Chapter 5

5.1 Given that the characteristic equation of the system is $s^3+2s^2+s+2=0$, try to evaluate the stability of the system.

5.2 The characteristic equation of a speed regulating system is $s^3+41.5s^2+517s+2.3\times10^4=0$. Try to evaluate the stability of the system.

5.3 The closed-loop characteristic equation of the system is $s^4+2s^3+3s^2+4s+5=0$. Try to evaluate the stability of the system.

5.4 The block diagram of a system is shown in Fig. A-40. Try to determine the range of K and ξ that can make the system stable.

5.5 The characteristic equation of a speed regulating system is $s^3+41.5s^2+517s+1670(1+K)=0$. Try to determine the range of parameter K that can make the system stable.

5.6 A unit feedback control system is shown in Fig. A-41. When $G_c(s)=1$ and $G_c(s)=\dfrac{K_p(s+1)}{s}$, is the corresponding closed-loop system stable? What are the stability conditions?

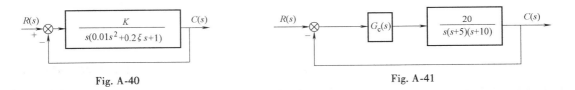

Fig. A-40   Fig. A-41

5.7 The open-loop transfer function of a unit feedback system is $G(s)=\dfrac{1}{s(Ts+1)}$. Figure out the amplitude crossing frequency and the inertia time constant T when the system satisfies the phase margin of 45°.

5.8 As is shown in Fig. A-42, a differential compensation control is adopted in the system. When the input signal is $r(t)=t$, the final steady-state error value of the system is required to be 0. Try to determine the value of the parameter τ_d.

5.9 The asymptotic logarithmic amplitude-frequency characteristic curve of a minimum phase system is shown in Fig. A-43. Try to figure out the closed-loop transfer function of the system and evaluate the stability of the system.

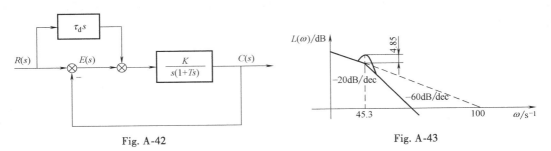

Fig. A-42   Fig. A-43

5.10 The asymptotic logarithmic amplitude-frequency characteristic curve of a minimum phase system is shown in Fig. A-44.

(1) Figure out the open-loop transfer function of the system.

(2) Calculate the steady-state error of the system when $r(t) = 0.5t^2$.

(3) Calculate the phase margin γ.

5.11 The open-loop transfer function of a system is $G(s) = 2KG_0(s)$. The open-loop frequency characteristics regarding $G_0(s)$ is shown in Fig. A-45, where the number of poles of the open-loop system in the right half of the s-plane is N_p. Try to determine the range of value of K that makes the closed-loop system stable.

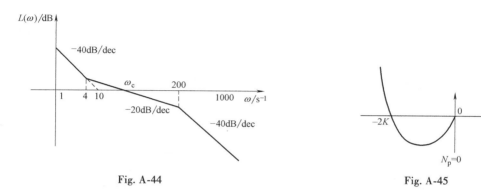

Fig. A-44　　　　　　　　　　　　Fig. A-45

5.12 The frequency characteristics $G_0(s)$ at $\omega = 0 \to +\infty$ is shown in Fig. A-46, where the number of open-loop poles in the right half of the s-plane is $N_p = 2$. Try to determine the value range of K that makes the system with the open-loop transfer function of $G(s) = 2KG_0(s)$ stable.

5.13 The open-loop transfer function of a unit feedback system is $G(s) = \dfrac{\alpha s + 1}{s^2}$. Try to determine the value of α when the phase margin of the system is $45°$.

5.14 The open-loop transfer function of a unit feedback system is $G(s) = \dfrac{s}{Ts+1}$. Try to determine the amplitude crossing frequency and the inertia time constant T of the system when the system is required to satisfy the phase margin of $45°$.

5.15 The open-loop transfer function of a system is $G(s) = 2KG_0(s)$. The open-loop frequency characteristics regarding $G_0(s)$ is shown in Fig. A-47, where the number of poles of the open-loop system in the right half of the s-plane is N_p. Try to determine the value range of K that makes the closed-loop system stable.

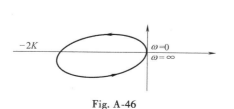 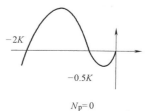

Fig. A-46　　　　　　　　　　　　Fig. A-47

5.16 The open-loop frequency characteristic $G_0(s)$ at $\omega = 0 \to +\infty$ is shown in Fig. A-48, where the number of open-loop poles in the right half of the s-plane is $N_p = 2$. Try to determine the value range of K that makes the system with the open-loop transfer function stable.

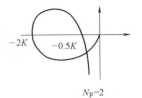

Fig. A-48

5.17 The open-loop transfer function of a unit feedback system is $G(s) = \dfrac{Ks}{\dfrac{\sqrt{3}}{3}s+1}$. Try to determine the amplitude crossing frequency and the gain coefficient K of the system when the system is required to satisfy the phase margin of $60°$.

5.18 The block diagram of a system is shown in Fig. A-49. If $\xi = 0.6$, the steady-state error of the system under the input of unit slope is $e_{ss} = 0.2$. Please calculate the value of K_f, the amplification factor K_a and the natural frequency of the system.

5.19 The open-loop logarithmic frequency characteristic of the system is shown in Fig. A-50.

(1) Figure out the open-loop transfer function of the system.

(2) Calculate the phase stability margin of the system and evaluate the stability of the closed-loop system.

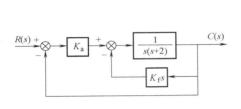

Fig. A-49

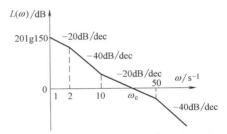

Fig. A-50

5.20 The open-loop transfer function of the system is $G(s) = \dfrac{K(0.2s+1)}{s^2(0.02s+1)}$. Determine the value of K if the phase margin of the system is required to be $45°$.

5.21 The closed-loop logarithmic amplitude-frequency characteristic of a unit feedback system is shown in Fig. A-51. The system is required to have the phase margin of $30°$. How many times should the open-loop gain be increased?

5.22 The open-loop logarithmic asymptote of a unit feedback system is shown in Fig. A-52.

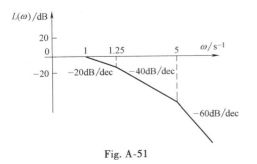

Fig. A-51

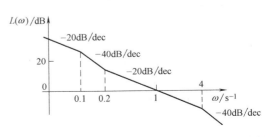

Fig. A-52

(1) Figure out the open-loop transfer function of the system.

(2) Evaluate the stability of the closed-loop system.

5.23 The open-loop transfer function of a unit feedback system is $G(s)=\dfrac{16}{s(s+1)(0.01s+1)}$. Try to determine the crossing frequency ω_c of the system, and calculate the phase margin of the system to evaluate its stability.

5.24 The open-loop transfer function of a unit feedback system is $G(s)=\dfrac{5}{s(s+1)(0.5s+1)}$. Try to determine the crossing frequency ω_c of the system, and calculate the phase margin of the system to evaluate its stability.

5.25 The block diagram of a system is shown in Fig. A-53, where $K>0$, $T>0$. the input signal of the system is set as $r(t)=t$, and the error is $e(t)=r(t)-c(t)$. In order to make the steady-state error $e_{ss}(t)=0$, what value should be taken for K_c?

5.26 The open-loop transfer function of a unit negative feedback system is $G(s)=\dfrac{K}{s(0.1s+1)(0.25s+1)}$.

(1) Figure out the value range of K that makes the system stable.

(2) Calculate the minimum steady-state error of the system with unit ramping signal.

5.27 The open-loop transfer function of a system is $G(s)=\dfrac{3s}{2s+1}$. Calculate the amplitude-frequency characteristics, phase-frequency characteristics and amplitude-crossing frequency of the system.

5.28 The open-loop transfer function of a system is $G(s)=\dfrac{15s}{5s+1}$. Calculate the amplitude-frequency characteristics, phase-frequency characteristics and amplitude-crossing frequency of the system.

5.29 The open-loop logarithmic amplitude-frequency characteristic curve of a minimum phase system is shown in Fig. A-54. Figure out the open-loop transfer function and the phase margin γ of the system.

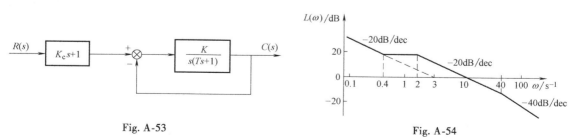

Fig. A-53   Fig. A-54

5.30 The open-loop transfer function of a unit negative feedback system is $G_K(s)=\dfrac{K}{(s+1)(s+2)(s+5)(s+10)}$.

(1) Figure out the value range of K that makes the system stable.

(2) Calculate the steady-state error of the system with unit step signal.

5.31 The open-loop logarithmic amplitude-frequency characteristic curve of a minimum phase system is shown in Fig. A-55. Try to figure out the open-loop transfer function and phase margin γ of the system.

5.32 The open-loop transfer function of a unit negative feedback system is $G_K(s) = \dfrac{K}{s(s+1)(s+2)(s+3)}$.

(1) Figure out the value range of K that makes the system stable.

(2) Calculate the steady-state error of the system with unit step signal.

5.33 The open-loop logarithmic amplitude-frequency characteristic curve of a minimum phase system is shown in Fig. A-56.

(1) Figure out the open-loop transfer function $G(s)$ of the system.

(2) Figure out the open-loop gain K of the system.

(3) Calculate the phase margin of the system.

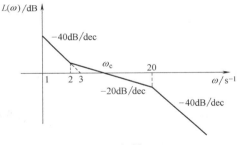

Fig. A-55

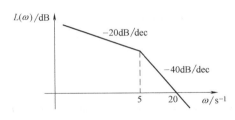

Fig. A-56

5.34 Fig. A-57 shows the open-loop logarithmic amplitude-frequency characteristic curve of a minimum phase system.

(1) Figure out the open-loop transfer function $G(s)$ of the system.

(2) Figure out the open-loop gain K of the system.

(3) Calculate the steady-state error with unit ramping input signal.

5.35 The open-loop logarithmic amplitude-frequency characteristics of a unit negative feedback system is shown in Fig. A-58.

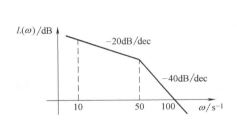

Fig. A-57

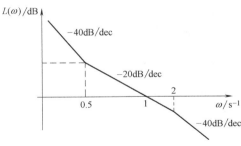

Fig. A-58

(1) Figure out the open-loop transfer function of the system.

(2) Figure out the open-loop gain K that makes the system stable.

(3) Calculate phase margin γ of the system.

5.36 The open-loop transfer function of a unit negative feedback system is $G_K(s) = \dfrac{K}{(s+1)(s+2)(s+3)(s+4)}$.

(1) Figure out the value range of K that makes the system stable.

(2) Calculate the steady-state error of the system with unit step signal.

5.37 The open-loop transfer function of a unit negative feedback system is $G_K(s) = \dfrac{K}{s(s+5)(s+50)}$.

(1) Figure out the value range of K that makes the system stable.

(2) Calculate the steady-state error of the system with unit ramp signal.

5.38 The open-loop transfer function of a unit feedback system is $G_K(s) = \dfrac{K}{(s+1)(s+3)(s+5)(s+7)}$.

(1) Figure out the value range of K that makes the system stable.

(2) Calculate the steady-state error of the system with unit step signal.

Chapter 6

6.1 The open-loop transfer function of a unit feedback system is $G_0(s) = \dfrac{K}{s(0.5s+1)}$. When the system is with the unit ramping input signal of the steady-state error is required to be $e_{ss} = 0.05$, and the performance indicators in the frequency domain are required that: amplitude margin $L_g(\omega) = 10\mathrm{dB}$, phase margin $\gamma = 50°$. Try to design the phase-lead compensation component of the system.

6.2 The open-loop transfer function of a unit feedback system is $G_0(s) = \dfrac{K}{s(0.2s+1)(0.5s+1)}$. It is required that $K_v = 20\mathrm{s}^{-1}$, the phase margin $\gamma \geqslant 35°$, and the amplitude margin $L_g(\omega) = 10\mathrm{dB}$. Try to figure out the transfer function of the cascade phase-lag compensation component.

6.3 The open-loop transfer function of a unit feedback system is $G_0(s) = \dfrac{K}{s(s+1)(0.125s+1)}$. It is required that $K_v = 20\mathrm{s}^{-1}$, the phase margin $\gamma = 50°$, the cutoff frequency $\omega_c \geqslant 2\mathrm{s}^{-1}$. Try to design a cascade lag-lead compensation component to meet the requirements.

6.4 The open-loop transfer function of a unit feedback system is $G_0(s) = \dfrac{K}{s(s+1)(0.125s+1)}$. The performance indicators are required as $K_v = 20\mathrm{s}^{-1}$, the phase margin $\gamma = 50°$, the cutoff frequency $\omega_c \geqslant 2\mathrm{s}^{-1}$. Try to design a PID compensation component to meet the requirements.

6.5 The open-loop transfer function of a unit feedback control system is $G_0(s) = \dfrac{20}{s(0.5s+1)}$.

The performance indicators are required as the phase margin $\gamma = 50°$, the amplitude margin $L_g(\omega) = 10\text{dB}$. When PD, PI, PID compensation components are used to compensate the system, figure out the transfer function of the corresponding compensation component.

6.6 The open-loop transfer function of a unit feedback system is $G_0(s) = \dfrac{K}{s(0.2s+1)(0.5s+1)}$. The system is compensate by a proportion-integral (PI) component of which the transfer function is $G_c(s) = \dfrac{K(Ts+1)}{s}$.

(1) Calculate the phase margin and amplitude margin of the system without compensation.

(2) The performance indicators are required as phase margin $\gamma = 35°$ and amplitude margin $L_g(\omega) = 10\text{dB}$. Figure out the transfer function of the compensation component.

Chapter 7

7.1 Describe the following transfer function of a system by using MATLAB.
$$G(s) = \frac{s+5}{s^4+2s^3+3s^2+4s+5}$$

7.2 Describe the following transfer function of a system by using MATLAB.
$$G(s) = \frac{6(s+5)}{(s^2+3s+1)^2(s+6)}$$

7.3 Build the zero-pole model of the system with following transfer function in MATLAB.
$$G(s) = 6\frac{(s+1.9294)(s+0.0353+0.9287j)}{(s+0.9567+1.2272j)(s-0.0433+0.6412j)}$$

7.4 Build the zero-pole model of the system with following transfer function in MATLAB.
$$G(s) = 6.8\frac{(s+2)(s+7)}{s(s+3-j2)(s+1.5)}$$

7.5 The transfer function of the system is $G(s) = \dfrac{25}{s^2+4s+25}$. Plot the unit step response curve of the system in MATLAB.

7.6 The transfer function of the system is $G(s) = \dfrac{3}{(s+1-3j)(s+1+3j)}$. Plot the unit step response curve of the system in MATLAB.

7.7 The transfer function of the system is $G(s) = \dfrac{1}{s^2+0.8s+1}$. Draw the Nyquist plot of the system in MATLAB.

7.8 The open-loop transfer function of a unit negative feedback system is $G(s) = \dfrac{10(s+1)}{s(s+7)}$. Draw the Bode plot of the system in MATLAB.

References
参考文献

[1] NISE N S. Control Systems Engineering [M]. 6th ed. Hoboken: John Wiley & Sons Inc, 2011.
[2] SOLOMON C W. Introduction to Dynamic and Control in Mechanical Engineering Systems [M]. Hoboken: John Wiley & Sons Inc, 2016.
[3] BISSELL C C. Control Engineering [M]. 2th ed. Boca Raton: CRC Press Inc, 2017.
[4] DORF R C, BISHOP R H. 现代控制系统:第十三版:英文 [M]. 北京:电子工业出版社, 2018.
[5] 曾励. 控制工程基础 [M]. 北京:机械工业出版社, 2012.
[6] 黄安贻. 机械控制工程基础 [M]. 武汉:武汉理工大学出版社, 2004.
[7] 孙晶. 控制工程基础:英文 [M]. 北京:科学出版社, 2017.
[8] 孔祥东, 姚成玉. 控制工程基础. 4版 [M]. 北京:机械工业出版社, 2019.
[9] 董景新, 赵长德, 郭美凤, 等. 控制工程基础 [M]. 4版. 北京:清华大学出版社, 2015.
[10] 罗庚合. 机电控制工程基础 [M]. 西安:西安电子科技大学出版社, 2016.
[11] 左健民. 机电控制工程基础 [M]. 北京:机械工业出版社, 2020.
[12] D'AZZO J J, HOUPIS C H, SHELDON S N. 基于 MATLAB 的线性控制系统分析与设计:原书第5版 [M]. 张武, 王玲芳, 孙鹏, 译. 北京:机械工业出版社, 2008.
[13] 魏巍. MATLAB 控制工程工具箱技术手册 [M]. 北京:国防工业出版社, 2004.
[14] 张若青, 罗学科, 王民. 控制工程基础及 MATLAB 实践 [M]. 北京:高等教育出版社, 2008.